Physical Metallurgy of Light Alloys and Composite Materials

Physical Metallurgy of Light Alloys and Composite Materials

Editor

Nikolay A. Belov

MDPI • Basel • Beijing • Wuhan • Barcelona • Belgrade • Manchester • Tokyo • Cluj • Tianjin

Editor
Nikolay A. Belov
National Research
and Technological
University "MISIS",
Moscow Russia

Editorial Office
MDPI
St. Alban-Anlage 66
4052 Basel, Switzerland

This is a reprint of articles from the Special Issue published online in the open access journal *Metals* (ISSN 2075-4701) (available at: https://www.mdpi.com/journal/metals/special_issues/physical_metallurgy_alloys_composite).

For citation purposes, cite each article independently as indicated on the article page online and as indicated below:

LastName, A.A.; LastName, B.B.; LastName, C.C. Article Title. *Journal Name* **Year**, *Volume Number*, Page Range.

ISBN 978-3-0365-8074-6 (Hbk)
ISBN 978-3-0365-8075-3 (PDF)

Contents

About the Editor

Nikolay A. Belov

Nikolay A. Belov is currently the Chief Scientist and Professor of Materials Science and Light Alloys at the National University of Science and Technology MISiS (former name "Moscow Institute of Steel and Alloys"), Russia. His areas of expertise include the physical metallurgy of aluminum alloys, including the development of new Al-alloys (both casting and wrought), the analysis of multicomponent phase diagrams, structure formation during solidification and its evolution during heat treating and plastic deformation. He has published 18 books and over 300 papers in scientific journals and conference proceedings and has been involved with over 50 patents and inventions.

metals

Article

A Comparative Investigation of Conventional and Hammering-Assisted Incremental Sheet Forming Processes for AA1050 H14 Sheets

Harshal Y. Shahare [1], Abhay Kumar Dubey [1], Pavan Kumar [1], Hailiang Yu [2], Alexander Pesin [3], Denis Pustovoytov [3] and Puneet Tandon [1,*]

[1] deLOGIC Lab, PDPM Indian Institute of Information Technology, Design and Manufacturing, Jabalpur 482005, India; 1823604@iiitdmj.ac.in (H.Y.S.); adubey@iiitdmj.ac.in (A.K.D.); pavan.kumar@iiitdmj.ac.in (P.K.)

[2] Light Alloys Research Institute, Central South University, Changsha 410083, China; yuhailiang@csu.edu.cn

[3] Laboratory of Mechanics of Gradient Nanomaterials Named after A.P. Zhilyaeva, Nosov Magnitogorsk State Technical University, 38, Lenin Prospect, 455000 Magnitogorsk, Russia; pesin@bk.ru (A.P.); pustovoitov_den@mail.ru (D.P.)

* Correspondence: ptandon@iiitdmj.ac.in; Tel.: +91-9425324240

Citation: Shahare, H.Y.; Dubey, A.K.; Kumar, P.; Yu, H.; Pesin, A.; Pustovoytov, D.; Tandon, P. A Comparative Investigation of Conventional and Hammering-Assisted Incremental Sheet Forming Processes for AA1050 H14 Sheets. *Metals* **2021**, *11*, 1862. https://doi.org/10.3390/met11111862

Academic Editors: José Valdemar Fernandes and Marta Oliveira

Received: 22 September 2021
Accepted: 17 November 2021
Published: 19 November 2021

Abstract: Incremental Sheet Forming (ISF) is emerging as one of the popular dieless forming processes for the small-sized batch production of sheet metal components. However, the parts formed by the ISF process suffer from poor surface finish, geometric inaccuracy, and non-uniform thinning, which leads to poor part characteristics. Hammering, on the other hand, plays an important role in relieving residual stresses, and thus enhances the material properties through a change in grain structure. A few studies based on shot peening, one of the types of hammering operation, revealed that shot peening can produce nanostructure surfaces with different characteristics. This paper introduces a novel process, named the Incremental Sheet Hammering (ISH) process, i.e., integration of incremental sheet forming (ISF) process and hammering to improve the efficacy of the ISF process. Controlled hammering in the ISF process causes an alternating motion at the tool-sheet interface in the local deformation zone. This motion leads to enhanced material flow and subsequent improvement in the surface finish. Typical toolpath strategies are incorporated to impart the tool movement. The mechanics of the process is further explored through explicit-dynamic numerical models and experimental investigations on 1 mm thick AA1050 sheets. The varying wall angle truncated cone (VWATC) and constant wall angle truncated cone (CWATC) test geometries are identified to compare the ISF and ISH processes. The results indicate that the formability is improved in terms of wall angle, forming depth and forming limits. Further, ISF and ISH processes are compared based on the numerical and experimental results. The indicative statistical analysis is performed which shows that the ISH process would lead to an overall 10.99% improvement in the quality of the parts primarily in the surface finish and forming forces.

Keywords: incremental sheet forming; hammering; aluminum alloy; incremental sheet hammering

1. Introduction

The incremental sheet forming (ISF) process is a flexible dieless manufacturing process [1]. The ISF process starts with the three-dimensional (3D) computer-aided design (CAD) model of the geometry of the component to be formed. The desired CAD model can be generated with any modelling software (e.g., SOLIDWORKS, CATIA, Siemens NX, PTC Creo, etc.). The required toolpath to incrementally form the desired geometry is evolved on the basis of the sliced 3D model. A forming tool is given the controlled motion along the generated toolpath and deforms the sheet into the desired shape [2]. The sheet is clamped at its outer periphery. Stresses are generated on the sheet due to the continuous motion of the forming tool, which leads to local plastic deformation on the sheet [3,4]. As the process does not require dies to manufacture the components, thus, it can be used in forming customized products at a lower cost in comparison to the conventional process. With an increase in the demand for customized products, the ISF process is drastically gaining popularity [5]. The ISF process finds its applications mainly in the automobile, aerospace, and medical implant industries. In the automobile industry, the headlight reflector, heat or vibration shield, solar oven, and silencer of the trucks are manufactured using the ISF process. Apart from this, the ISF process also finds its application in the medical implant industry, where it mainly helps to form facial and cranial implants. Even though the ISF process has numerous advantages over the conventional forming process, it has some limitations while forming materials with poor forming characteristics such as geometric inaccuracy, non-uniform thinning, poor surface finish, higher processing time than conventional forming processes, and difficulties in forming geometries with 90° (or more than 90°) wall angles. Formability of the material is the degree of deformation during the forming operation without any unwanted flaws like tearing, buckling, necking, or any other defect [6]. The different forms of the ISF process are studied to overcome the flaws that arise during the ISF process.

The ISF process is broadly classified according to the number of contact points during the forming process, i.e., single point incremental forming (SPIF) process and two-point incremental forming (TPIF) process. In the SPIF process, there is only one forming tool, which is controllable, whereas in the TPIF process, there is an additional supporting tool with the master tool to control the flow of the deforming material [7,8]. The TPIF process is further extended with the help of a partial die which limits the neck formation during the process [9]. Researchers had worked on heat-assisted incremental forming techniques to improve formability and geometrical accuracy [10]. Further, an ultrasonic-assisted incremental forming setup had been devised to investigate the formability and the forming forces arising due to ultrasonic energy infused during the process [11]. Vihtonen et al. [12] compared the robot-assisted incremental forming by pressing (RAIFP) and robot-assisted incremental forming by hammering (RAIFH) based on the strain measurement and found that the formability of the material can be enhanced with the help of hammering. Luo et al. [13,14] designed and developed a new sheet metal forming system based on incremental punching. They have developed a mechanics model, which helps to predict the final shape of the component using the minimum energy principle. One-step FEM is developed to predict the stress and strain distribution of the component using inverse finite element modelling. Further, they have proposed that the incremental sheet metal punching (ISMP) is effective for the rapid prototyping of sheet metal components, whereas it is effective in developing free form sheet geometries without bottom support. Further, Wang et al. [15] performed incremental sheet punching based on the sinusoidal toolpath and the results indicated that the wavelength and amplitude of the sinusoidal toolpath affect the material formability, surface quality, and fatigue life. The dimensional accuracy was investigated in the incremental sheet metal hammering process and the results revealed that the springback and surface roughness can be minimized by increasing the punch diameter [16]. Though limited research articles are available on the effect of hammering in the incremental sheet forming, the scope of the work is not limited, as shot peening, one of the types of hammering process, had revealed that the material characteristics can be improved with the help of the controlled impact on the material surface [17].

The present study aims to integrate the hammering operation with the incremental sheet forming process to enhance the material properties of the sheet during the forming. The process uses a working principle similar to the ISF process, i.e., the sheet is clamped at its ends and is deformed into desired geometry with the help of forming tool. The forming tool is provided with a hammering induced toolpath. As the process is applied for the sheet materials, thus, it is also called the incremental sheet hammering (ISH) process. Figure 1 shows the schematics of the ISH process. The present work comprises experimental and numerical investigations of the ISF and ISH processes on the AA1050 sheet and a comparative conclusion has been drawn based on the analysis.

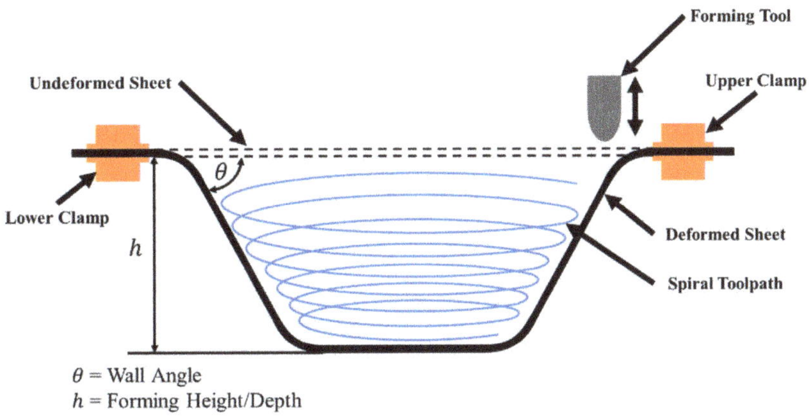

Figure 1. Schematics of incremental sheet hammering (ISH) process.

2. Deformation Methods and Methodology

2.1. *Deformation Behaviour*

To understand the deformation behavior of the hammering integrated with the ISF process, the deformation of the workpiece needs to be studied based on the linear tool movement due to the hammering operation. Figure 2 shows the schematics of deformation of the workpiece with respect to the hammering induced linear tool movement. The tool moves from one position to another position in certain steps. Initially, the tool aligns itself to the first hammering position where it deforms the workpiece with predefined amplitude. In the next step, the tool retracts to its original position and then moves to the next hammering position and similarly deforms the workpiece.

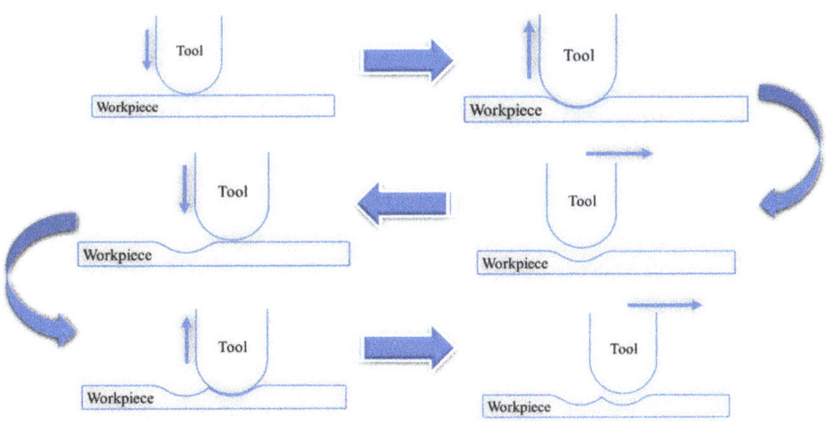

Figure 2. Schematics of deformation of the workpiece based on linear tool movement with hammering.

When the tool deforms two successive hammering positions, it forms an overlapping region, as shown in Figure 3a. The overlapping region affects the surface quality of the formed component, i.e., if the overlapping region is bigger, the surface quality of the deformed product is poor. Further, the overlapping region dimension adversely affects the production time. Thus, the overlapping region should be chosen such that the surface quality of the product is maintained, and the production time is minimized [18].

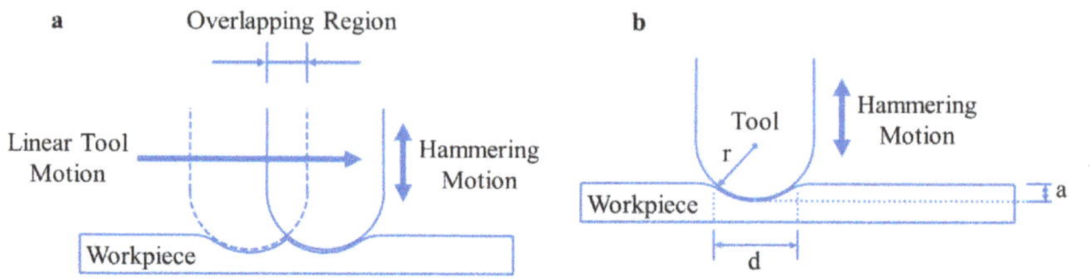

Figure 3. (**a**) Schematics of the overlapping region caused due to linear tool movement with hammering (**b**) Schematics of the basic hammering mechanics.

Figure 3b shows the basic hammering mechanics in which a hemispherical tool with radius (r, mm) impacts the workpiece with predetermined amplitude (a, mm). 'd' is the maximum length of the deformed area of the workpiece (mm), which further helps in deducing the frequency of the impact (f, Hz), when the tool moves with certain feed rate (V, mm/s). The frequency of the impact is calculated using Equations (1) and (2).

$$f = \frac{V}{d} \tag{1}$$

$$d = 2\sqrt{2ra - a^2} \tag{2}$$

Equations (1) and (2) help to investigate the frequency of the impact during the spiral movement of the tool deforming the workpiece as shown in Figure 3b. Further, the indentation marks and the overlapping region can be controlled with the help of controlling the input process parameters such as predetermined amplitude and feed rate of the tool.

2.2. Methodology

The formability was studied for both the ISF and ISH processes with the aluminum alloy (AA1050) in terms of the wall angle, forming depth, and forming limits of the formed geometries. Hussain et al. [19] had proposed that the number of experiments would increase to study the formability of the component. Thus, to minimize the number of experiments to determine the formability of a material, it was advised to form a varying wall angle truncated cone (VWATC). The VWATC was used as the initial test geometry, with wall angle, varying from 0 degrees on the top to 90° at the bottom of the cone, as shown in Figure 4. In the incremental forming process, the tool motion was controlled with the help of numerical control (NC) codes of the desired component. Generally, the NC codes were obtained by generating a spiral toolpath of the required geometry in CAD/CAM software [16]. The toolpath for the VWATC was prepared with the help of an in-house developed generalized python code (kindly refer to Appendix A), which is independent of any commercial CAD/CAM software. The input of the generalized python code for VWATC was the major and minor diameters of the geometry, the height of the VWATC, number of layers or step depth and number of points in one spiral contour. Figure 5 shows the toolpath trajectory for the ISF and ISH processes. The enlarged section in Figure 5 shows the tool motion during both processes.

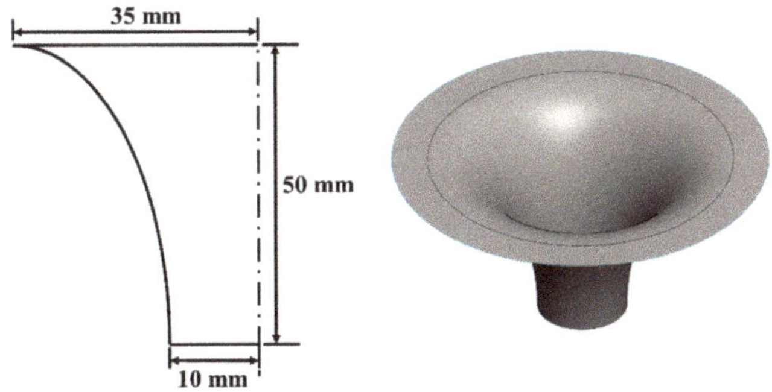

Figure 4. Varying wall angle truncated cone (VWATC) test geometry.

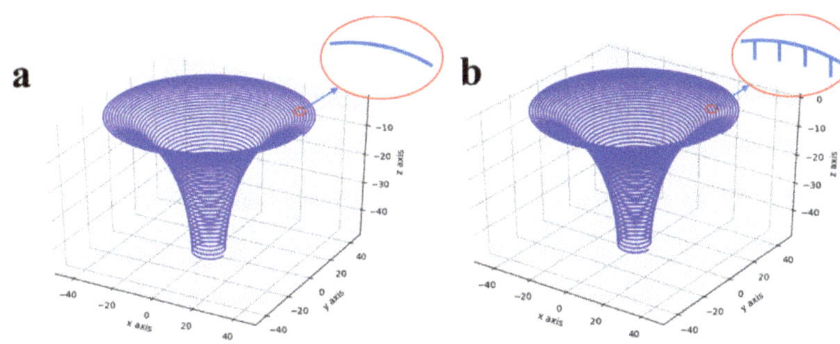

Figure 5. Toolpath for the VWATC for (**a**) ISF process and (**b**) ISH process.

During forming of the VWATC, the sheet thickness decreases continuously with increasing wall angle, which ultimately leads to rupture of the component at the point of minimum thickness. The angle at which the VWATC fractures was considered as the maximum achievable wall angle for the sheet material. The present work initiated with the forming of VWATC using the toolpath strategies for ISF and ISH processes as shown in Figure 5. Further, the wall angle achieved through both processes was investigated through experimentation. The VWATC was formed till the point of fracture to determine the forming height achieved through both processes. The formed height of the VWATC was used to determine the wall angle achieved for the respective components formed through ISF and ISH processes. The tangent at the point of fracture revealed the wall angle achieved in the components formed through ISF and ISH processes. To ensure correctness in the results, three sets of VWATC components were formed through ISF and ISH processes individually. Further, the mean values of the wall angle achieved through ISF and ISH processes were compared to find the minimum value of the achieved wall angle. The minimum wall angle observed after comparison was taken as an input wall angle for the constant wall angle truncated cone (CWATC) geometry. The minimum wall angle found through comparison was 78° and it was considered as the wall angle for CWATC. The schematic of the CWATC is shown with the help of Figure 6. To form the CWATC through ISF and ISH processes, the toolpaths were developed using the similar generalized python code which was discussed earlier. The results obtained in terms of sheet thickness, depth achieved, surface roughness, wall angle, and forming limits were compared for the two processes.

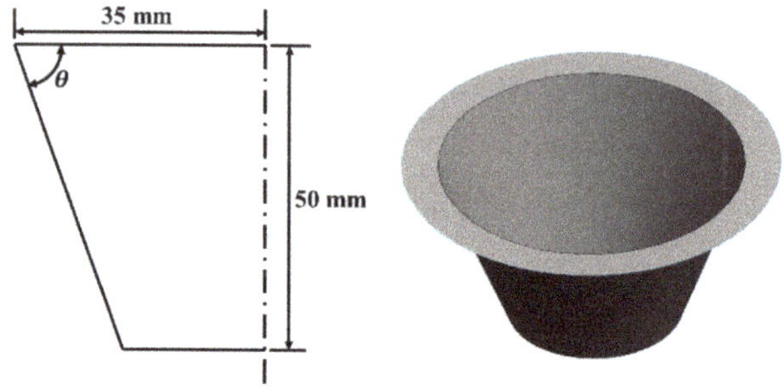

Figure 6. Constant wall angle truncated cone (CWATC) test geometry.

3. Numerical Simulation and Experimental Work

3.1. Material Properties

The ISF process is more widely adopted in the automobile and aerospace industries where lightweight materials are preferred. Aluminum and its alloys have gained tremendous popularity in recent decades in aerospace and automobile applications. Keeping these factors in mind, a commercial aluminum alloy AA1050 was used for the current research purpose. A sheet of 1 mm thickness was used during the experimentation. The tensile test was performed on the mechanical testing equipment GLEEBLE 350 (GLEEBLE—A VPG Brand, Poestenkill, NY, United States) to obtain the mechanical properties and elastoplastic response of AA1050 at room temperature. Table 1 depicts the material properties of aluminum alloy AA1050 at room temperature, which is responsible for the material behaviour. Figure 7 shows the flow stress-strain curves for different strain rates at room temperature obtained through mechanical testing.

Table 1. Material properties of aluminum alloy AA1050.

S. No.	Property	Value
1	Young's modulus	71,000 MPa
2	Poisson's ratio	0.33
3	Density	2700 kg/m^3

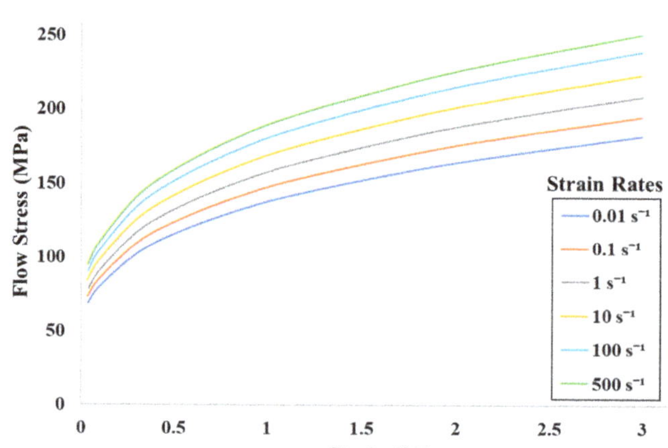

Figure 7. Flow stress-strain curves at different strain rates of aluminum alloy AA1050.

3.2. Numerical Simulation

The numerical simulations of both the processes were carried out in the ABAQUS/CAE (SIMULIA™ by Dassault Systèmes®, Vélizy-Villacoublay, France) version 2017 platform with a built-in explicit solver. The explicit approach was selected because it is computationally more efficient. Figure 8 shows the meshed model used during the numerical simulation. The workpiece material, i.e., sheet blank with a thickness of 1 mm, was modelled. The ratio of the material thickness (h) to the length of the workpiece (L) was less than 0.3 which allows usage of shell elements instead of solid elements [20]. Therefore, the sheet was meshed using the four-node square shell element S4R, having a mesh size of 1 mm. Chung et al. [21] had reported that the shell elements are not able to properly predict the plastic deformation in the thickness direction. Li et al. [22] had suggested that the use of shell elements lead to an inaccurate evaluation of transverse shear strain when the bending effect is important. In the present work, the numerical simulation results are used only to access qualitative differences between the two processes, thus, the numerical simulation was carried out with explicit solver and shell elements due to their computational efficiency. To satisfy the experimental conditions, the sheet was fixed at its outer boundary with the help of upper and lower clamps. The upper and lower clamps were meshed as discrete rigid elements as these fixtures were assumed to be non-deformable. The hemispherical tool of diameter 10 mm was modelled and meshed with the help of discrete rigid shell elements, which moves according to the toolpath strategies developed for both processes. The size of the sheet blank was 110 mm × 110 mm. The simulation of the incremental forming process has various nonlinearities due to the nature of the process. Furthermore, the incremental forming process is 3D in nature without any symmetry plane. Typically, a significant number of elements must be employed, and the tool must follow a longer trajectory, which makes the finite element analysis difficult and time-consuming. A high-velocity artificial tool was found to be suitable for the simulation of the incremental forming process [23]. Thus, the artificially increased feed rate was taken as 2000 mm/s to reduce the computational time maintaining quasi-static conditions. The computation time depends on many factors like element size, analysis procedure, i.e., explicit or implicit and the machine specifications. The total step time for forming a CWATC was 4.0041 s with the ISF process, while it took 4.7169 s in the case of the ISH process. The hammering operation was incorporated in the toolpath strategy of the ISH process; thus, the computation time and production time was higher in the case of the ISH process than the ISF process. Further, the time for forming a component in the ISH process depends on the overlapping region and the amount of predefined amplitude. In the present work, the overlapping region in the case of the ISH process was 25%, i.e., 1/4th of the trailing deformed area overlapped with the leading area. The increasing amplitude of hammering results in the poor surface quality of the component and it eventually leads to early rupture of the component [15]. Therefore, the amplitude of hammering was considered as 0.05 mm. The contact condition and the type of interaction of the deforming tool with the sheet during processing has a great influence on the redistribution of the material of the surface layer [24,25]. The interaction between tool and sheet was described using master and slave algorithm with the coefficient of friction taken as 0.05 [26]. The results obtained through the numerical simulation are further compared with experimental results and are discussed later in the results section.

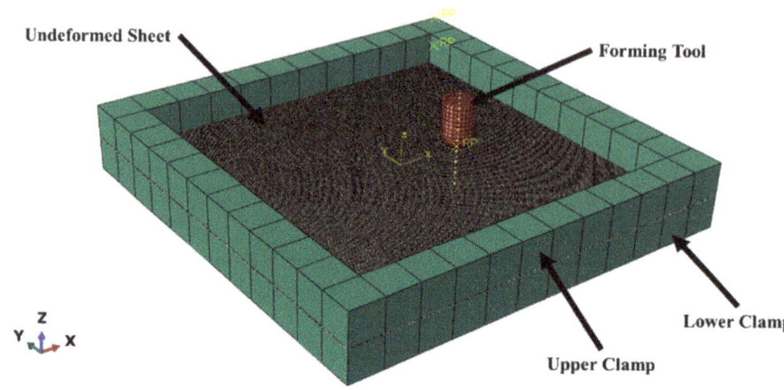

Figure 8. Numerical model used during simulation of ISF and ISH processes.

3.3. Experimental Setup

The experimental work was mainly carried out to validate the simulation results. Figure 9 shows the experimental setup of the incremental sheet forming process. It mainly consists of a CNC forming rig, sheet, forming tool, clamping plate, and fixture plate. Sheet blank was clamped on the fixture plate with the help of a clamping plate through threaded bolts. During the experimentation, a stainless-steel hemispherical tool of 10 mm diameter was used for all the experiments. The tool can move in the X, Y, and Z directions with the help of numerical control at a predetermined feed rate, i.e., 800 mm/min.

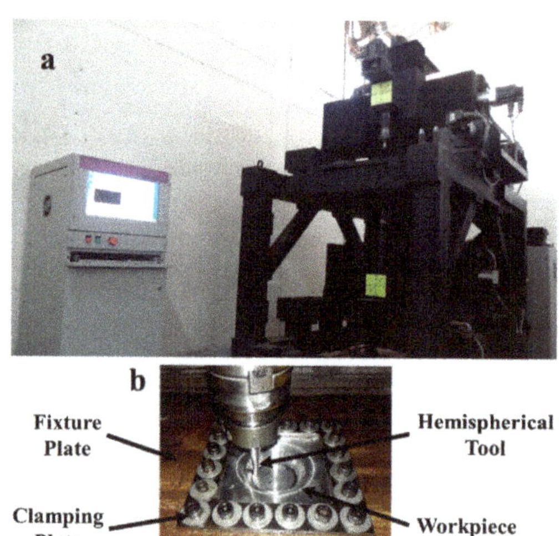

Figure 9. Experimental setup consisting of (**a**) dieless manufacturing center (**b**) forming setup.

4. Results and Discussions

This section discusses the comparative results of both the processes in terms of formability, thickness distribution, geometrical accuracy, and surface roughness.

4.1. Results Based on VWATC Test Geometry

The experiments performed to form the VWATC provides the maximum wall angle for both processes. Further, it compares the formability of both the processes in terms of maximum forming depth while forming the VWATC. The maximum forming depth of the fracture depth reveals the maximum wall angle that can be achieved for respective processes. Figure 10 shows the VWATC formed through ISF and ISH processes. Table 2 presents the maximum forming depth and maximum wall angle observed through VWATC during experimentation for both processes. The maximum forming depths observed for the ISF and ISH processes were 30.71 mm and 32.31 mm, respectively. The wall angles at these depths were observed to be 78.19° and 79.29°, respectively, for the ISF and ISH processes. The experimentation carried out to form the VWATC through both processes reveals that the higher forming depth was achieved with the ISH process compared to the ISF process. This leads to a higher wall angle achieved in the case of the ISH process. Thus, it authenticates that higher formability can be achieved for the ISH process.

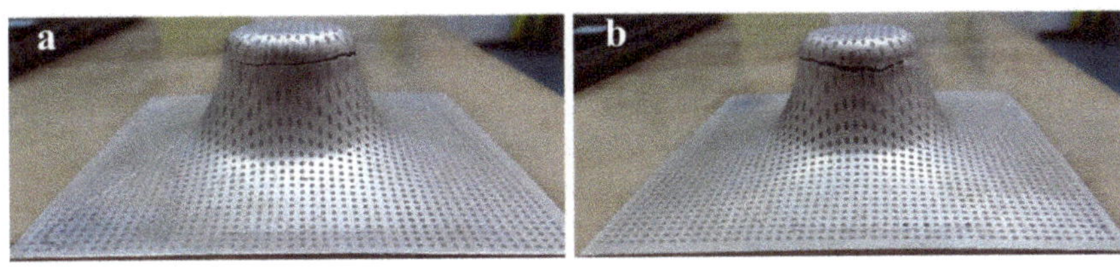

Figure 10. (**a**) VWATC formed through ISF process and (**b**) VWATC formed through ISH process.

Table 2. Observations from VWATC test geometry.

Process	Depth Achieved	Wall Angle Achieved
ISF	30.71 mm	78.19°
ISH	32.31 mm	79.29°

4.2. Results Based on the CWATC Test Geometry

The maximum wall angle achieved through experimentation of VWATC was 78.19° for the ISF process. Therefore, 78° CWATC were fabricated through both the processes to further validate that the higher formability associated with the ISH process. The forming time was 16.5 min for components formed through ISF process and it was 18.7 min for components formed through ISH process. Figure 11 shows the CWATC formed through ISF and ISH processes. The maximum forming depth achieved for both the processes is presented in Table 3.

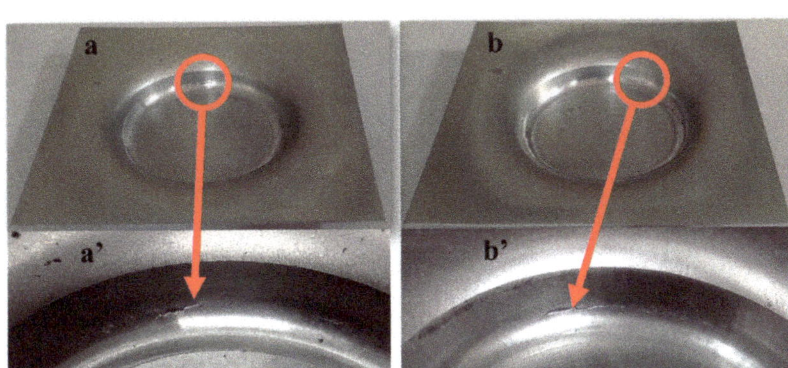

Figure 11. (**a**) CWATC formed through ISF process, (**a'**) Scaled fracture section in CWATC formed through ISF process (**b**) CWATC formed through ISH process, (**b'**) Scaled fracture section in CWATC formed through ISH process.

4.2.1. Geometrical Parameters

The experimentally formed components were scanned with the help of the ATOS Core 200 (GOM – a ZEISS company, Brunswick, Germany) scanner for the evaluation of geometrical accuracy, thickness distribution, and forming limits. The scanned components were evaluated on the GOM Inspect platform to study the geometric aspects of the component. Figure 12a,b shows the material thickness distribution in the component formed with ISF and ISH processes, respectively, whereas the scanned parts were sliced along the section lines to investigate the material thickness distribution. Figure 13 shows the material thickness distribution on any one side of the sliced section, where the thickness was varying at a similar position for both the formed components. Figure 13 shows that the slope of the thickness distribution is more in the case of components formed through the ISH process as compared to the ISF process. This indicates that the material availability during forming of the component was low in the ISF process as compared to the ISH process. Thus, a more uniform material distribution was observed in the component formed through the ISH process. The minimum value of sheet thickness was achieved in the case of components formed through the ISH process which leads to the delayed fracture of the material as compared to the components formed through the ISF process. It should be noted that the deformation in ISF is a combination of stretching, bending, and shearing [27]. The friction between the tool and the sheet is a major factor that causes the through-the-thickness-shear deformation in the ISF process [28], whereas in the ISH process, the friction between the tool and sheet is minimal because of the intermittent tool-sheet contact. Thus, it could be concluded that the hammering dominates the deformation in bending and stretching of the sheet. The numerical simulation was performed till the fracture depth for both the processes to compare the material thickness distribution for CWATC. Figure 14 shows the material thickness distribution obtained through the numerical simulation. The numerical simulation predicted the thickness distribution with the minimum thickness to be 0.2668 mm and 0.2672 mm for the ISF and ISH processes, respectively.

Table 3 shows the wall angle and depth achieved for the components formed through the ISF and ISH processes. The result obtained for the wall angle achieved was more favorable for the components formed through the ISF process than the ISH process, which indicates that the springback is more in the case of the ISH process than the ISF process. The forming depth obtained supports the formability remarks made in the case of VWATC. The effect of springback is the subject of further research for the ISH process.

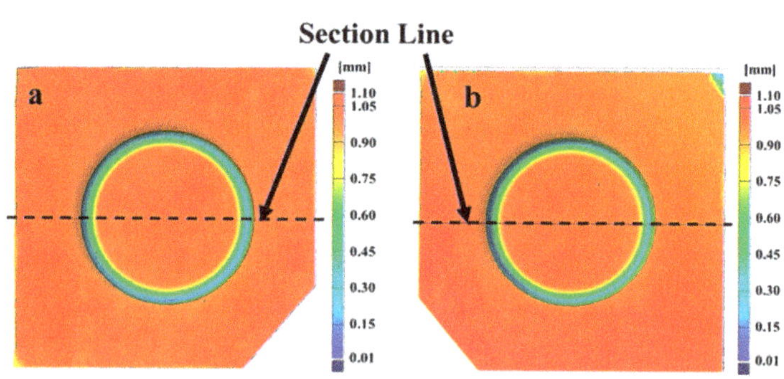

Figure 12. 3D scanned CWATC components formed with (**a**) ISF process, (**b**) ISH process.

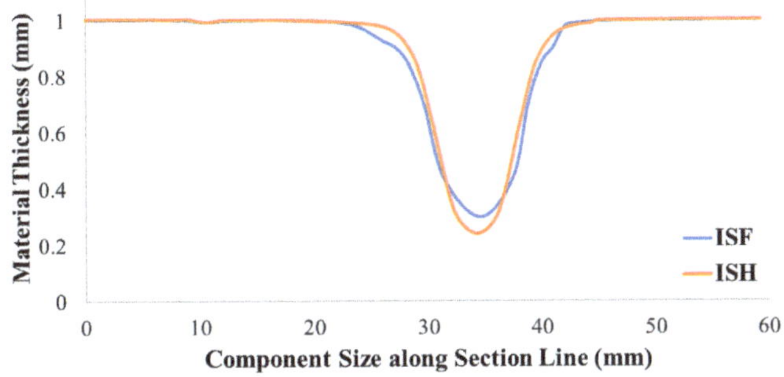

Figure 13. Material thickness distribution along the sliced section in the formed CWATC component.

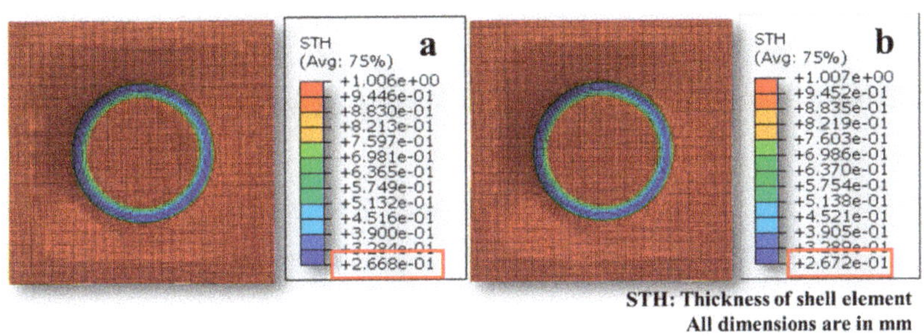

Figure 14. Material thickness distribution obtained through numerical simulation for the (**a**) ISF process, (**b**) ISH process.

Table 3. Wall angle and depth achieved in ISF and ISH processes for 78° wall angle.

Process	Depth Achieved	Wall Angle Achieved
ISF	13.62 mm	73.62°
ISH	13.80 mm	73.59°

4.2.2. Forming Limit Diagram

To evaluate the formability of ISF and ISH processes the forming limits were also evaluated through surface strains developed in the formed components. To measure the strains in the experimentally formed components, circular grids having the size of 1 mm diameter and center-to-center distance of 2 mm, were electrochemically marked on the surface of the sheet blank before experimentation. The experimentally formed components were scanned using an ATOS Core 200 scanner. ARGUS (GOM—a ZEISS company, Brunswick, Germany) digital image correlation (DIC) software was used to measure the surface strains. The distorted circles after forming help to calculate the major (maximum) and minor (minimum) surface strains using the following equations:

$$\varepsilon_{major} = ln\left(\frac{d_{major}}{d_o}\right) \tag{3}$$

$$\varepsilon_{minor} = ln\left(\frac{d_{minor}}{d_o}\right) \tag{4}$$

In the above equations, ε_{major} and ε_{minor} are the major and minor strains, while d_{major} and d_{minor} are the major and minor diameters of the ellipse formed. d_o is the diameter of the circle etched on the sheet before forming. Figure 15 shows the comparison of the forming limit diagrams (FLD) for both processes. The higher values of the strain were observed in the components formed through the ISH process compared to the ISF process. The maximum value of the major strain observed through ISF and ISH processes are 1.54 and 1.78 respectively. This indicates that the higher formability associated with the ISH process compared to the ISF process. Further, as visible from FLD that the strain data are lying in the first quadrant of the FLD indicating the plane strain condition. This implies that the conical objects are subjected to the plane strain condition during forming. The same observations are also reported in the literature [29,30].

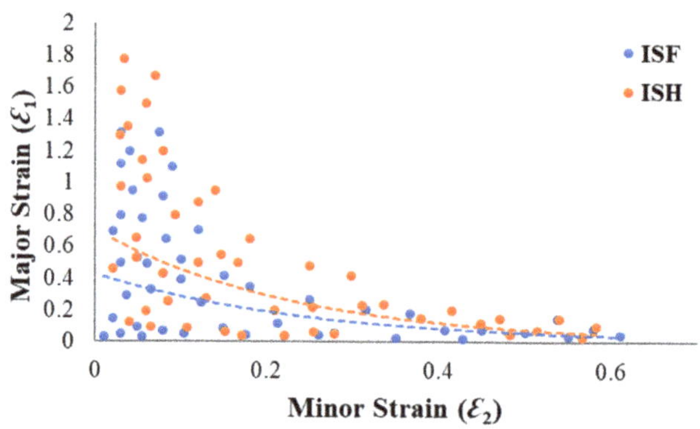

Figure 15. Comparison of strain values for the components formed through ISF and ISH processes.

4.2.3. Surface Roughness

The surface roughness of the formed components was evaluated using the Mitutoyo Surftest SJ-500 (Mitutoyo, Neuss, Germany). The resolution of the device was 0.0001 μm with a measurement accuracy of ±10%. The repeatability of the equipment was <6%. The lateral surface of the formed components was identified as the testing surface. Each test specimen was measured three times to assure the correctness of the results. Figure 16 and Table 4 presents the average roughness values (R_a) and maximum average roughness (R_{amax}) values obtained for components formed through ISF and ISH processes. It was observed that the average roughness value for components formed through the ISF process was 0.2860 μm, while it was lower, i.e., 0.2401 μm, for the ISH process. The maximum average roughness value was also lower for the ISH process, i.e., 0.2903 μm, as compared to the ISF process, i.e., 0.4492 μm. The range of mean roughness depth (R_z) for components formed through the ISF process is 0.8676–1.3316 μm, whereas it is 1.0839–1.4974 μm for the ISH process. In the ISH process, the tool applies compressive force through the hammering motion, which leaves the punched marks over the formed sheet surface. Thus, the range of mean roughness depth (R_z) is higher in the ISH process as compared to the ISF process.

In the ISF process, the deformation behavior is a combination of bending, stretching, and shearing, where the tool stretches the sheet along the predefined forming path. The stretching behavior along the forming path in ISF leads to large scale waviness resulting in large surface strains. Further, it resulted in roughness over the formed area; whereas, in the case of the ISH process, the intermittent hammering restricts stretching over the sheet and generates compressive strain on the punched area. This curtails the stretch marks on the formed area of the sheet. Thus, it can be concluded that the surface quality produced by the ISH process was better than that with the ISF process.

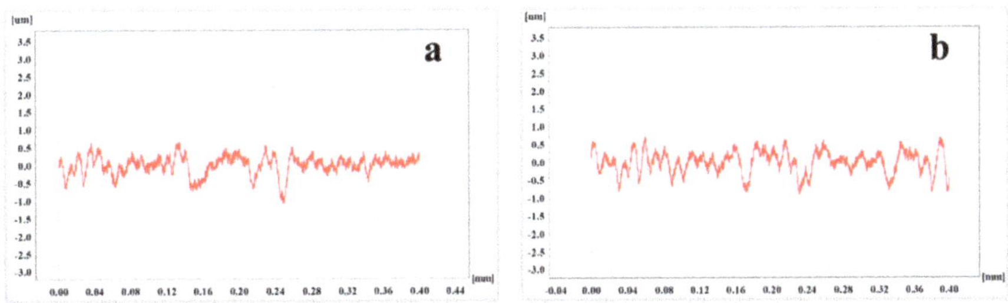

Figure 16. Surface roughness profiles of components formed through (**a**) ISF process, (**b**) ISH process.

Table 4. Surface roughness in the components formed with ISF and ISH processes.

Process	Average Roughness (R_a, µm)	Maximum Average Roughness (R_{amax}, µm)	Mean Roughness Depth (R_z, µm)
ISF	0.2860	0.4492	0.867–1.3316
ISH	0.2401	0.2903	1.0839–1.4974

4.2.4. Forming Forces

The forming force is an important parameter in the forming processes. The forming forces were studied under the same process parameters during simulations. The maximum forming forces in the horizontal direction (X and Y direction) and vertical direction (Z direction) observed during the simulation of ISF and ISF processes are presented in Table 5. The forming forces along Z-direction are 1350 N and 1225 N in ISF and ISH processes respectively. As the forming force in the Z direction is most significant during the incremental forming processes, therefore, the forming force in the Z direction for both the process have been plotted. Figure 17 shows the comparative forming forces in the Z direction observed during the simulation of ISF and ISH processes. As visible from the figure the trend of forming forces for ISH and ISF processes was increasing throughout the process; although, in the case of the ISH process, the force gets reduced after a certain forming depth compared to the ISF process. In the ISF process, the tool was in continuous contact with the sheet throughout the process. It results in a larger forming force as the deformation of the sheet must overcome the constraint impact of the entire sheet on the deformation area. In the ISH process, initially, the tool aligns itself to the first hammering position where it deforms the workpiece with predefined amplitude. In the next step, the tool retracts to its original position and then moves to the next hammering position and similarly deforms the workpiece. The same phenomenon was explained in Section 2.1 and shown with the help of Figure 2. In addition, when the tool deforms the sheet in two successive hammering positions, it forms an overlapping region, as mentioned earlier and shown in Figure 3a. In the present work, the overlapping region in the case of the ISH process was 25%, i.e., 1/4th of the trailing deformed area overlapped with the leading area. Further, the hammering amplitude considered in the present work is very small, i.e., 0.05 mm, which does not allow the tool to leave contact with the sheet. Due to the overlapping region, lower hammering amplitude, and spiral toolpath used in the present investigation, the force value does not fluctuate greatly; as the tool and sheet have intermittent contact, which leads to shorter deformation time, and it further leads to a reduction in forming forces. Thus, it signifies that the integration of hammering with the ISF process leads to the reduction of the forming forces compared to the ISF process.

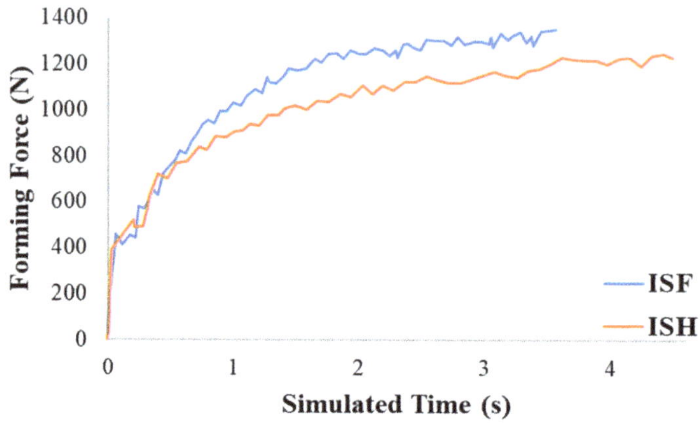

Figure 17. Forming forces in Z direction during ISF and ISH processes.

Table 5. Forming forces in the components formed with ISF and ISH processes.

Process	ISF	ISH
Forces in X-direction	1031 N	861 N
Forces in Y-direction	913 N	803 N
Forces in Z-direction	1350 N	1225 N

4.3. Statistical Analysis

An indicative statistical analysis is performed to understand the experimental results obtained for both the processes. Figure 18 shows the percentage change in the values obtained with the ISH process in comparison to that with the ISF process. The statistical analysis helps in understanding the comparison of the ISF and ISH processes.

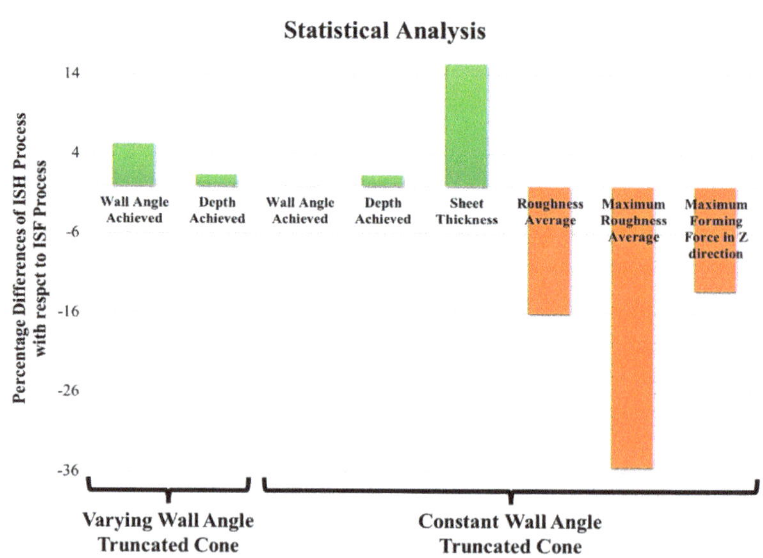

Figure 18. Experimental statistical analysis of the ISH process with respect to the ISF process.

The statistical analysis shows the differences mainly in terms of wall angle, forming depth, and surface quality of the components observed for the ISF and ISH processes.

The forming depth and wall angle while forming the VWATC through the ISH process were found to be 1.5% and 5.2% higher, respectively, compared to the ISF process. These initial results are the foundations for the CWATC components for further study. When the CWATCs of the same wall angle were experimentally formed through both the processes, only 0.1% difference is observed in the ISH process compared to the ISF process, whereas the achieved depth is observed to be higher by 1.32% in the components formed through the ISH process. The major differences are observed in the parameters such as sheet thickness, surface quality and forming forces. The thickness distribution was compared for both the processes at a forming depth of 11.59 mm because this is the maximum depth achieved during the ISF process. It was found that the minimum sheet thickness was 15.37% higher for the ISH process compared to the ISF process. The surface quality was observed to be better through visual inspection of the components formed through the ISH process. The average roughness and maximum average roughness were found to be 16% and 35.37% higher, respectively, in the case of components formed through the ISF process compared to the ISH process. The maximum values of forming forces in the Z direction were found to be 13.16% lower for the ISH process. Further, it was observed that the ISH process would lead to an overall 10.99% improvement in the quality of the parts, where the majority of the improvement is contributed by surface finish and forming forces.

5. Conclusions

This work presents the integration of the hammering operation with the incremental sheet forming process. The differences in terms of formability are investigated with the help of a VWATC test geometry. Further, the comparative experimental results of the CWATC investigate the geometrical aspects of both processes. It is observed that the introduction of the hammering effect during the ISF process imparts significant improvements, such as increased fracture depth, better surface roughness, and uniform thickness distribution, in the formed components. The concluding remarks based on the statistical analysis can be summarized as:

- The ISH process is an attempt to combine the effect of shot peening with the incremental sheet forming process. It would possibly result in the prevention of microcrack propagation because of the application of compressive forces which can delay the fracture initiation. The obtained results showcase that the fracture depth and wall angle achieved are more in the case of the ISH process compared to that in the ISF process.
- The comparison of both the processes through the strain values shows the higher formability associated with the ISH process compared to the ISF process. Further, it was also noticed that the conical objects are subjected to the plane strain condition.
- In ISF, the deformation mode of a material is a combination of bending, stretching, and shearing. Material deformation in ISH is dominated by direct bending and stretching which limits the reduction in workpiece thickness drastically. As a result of this, the ISH process exhibits a more uniform material thickness distribution as compared to the ISF process.
- The component formed through the ISH process reports better surface quality than the components formed through the ISF process.
- The reduced forming forces were attained while incorporating the hammering in the incremental sheet forming process.

The results in terms of wall angle achieved and fracture depth show the percentage differences of less than 5%, but the variation in sheet thickness and surface roughness shows promising results with more than 5% differences. The indicative statistical analysis shows that the ISH process would lead to an overall 10.99% improvement in the quality of the parts, primarily in the surface finish and forming forces. Further, the experiments are performed with one set of hammering parameters; thus, the proper choice and optimization of process parameters is a subject of further extensive investigation.

Author Contributions: H.Y.S., A.K.D. and P.K. conducted the numerical modelling, performed experiments, and wrote the first draft of the manuscript; P.T., A.P. and H.Y. provided the concept; D.P. and A.P. provided the testing data for numerical simulation; P.T. edited the draft of the manuscript and designed the overall simulation and the experimentation. All authors have read and agreed to the published version of the manuscript.

Funding: The study was carried out within the framework of the implementation of the Resolution of the Government of the Russian Federation of 9 April 2010, No. 220 (Contract No. 075-15-2019-869 from 12 May 2019) and by a grant of the Russian Science Foundation (project No. 20-69-46042 of 20 May 2020).

Institutional Review Board Statement: Not applicable.

Informed Consent Statement: Not applicable.

Data Availability Statement: The data sets used or analyzed during the current study are available with the corresponding author upon reasonable request.

Acknowledgments: This research is supported by the Department of Science and Technology (DST), India (Project No.: INT/RUS/RFBR/P-349) and Russian Foundation for Basic Research (RFBR), Russia. The authors wish to acknowledge the funding agency of both countries.

Conflicts of Interest: The authors declare no conflict of interest.

Appendix A

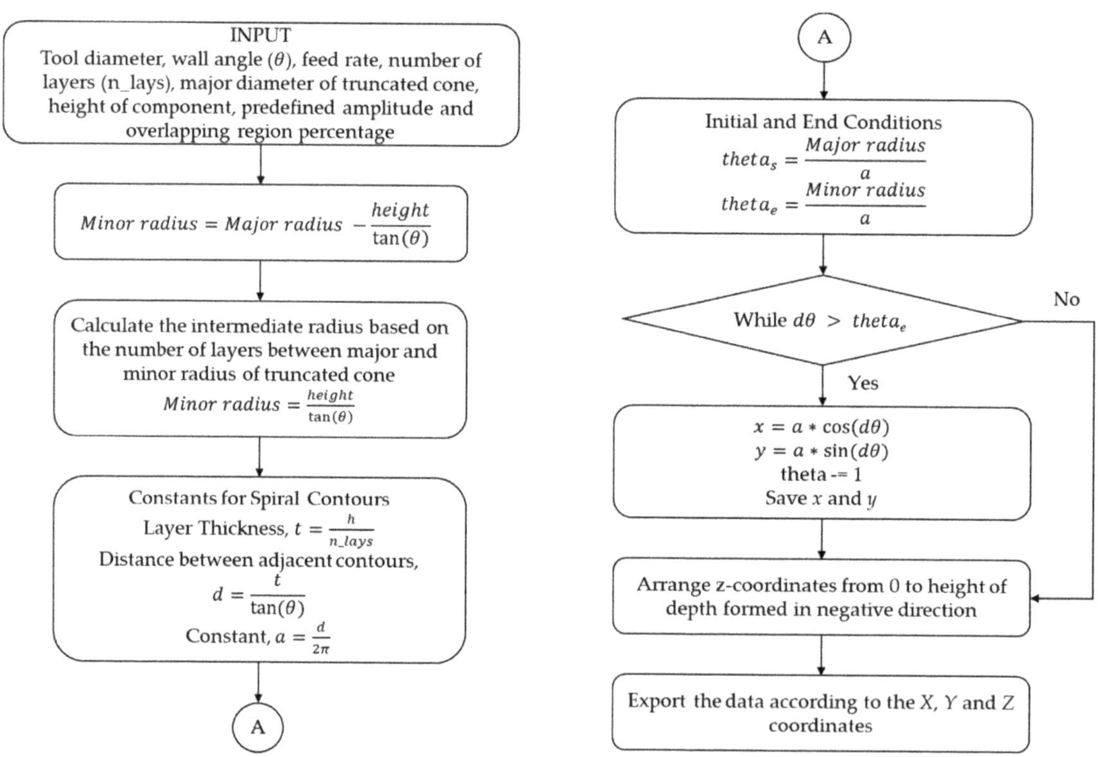

Figure A1. Flowchart of the generalized python code for CWATC.

References

1. Amino, M.; Mizoguchi, M.; Terauchi, Y.; Maki, T. Current status of "Dieless" Amino's incremental forming. *Procedia Eng.* **2014**, *81*, 54–62. [CrossRef]
2. Park, J.J.; Kim, Y.H. Fundamental studies on the incremental sheet metal forming technique. *J. Mater. Process. Technol.* **2003**, *140*, 447–453. [CrossRef]
3. Shrivastava, P.; Tandon, P. Microstructure and texture based analysis of forming behavior and deformation mechanism of AA1050 sheet during Single Point Incremental Forming. *J. Mater. Process. Technol.* **2019**, *266*, 292–310. [CrossRef]
4. Jeswiet, J.; Micari, F.; Hirt, G.; Bramley, A.; Duflou, J.; Allwood, J. Asymmetric single point incremental forming of sheet metal. *CIRP Ann.* **2005**, *54*, 88–114. [CrossRef]
5. Gupta, P.; Jeswiet, J. Manufacture of an aerospace component by single point incremental forming. *Procedia Manuf.* **2019**, *29*, 112–119. [CrossRef]
6. Kang, D.M.; Manabe, K.I. Improvement on the Formability of Magnesium Alloy Sheet by Heating and Cooling Method. *J. Korea Soc. Power Syst. Eng.* **2005**, *9*, 66–70.
7. Martins, P.A.F.; Bay, N.; Skjoedt, M.; Silva, M.B. Theory of single point incremental forming. *CIRP Ann.* **2008**, *57*, 247–252. [CrossRef]
8. Meier, H.; Smukala, V.; Dewald, O.; Zhang, J. Two point incremental forming with two moving forming tools. *Key Eng. Mater.* **2007**, *344*, 599–605. [CrossRef]
9. Silva, M.B.; Martins, P.A.F. Two-point incremental forming with partial die: Theory and experimentation. *J. Mater. Eng. Perform.* **2013**, *22*, 1018–1027. [CrossRef]
10. Liu, Z. Heat-assisted incremental sheet forming: A state-of-the-art review. *Int. J. Adv. Manuf. Technol.* **2018**, *98*, 2987–3003. [CrossRef]
11. Amini, S.; Gollo, A.H.; Paktinat, H. An investigation of conventional and ultrasonic-assisted incremental forming of annealed AA1050 sheet. *Int. J. Adv. Manuf. Technol.* **2017**, *90*, 1569–1578. [CrossRef]
12. Vihtonen, L.; Puzik, A.; Katajarinne, T. Comparing two robot assisted incremental forming methods: Incremental forming by pressing and incremental hammering. *Int. J. Mater. Form.* **2008**, *1*, 1207–1210. [CrossRef]
13. Luo, Y.; He, K.; Du, R. A new sheet metal forming system based on the incremental Punching, part 1: Modeling and simulation. *Int. J. Adv. Manuf. Technol.* **2010**, *51*, 481–491. [CrossRef]
14. Luo, Y.; He, K.; Du, R. A new sheet metal forming system based on incremental punching, part 2: Machine building and experiment results. *Int. J. Adv. Manuf. Technol.* **2010**, *51*, 493–506. [CrossRef]
15. Wang, J.; Wang, X.; Wang, Y.; Wang, M. Effects of Forming Parameters on Fatigue Life in Incremental Sheet Punching. *Materials* **2021**, *14*, 2308. [CrossRef] [PubMed]
16. Asgari, A.; Sedighi, M.; Riahi, M. Investigation of dimensional accuracy in incremental sheet metal hammering process: A parametric study. *Mech. Ind.* **2015**, *16*, 308. [CrossRef]
17. Bagheri, S.; Guagliano, M. Review of shot peening processes to obtain nanocrystalline surfaces in metal alloys. *Surf. Eng.* **2009**, *25*, 3–14. [CrossRef]
18. Schubnell, J.; Eichheimer, C.; Ernould, C.; Maciolek, A.; Rebelo-Kornmeier, J.; Farajian, M. The influence of coverage for high frequency mechanical impact treatment of different steel grades. *J. Mater. Process. Technol.* **2020**, *277*, 116437. [CrossRef]
19. Hussain, G.; Gao, L. A novel method to test the thinning limits of sheet metals in negative incremental forming. *Int. J. Mach. Tools Manuf.* **2007**, *47*, 419–435. [CrossRef]
20. Akin, J.E. *Finite Element Analysis Concepts: Via SolidWorks*; World Scienfic: Hackensack, NJ, USA, 2010.
21. Chung, W.; Kim, B.; Lee, S.; Ryu, H.; Joun, M. Finite element simulation of plate or sheet metal forming processes using tetrahedral MINI-elements. *J. Mech. Sci. Technol.* **2014**, *28*, 237–243. [CrossRef]
22. Li, K.P.; Carden, W.P.; Wagoner, R.H. Simulation of springback. *Int. J. Mech. Sci.* **2002**, *44*, 103–122. [CrossRef]
23. Han, F.; Mo, J.H. Numerical simulation and experimental investigation of incremental sheet forming process. *J. Cent. South Univ. Technol.* **2008**, *15*, 581–587. [CrossRef]
24. Zaydes, S.A. New surface plastic deformation techniques in the manufacture of machine parts. *Vestnik Magnitogorskogo Gosudarstvennogo Tekhnicheskogo Universiteta im. G.I. Nosova [Vestnik of Nosov Magnitogorsk State Technical University]* **2018**, *16*, 129–139. [CrossRef]
25. Belevskiy, L.S.; Belevskaya, I.V.; Efimova, Y.Y. Study of the process of surface cold working with punching blocks. *Vestnik Magnitogorskogo Gosudarstvennogo Tekhnicheskogo Universiteta im. G.I. Nosova [Vestnik of Nosov Magnitogorsk State Technical University]* **2016**, *14*, 63–68. [CrossRef]
26. Mohammadi, A.; Vanhove, H.; Van Bael, A.; Duflou, J.R. Towards accuracy improvement in single point incremental forming of shallow parts formed under laser assisted conditions. *Int. J. Mater. Form.* **2016**, *9*, 339–351. [CrossRef]
27. Jackson, K.; Allwood, J. The mechanics of incremental sheet forming. *J. Mater. Process. Technol.* **2009**, *209*, 1158–1174. [CrossRef]
28. Lu, B.; Fang, Y.; Xu, D.K.; Chen, J.; Ou, H.; Moser, N.H.; Cao, J. Mechanism investigation of friction-related effects in single point incremental forming using a developed oblique roller-ball tool. *Int. J. Mach. Tools Manuf.* **2014**, *85*, 14–29. [CrossRef]
29. Kumar, P.; Tandon, P. Process capabilities of commercially pure titanium grade 2 formed through warm incremental sheet forming. *Proc. Inst. Mech. Eng. Part B J. Eng. Manuf.* **2021**, in press. [CrossRef]
30. Said, L.B.; Mars, J.; Wali, M.; Dammak, F. Numerical prediction of the ductile damage in single point incremental forming process. *Int. J. Mech. Sci.* **2017**, *131*, 546–558. [CrossRef]

metals

MDPI

Article

Influence of the Small Sc and Zr Additions on the As-Cast Microstructure of Al–Mg–Si Alloys with Excess Silicon

Evgenii Aryshenskii [1], Maksim Lapshov [1], Juergen Hirsch [1], Sergey Konovalov [1,*], Viacheslav Bazhenov [2], Alexander Drits [1] and Denis Zaitsev [3]

1 Department of Metals Technology and Aviation Materials, Samara National Research University, Moskovskoye Shosse 34, 443086 Samara, Russia; ar-evgenii@yandex.ru (E.A.); lapshov.m.syz@gmail.com (M.L.); juergen.hirsch@hydro.com (J.H.); Alexander.Drits@arconic.com (A.D.)
2 Casting Department, National University of Science and Technology (MISIS), Leninskiy Prospekt 4, 119049 Moscow, Russia; v.e.bagenov@gmail.com
3 Laboratory of Metallophysical Research, All-Russian Scientific Research Institute of Aviation Materials, Radio Street 17, 105005 Moscow, Russia; sale@isp.viam.ru
* Correspondence: ksv@ssau.ru; Tel.: +7-846-2674-640

Abstract: This research is devoted to the study effects of complex alloying of Al-0.3 wt. % Mg-1 wt. % Si and Al-0.5 wt. % Mg-1.3 wt. % Si alloys by small additions of Sc and Zr on the microstructure in the as-cast condition. The effect of small additions of these elements on the microhardness, electrical conductivity, grain size and phase composition of the indicated alloy systems was studied. The methods of optical and electron microscopy were used for the study. Moreover, the phase composition was calculated using the Thermo-Calc software package. The study showed a strong effect of the chemical composition of investigated alloys on the size of the grains, which, with a certain combination of additives, can decrease several times. Grain refinement occurs both due to supercooling and formation of primary Al_3Sc particles in the liquid phase. Alloys based on Al-0.5 wt. % Mg-1.3 wt. % Si are more prone to the formation of this phase since a lower concentration of Sc is required for it to occur. In addition, more silicon interacts with other elements. At the same time, Al-0.3 wt. % Mg-1 wt. % Si requires lower temperature for complete dissolution of Mg_2Si, which can contribute to more efficient heat treatment, which includes reducing the number of steps. TEM data show that during ingot cooling $(AlSi)_3ScZr$ dispersoid precipitates. This dispersoid could precipitate as coherent and semi-coherent particles with $L1_2$ structure as well as needle-shaped particles. The precipitation of coherent and semi-coherent particles during cooling of the ingot indicates that they can be obtained during subsequent multistage heat treatment. In addition, in the $Al_{0.5}Mg_{1.3}Si_{0.3}Sc$ alloy, metastable β″ (Mg_5Si_6) are precipitated.

Keywords: aluminum alloys; phase composition; casting; Sc and Zr small addition; grain refinement; dispersoid

Citation: Aryshenskii, E.; Lapshov, M.; Hirsch, J.; Konovalov, S.; Bazhenov, V.; Drits, A.; Zaitsev, D. Influence of the Small Sc and Zr Additions on the As-Cast Microstructure of Al–Mg–Si Alloys with Excess Silicon. *Metals* **2021**, *11*, 1797. https://doi.org/10.3390/met11111797

Academic Editor: Elisabetta Gariboldi

Received: 14 October 2021
Accepted: 5 November 2021
Published: 8 November 2021

1. Introduction

Aluminum alloys are widely used in modern industry due to their combination of low density and high ductility [1–10]. Nowadays, aluminum alloys of the 6XXX series (Al–Mg–Si) are gaining popularity [11–18]. This series has an average strength that can be increased by heat treatment, namely by metastable β″ (Mg_5Si_6) particles coherent to the aluminum matrix.

Further improvement of the properties of this alloy series is possible due to small additions of Sc and Zr. These elements are poorly soluble in the aluminum matrix and form coherent strengthening $Al_3(ScZr)$ nanoparticles [19–21]. In addition, these additives contribute to grain refinement during casting [19–21], and also contribute to the control of the recrystallization process [21] which also increases the mechanical properties. Alloying with zirconium and scandium is especially common for 5XXX group alloys, i.e., with a high content of magnesium [22,23], however, it is effectively used also in other aluminum alloy systems [24]. In recent years, active research on the use of these additives in 6XXX series alloys has been started [25–28]. In the future, it is possible to achieve a significant increase in strength due to the combined effect of grain refinement, and the simultaneous usage of strengthening particles $Al_3(ScZr)$ and β″ (Mg_5Si_6). At the same time, Sc alloying of this group is a rather difficult task. The reason is that silicon addition changes the kinetics and chemical composition of scandium-containing dispersoids. As a result it provokes the appearance of an undesirable, $AlSc_2Si_2$ phase [25]. Because of this, some sources have noted a negative effect of scandium on the properties of the Al–Mg–Si alloys [25,29]. In order to avoid $AlSc_2Si_2$, it is necessary to carefully select the chemical composition (mainly by changing the amount of zirconium) and control the casting and heat treatment modes. In this case, due to joint scandium-zirconium alloying, it is possible to improve the properties of the 6XXX group alloys even if the ratio of the wt. fractions is Mg/Si < 1.73 [25,30,31]. In alloys of the Al–Mg–Si system with ratio Mg/Si < 1.73, not all silicon combines with magnesium into the Mg_2Si phase (or its metastable modifications such as β″) and there is an excess of it. It should be noted that, on the one hand, the increased silicon content enhances the properties in this group of alloys, due to the more evenly distributed and more finely dispersed strengthening particles β″ [17]. However, on the other hand, with the addition of scandium, excess silicon will increase the tendency to form $AlSc_2Si_2$. At the same time, the study of the processes occurring during casting and heat treatment of 6XXX group alloys with scandium-zirconium additives made it possible to avoid the undesirable phase and, in general, to improve the properties at a ratio Mg/Si > 1.73 in AA6082 [26,30,32]. The possibility of further decreasing this ratio is promising since it will allow scandium and zirconium to be added to alloys such as 6016 where the excess of silicon is extremely large. However, it requires a thorough study on the effect of casting and thermomechanical treatment on the microstructure [26,30]. This article is devoted to the study of microstructure formation during casting of the 6XXX group alloys with a high silicon content and complex scandium and zirconium additives, especially in the range of element concentrations for which this problem has not yet be investigated.

2. Experimental Methods

2.1. Material Preparation

Two alloys with a low Mg/Si ratio: 0.3 for $Al_{0.3}Mg_1Si$ and 0.38 for $Al_{0.5}Mg_{1.3}Si$ were chosen for the study. Both alloys were alloyed with small additions of Zr and Sc in the following combinations: Al-$_{0.3}$Mg-$_1$Si-$_X$Sc-$_Y$Zr and Al-$_{0.5}$Mg-$_{1.3}$Si-$_X$Sc-$_Y$Zr alloys, where X = 0, 0.1, 0.2, 0.3 wt. % and Y = 0, 0.05, 0.10, 0.15 wt. %. Its chemical composition are shown in Table 1.

Melting of alloys was carried out in a medium-frequency induction furnace using a graphite crucible, the melt weight was 4–5 kg. The weight of the cast ingot was 3 kg. The following materials were used as a charge for the alloy: aluminum (purity 99.8%), magnesium (99.9%), Al-2%Sc master alloy, Al-12%Si master alloy, and Al-5%Zr master alloy. The casting temperature was 720–740 °C. Before pouring the molten metal into the casting mold, it was refined with a carnallite flux introduced at the rate of 5 g per 1 kg of charge. After that, the metal was poured into a steel mold at a uniform pouring time of 40 s. The obtained ingot size was 135 mm × 30 mm × 200 mm and its cooling speed was 2 °C/s. After solidification, the ingot was removed from the mold and cooled in water.

2.2. Material Characterization

Micro-hardness tests were carried out on a digital stationary hardness tester according to the micro-Vickers method on the HV-1000 model (test forces 0.2452 N). In order to exclude the influence of grain boundaries and large intermetallic compounds, measurements were made in the grain center. Measurements of microhardness by indentation were carried out in accordance with GOST 9450-76 [33].

Table 1. Chemical composition of the investigated alloys.

Sample Number № 1	Chemical Composition					Sample Number № 2	Chemical Composition				
	Al	Mg	Si	Sc	Zr		Al	Mg	Si	Sc	Zr
1	base	0.3	1.0	0	0	1	base	0.5	1.3	0	0
2	base	0.3	1.0	0	0.05	2	base	0.5	1.3	0	0.05
3	base	0.3	1.0	0	0.1	3	base	0.5	1.3	0	0.1
4	base	0.3	1.0	0	0.15	4	base	0.5	1.3	0	0.15
5	base	0.3	1.0	0.1	0	5	base	0.5	1.3	0	0
6	base	0.3	1.0	0.1	0.05	6	base	0.5	1.3	0.1	0
7	base	0.3	1.0	0.1	0.1	7	base	0.5	1.3	0.1	0.05
8	base	0.3	1.0	0.1	0.15	8	base	0.5	1.3	0.1	0.1
9	base	0.3	1.0	0.2	0	9	base	0.5	1.3	0.1	0.15
10	base	0.3	1.0	0.2	0.05	10	base	0.5	1.3	0.2	0.05
11	base	0.3	1.0	0.2	0.1	11	base	0.5	1.3	0.2	0.1
12	base	0.3	1.0	0.2	0.15	12	base	0.5	1.3	0.2	0.15
13	base	0.3	1.0	0.3	0	13	base	0.5	1.3	0.3	0
14	base	0.3	1.0	0.3	0.05	14	base	0.5	1.3	0.3	0.05
15	base	0.3	1.0	0.3	0.1	15	base	0.5	1.3	0.3	0.1
16	base	0.3	1.0	0.3	0.15	16	base	0.5	1.3	0.3	0.15

The grain structure of the samples was examined using a Carl Zeiss Axiovert-40 MAT optical microscope (Carl Zeiss AG, Oberkochen, Germany). The preparation of microsections included cutting out samples, mechanical grinding, polishing, as well as electropolishing in a fluoroboric electrolyte of the following composition: boric acid—11 g, hydrofluoric acid—30 mL, distilled water—2200 mL. For each sample, the mean grain size was measured by the linear intercept method.

Intermetallic particles were investigated using a JEOL 6390A SEM. The sample preparation technique consisted of mechanical grinding, polishing and electropolishing. Electropolishing was carried out at a temperature of 85–110 °C and a voltage of 10–30 V in an electrolyte of the following composition: 500 mL of H_3PO_4; 300 mL H_2SO_4; 50 g CrO_3; 50 mL H_2O. Study of the chemical composition of structural components by energy dispersive spectroscopy using the X-Max 80T detector in the energy range 0–10 keV (the energy resolution of the detector is 122 eV).

For effective phase counting, the image was made with maximum contrast until the appearance of "noise" points. The ImageJ program (1.52 u, National Institutes of Health, Wayne Rasband (NIH), Bethesda, MD, USA) was used to process the image to the desired contrast of the aluminum matrix and secondary phases.

The study was carried out on the Tecnai G2 F20 S-TWIN TMP transmission electron microscope with a thermal-field cathode at accelerating voltage of 200 kV. The chemical composition of the fine dispersion participles was studied by energy-dispersive spectroscopy (EDS) method using the X-Max 80T detector in the energy range 0–10 keV. The energy resolution of the detector is 122 eV.

The sample preparation for the study was carried out in the following sequence: cutting out samples with a diameter of 3 mm and a thickness of 0.3 mm on the Art's electric discharge machine, mechanical thinning on SiC paper from both sides, electrolytic polishing on a Tenupol-5 installation in a nitrogen-alcohol electrolyte. Polishing mode: −38 °C, 16 V.

2.3. Phase Diagram Calculation

The polythermal sections of phase diagrams was calculated using the Thermo-Calc 2016a software. Thermodynamic database TCAL4 (Thermo-Calc Software Al-based alloy database, Stockholm, Sweden, Version 4.0) was used [34].

3. Results

3.1. Grain Refinement

The results of the grain structure investigation showed very large dendritic grains, the sizes of which were several millimeters (Figure 1a), in the $Al_{0.3}Mg_1Si$ alloy without the addition of any transition elements. This is generally characteristic of low-alloyed alloys [35]. Upon the addition of Sc (Figure 1b) and Zr (Figure 1c), the size of the grain structure continues to decrease but not significantly. The scandium concentration is 0.6% which is required to achieve a hypereutectic composition and the formation of primary Al_3Sc particle in the liquid [21]. Therefore, in $Al_{0.3}Mg_1Si_{0.3}Sc$ alloy grain refinement occurs due to supercooling at the boundaries of the nuclei and the liquid phase [36]. The concentration of zirconium $Al_{0.3}Mg_1Si_{0.15}Zr$ reaches a peretectic point of 0.11% [36]. However, significant grain refinement is not observed. It is explained by the fact, that according to [36] at the given concentration, the primary particles of Al_3Zr only begin to form in the liquid and a significant grain-refinement effect is observed at a concentration of 0.2%.The concentration of 0.3 Sc and 0.15 Zr is sufficient for the hypereutectic composition and the formation of primary Al_3Sc particles in the liquid phase, which leads to a significant appearance of a fine-grained equiaxed structure during casting (Figure 1d).

Figure 1. Grain structure after casting (**a**) $Al_{0.3}Mg_1Si$, (**b**) $Al_{0.3}Mg_1Si_{0.3}Sc$, (**c**) $Al_{0.3}Mg_1Si_{0.15}Zr$, (**d**) $Al_{0.3}Mg_1Si_{0.3}Sc_{0.15}Zr$.

A dendritic structure is observed in the $Al_{0.5}Mg_{1.3}Si$ alloy (Figure 2a), as well as in $Al_{0.3}Mg_{1}Si$. However, the grain size in $Al_{0.5}Mg_{1.3}Si$ is two times smaller. This effect is caused by an increase in the content of Mg and Si, which also leads to supercooling between the nuclei and the liquid phase. Additions of scandium (Figure 2b) or zirconium (Figure 2c) alone only slightly affect the grain size. Their combined use leads to the appearance of fine equiaxed grains which, however, only partially replace the dendritic structure (Figure 2d). The decrease in alloying efficiency will be explained later.

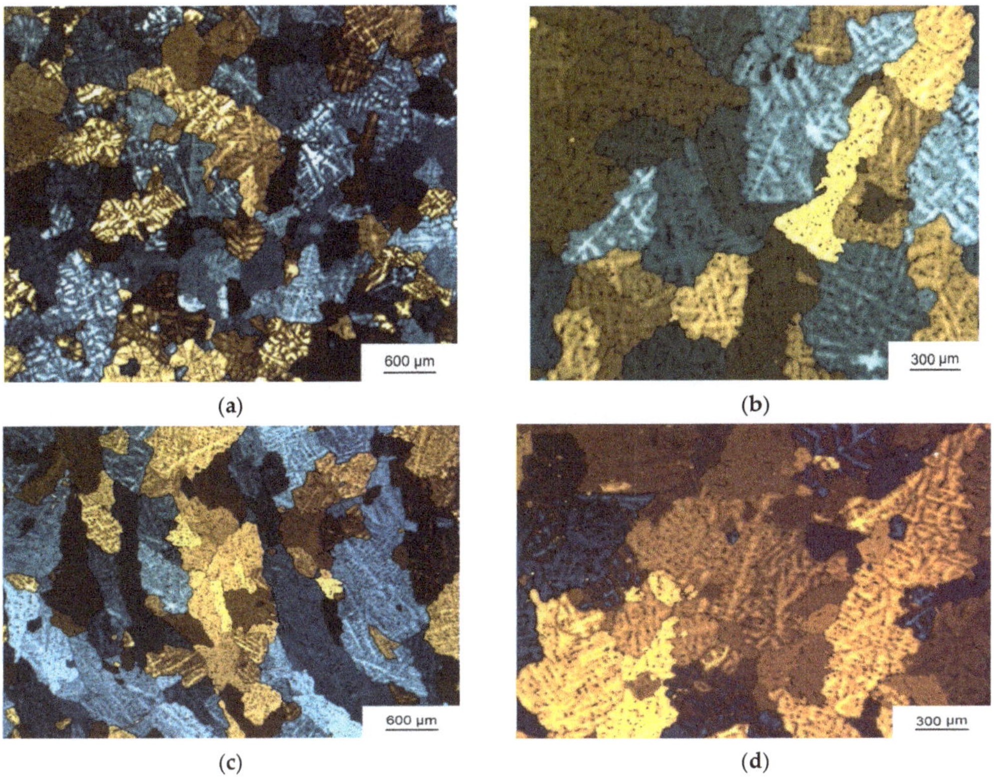

Figure 2. Grain structure after casting (**a**) $Al_{0.5}Mg_{1.3}Si$ (**b**) $Al_{0.5}Mg_{1.3}Si_{0.3}Sc$, (**c**) $Al_{0.5}Mg_{1.3}Si_{0.3}Zr$, (**d**) $Al_{0.5}Mg_{1.3}Si_{0.3}Sc_{0.15}Zr$.

Thus, a gradual grain refinement occurs during complex alloying with small additions of zirconium and scandium. The decrease in the grain size is explained, first, by the fact that the growing content of scandium and zirconium causes supercooling between the liquid and the surface of the nuclei formed during solidification thereby facilitating the passing of the latter [36]. This also explains the fact that the grain in $Al_{0.5}Mg_{1.3}Si$ (Figure 3a) is finer than in $Al_{0.3}Mg_{1}Si$ (Figure 3b) after solidification, since the total number of alloying elements increases. With a further increase in the Sc and Zr contents, an almost 10-fold decrease in grain size occurs, mainly because the eutectic composition for Al-Sc-Zr system is reached [37]. As already mentioned, due to the high concentration of Zr and Sc in the $Al_{0.5}Mg_{1.3}Si$ alloys, the grain refinement is more effective than in $Al_{0.3}Mg_{1}Si$. An explanation of this fact will be given later. Note, however, that no appreciable refinement occurs with an increase in Zr or Sc alone. This is due to the fact that a scandium content of 0.6% is required for the hypereutectic composition and significant grain refinement. At the same time, an increase in Zr content to 0.15% reduces the scandium content which is necessary for the hypereutectic composition to 0.2%. It should be noted that the amount of zirconium required for the overflow reaction of the zirconium content should be about

0.26%. Therefore, it is not efficient for a significant refinement of the grain structure during casting in the concentration range studied here.

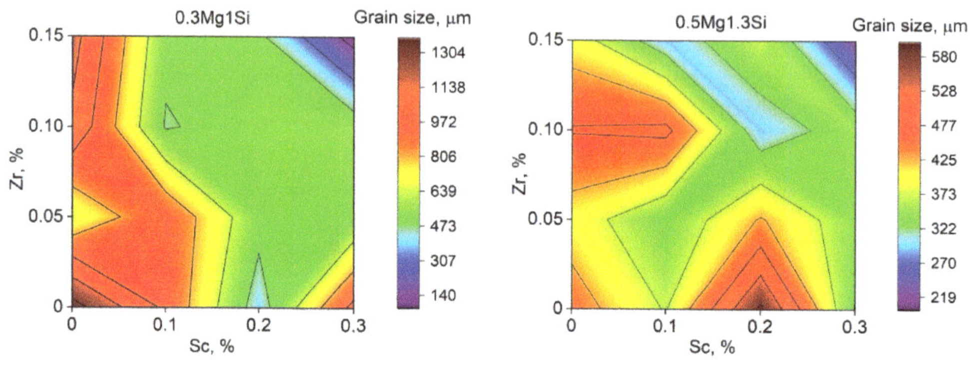

Figure 3. Grain size at complex alloying with small Sc and Zr additives (**a**) in alloys based on $Al_{0.3}Mg_1Si$, (**b**) on $AL_{0.5}Mg_{1.3}Si$.

3.2. Microhardness Behavior

The results of microhardness measurements are very interesting. It is generally higher (Figure 4a) in an alloy with a higher silicon content than in an alloy with a low silicon content (Figure 4b). This can be explained by the fact that alloy $Al_{0.3}Mg_1Si$ contains more particles of the type $(AlSi)_3ScZr$ and/or $AlSc_2Si_2$ than $Al_{0.5}Mg_{1.3}Si$. They can appear due to the fact that the cooling process during casting requires a certain time, which is sufficient for the precipitation of a certain amount of these particles from the supersaturated solid solution [26]. An increased ratio of magnesium and silicon will accelerate the precipitation of these particles from the supersaturated solid solution and thereby increase the microhardness. At the same time, an increase in the zirconium content will reduce the ability to form them upon cooling, since this element reduces the diffusion of scandium [19]. As for the $Al_{0.5}Mg_{1.3}Si$ alloy (Figure 4b), it can be assumed that grain refinement, associated with an increase in the proportion of zirconium and scandium, will have a stronger effect than a decrease in the diffusion of the latter. As mentioned earlier, due to the larger amount of Mg and Si, the refinement of the grain structure will be much stronger. Therefore, the length of grain boundaries is longer. The latter, in turn, serve as an additional source in the formation of secondary metastable particles, contributing to an increase in their number [38]. Note that the effect of the grain structure size can be excluded, since this indicator begins to exert a significant effect on the microhardness when reaching submicron sizes. Even in alloys with the finest grains its size is several microns.

3.3. Phase Diagram Calculation Result

Figure 5a,b shows polythermal sections of the systems Al-0.3%Mg-1%Si-0.15%Zr-(0–0.4)%Sc (wt. %) and Al-0.5%Mg-1.3%Si-0.15%Zr-(0–0.4)%Sc. Note that alloys with a combined content of scandium and zirconium seem to be the most promising. Firstly, they allow to refine the grain structure which will positively affect the mechanical properties and, secondly, slow down the diffusion of scandium and foster the formation of metastable nanoparticles $(AlSi)_3ScZr$ coherent to aluminum matrix, instead of harmful $AlSc_2Si_2$. Therefore, polythermal sections with the highest zirconium content are considered. On the whole, phase transformations in both groups of alloys (with different Mg/Si ratios) have common laws. Slightly below 700 °C, primary Al_3Zr particles begin to precipitate from the liquid. Three phases L + Al_3Zr + (Al) are simultaneously present in a narrow range between 653–650 °C in $Al_{0.3}Mg_1Si$ (Figure 5a) and 650–647 °C in $Al_{0.5}Mg_{1.3}Si$ (Figure 6b). After that, Al_3Zr dissolves and the ZrSi phase appears instead. With Sc content of up to 0.23%

in $Al_{0.3}Mg_1Si$ (Figure 5a) and up to 0.26% in $Al_{0.5}Mg_{1.3}Si$ (Figure 5b), the ZrSi phase is transformed into Zr_2Si. At zero scandium content, upon reaching a temperature of 601 °C in $Al_{0.3}Mg_1Si$ (Figure 5a) and 578 °C in $Al_{0.5}Mg_{1.3}Si$ (Figure 5b), the metal finally solidifies. After that an aluminum solid solution and the $ZrSi_2$ phase can be observed in the metal. With an increased concentration of scandium, the solidus temperature gradually decreases up to 0.14% and 0.07% of Sc for $Al_{0.3}Mg_1Si$ and $Al_{0.5}Mg_{1.3}Si$ alloys respectively and at higher concentration of Sc the solidus temperatures is slightly increased. When the Sc concentration reaches 0.21% in $Al_{0.3}Mg_1Si$ (Figure 5a) and 0.25% $Al_{0.5}Mg_{1.3}Si$ (Figure 5b), L + (Al) + $ZrSi_2$ first transforms into L + (Al) + $ZrSi_2$ + Al_3Sc and then finally is solidified as (Al) + $ZrSi_2$ + Al_3Sc. When scandium reaches 0.24% in the $Al_{0.3}Mg_1Si$ alloys (Figure 6a) and 0.27% in $Al_{0.5}Mg_{1.3}Si$ (Figure 6b), solidification occurs according to the following scheme: L + (Al) + $ZrSi_2$ → L + (Al) + $ZrSi_2$ + Al_3Sc → L + (Al) + $ZrSi_2$ + SiSc → (Al) + $ZrSi_2$ + Al_3Sc. Upon reaching Sc 0.24% and 0.28% in the alloys $Al_{0.3}Mg_1Si$ (Figure 5a) and $Al_{0.5}Mg_{1.3}Si$ (Figure 5b), solidification occurs along an even more complicated path: L + (Al) + ZrSi → L + (Al) + ZrSi + Al_3Sc → L + (Al) + $ZrSi_2$ + Al_3Sc → L + (Al) + $ZrSi_2$ + SiSc → (Al) + $ZrSi_2$ + SiSc. In addition, in a short interval between 0.23 ÷ 0.24% for $Al_{0.3}Mg_1Si$ and 0.27 ÷ 0.28%, solidification occurs according to the following scheme: L + (Al) + ZrSi → L + (Al) + $ZrSi_2$ + Al_3Sc → L + (Al) + $ZrSi_2$ + SiSc → (Al) + $ZrSi_2$ + SiSc.

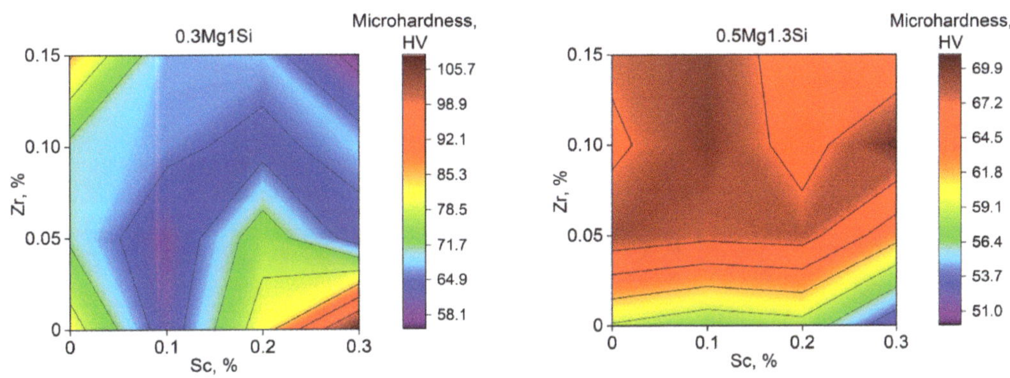

(a) (b)

Figure 4. Microhardness with complex alloying with small additives Sc and Zr (**a**) in alloys based on $Al_{0.3}Mg_1Si$, (**b**) on $Al_{0.5}Mg_{1.3}Si$.

It should be noted that the appearance of primary particles in the liquid phase will facilitate grain refinement but not all of them will be equally effective. Primary Al_3Zr, and especially Al_3Sc, are effective inoculators due to the good correspondence between them and the aluminum matrix. There is no such information about $ZrSi_2$ and SiSc particles. However, it can be assumed that their effectiveness is rather low.

Note that in the $Al_{0.3}Mg_1Si_{0.3}Sc_{0.15}Zr$ alloy the combined grain refinement effect of zirconium and scandium does not differ from that observed in pure aluminum and alloys with a high magnesium content [37]. However, in the $Al_{0.5}Mg_{1.3}Si_{0.3}Sc_{0.15}Zr$ alloy at concentrations sufficient to obtain a fine grain structure in pure aluminum and alloys with high magnesium content, a large number of dendritic grains are still observed. A first explanation for this is that Al_3Sc and Al_3Zr particles are not present at all stages of solidification. They will have a slightly weaker effect than in high-magnesium alloys. It should also be noted that Al_3Sc particles, which contribute to efficient grain refinement, act even more efficiently in $Al_{0.3}Mg_1Si$ alloys because their formation requires a lower scandium content than in $Al_{0.5}Mg_{1.3}Si$. This explains the fact that the grain is larger in this group of alloys with a low content of Zr and Sc than in $Al_{0.5}Mg_{1.3}Si$ with a similar content of these elements. However, the grain in $Al_{0.3}Mg1Si$ is refined better with an increase in their concentration. At a low concentration of Sc and Zr, the reason for the finer grain in the case of $Al_{0.5}Mg_{1.3}Si$ is the aforementioned supercooling at the interface between the nuclei and the liquid phase (arising from the dissolved Mg and Si). However, due to the lower Sc concentration required for the formation of Al_3Sc particles during metal solidification, the refinement of the grain structure at high scandium concentrations is more efficient in $Al_{0.3}Mg_1Si$.

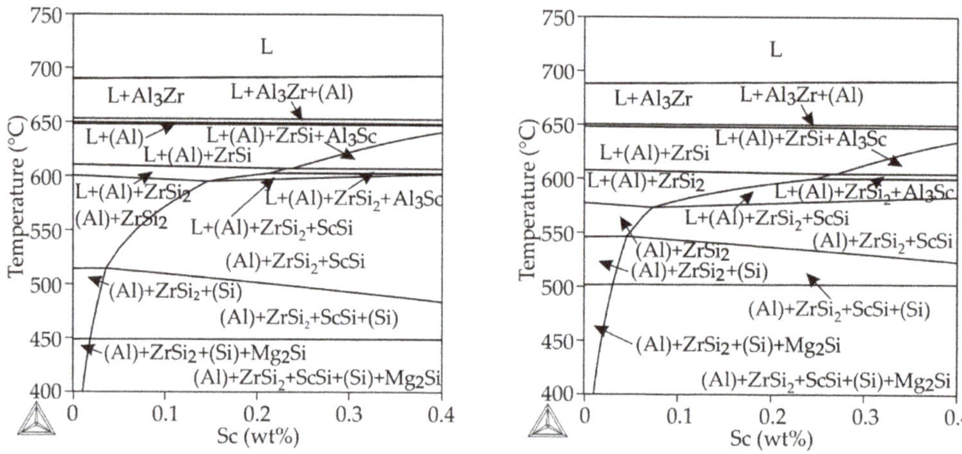

(a) (b)

Figure 5. Polythermal sections (**a**) Al-0.3%Mg-1%Si-0.15%Zr-(0–0.4)% Sc (wt. %) and (**b**) Al-0.5%Mg-1.3%Si-0.15%Zr-(0–0.4)%Sc (wt. %) of the Al–Mg–Si–Zr–Sc phase diagram calculated using Thermo-Calc software.

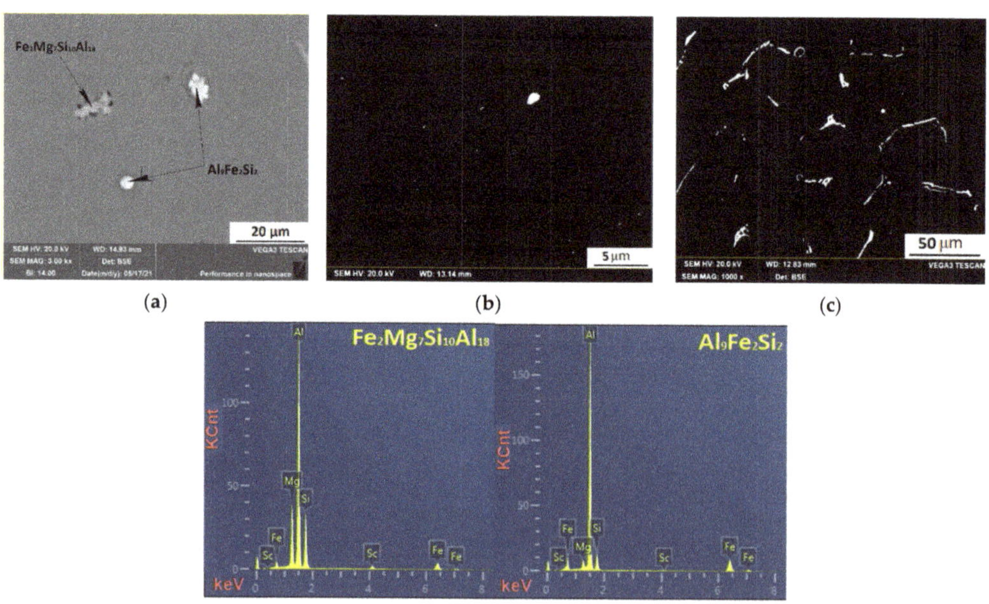

(a) (b) (c)

(d)

Figure 6. $Al_{0.3}Mg_1Si_{0.3}Sc$ main types of intermetallic compounds (**a**); small dispersoids (**b**); large particles (**c**); result of the energy-dispersive spectroscopy (EDS) investigation of chemical composition of the intermetallic compounds (**d**).

After solidification in both types of alloy, a similar chain of phase transformations is observed. At a low concentration of scandium, the following phase transformations will occur: (Al) + $ZrSi_2$ → (Al) + $ZrSi_2$ + (Si) → (Al) + $ZrSi_2$ + (Si) + Mg_2Si. At a higher concentration of scandium, the following transformations will be observed: (Al) + $ZrSi_2$ + SiSc, (Al) + $ZrSi_2$ + SiSc + (Si); (Al) + $ZrSi_2$ + SiSc + (Si) + Mg_2Si. It may seem that the $Al_{0.5}Mg_{1.3}Si$ system is more promising from the point of view of further use, since it has a larger Mg/Si relationship than $Al_{0.3}Mg_1Si$, which decreases the risk of the appearance of harmful particles of the $AlSc_2Si_2$-type. However, a lower dissolution temperature of Mg_2Si particles in $Al_{0.3}Mg_1$ Si alloys allows one to avoid additional heat treatment. To date, multistage heat treatment has been required to obtain two hardening phases $(AlSi)_3ZrSc$ and β″ (Mg_5Si_6) [26]. It consists in homogenization in order to maximize dissolution of Sc, Mg, Si and Zr into a supersaturated solid solution which includes eliminating large non-equilibrium particles. Three-stage annealing is applied after homogenization: The first step is the formation of $(AlSi)_3Sc$ particles. Then $(AlSi)_3ScZr$ particles are formed on their basis. The third stage consists of the dissolution of phases close to Mg_2Si, which at the fourth stage can be isolated in the form of a strengthening β″ (Mg_5Si_6) phase. However, at a low temperature of Mg_2Si solubility, the fourth stage of annealing can be avoided, since all the necessary processes will occur at the third stage. Note that, according to the calculations performed in the current Thermo-Calc version, only equilibrium primary particles Al_3Zr are present in the alloys studied. The appearance of Al_3Zr as solid state equilibrium particle which has D023-structure was not predicted by this calculation. This, however, does not exclude the emerging of metastable particles which have $L1_2$ structure and composition close to $(AlSi)_3Zr$.

3.4. Investigation of the Intermetallic Participles by Scanning Electron Microscopy (SEM) Method

The $Al_{0.3}Mg_1Si_{0.3}Sc$ alloy contains two types of particle; the first one can be described as being like $Al_3Fe_2Si_2$. The second can be interpreted as being like quaternary π phase $Fe_2Mg_7Si_{10}Al_{18}$ [39] which in this alloy can form due to there being no equilibrium solidification. Primary intermetallic compounds have both close to round (smaller) and needle-like shapes. Their average size is 12.4 microns. This alloy has an increased amount of fine particles which, among other things, explains its high microhardness (Figure 4a). Considering that no traces of scandium were found in large intermetallic particles with a fraction of its content in the solid solution of only 0.11%, it can be assumed with a high degree of probability that it is in the form of fine particles, the fraction of which is 0.14%. It should be noted that there is a high proportion of magnesium in the supersaturated solid solution. Therefore, it is insufficient for the formation of large intermetallic particles of the Mg_2Si type. This chemical composition, as predicted in [25], is extremely unfavorable for the production of strengthening Al_3Sc particles, since scandium actively interacts with silicon already during cooling after casting. In this case, Mg is used rather rarely. This means, firstly, the need for a prolonged homogenization (completed by quenching) to dissolve excess Sc and Si at high temperatures. In addition, with a high degree of probability the subsequent heat treatment and aging at 300 °C will not be accompanied by the precipitation of Mg_2Si-type particles. As a result, silicon will interact only with scandium, which will favor the predominant growth of unwanted $AlSc_2Si_2$ particles.

Figure 7 shows the results of electron microscopy for the alloy $Al_{0.3}Mg_1Si_{0.3}Sc_{0.15}Zr$. It contains a number of large primary intermetallic such as $Al_9Fe_2Si_2$, (Al, Si)$_3$Sc, $Al_3Sc_{0.6}Zr_{0.4}$ with an average size of 11.6 microns. It also contains particles with a composition close to quaternary π phase $Fe_2Mg_7Si_{10}Al_{18}$. In this alloy, a fairly large amount of Si and Sc is in solid solution. This suggests that, for a given chemical composition, Si and Sc do not interact so intensely with each other. The presence of particles of both the (Al, Si)$_3$Sc and $Al_3Sc_{0.6}Zr_{0.4}$ types indicates that Sc, upon further heat treatment, will react both with silicon, forming $AlSc_2Si_2$ particles, and with zirconium, forming $Al_3Sc_{0.6}Zr_{0.4}$ particles. In this case magnesium also "absorbs" part of the silicon, contributing to the formation of Mg_2Si-type particles. Taking into account the results of Thermo-Calc calculations (Figure 5a), it can also be assumed that various Zr and Si compounds will occur during heat treatment. Thus, this alloy seems to be very promising for further study, since excessive silicon will interact both with magnesium and zirconium, as well as various impurities. With proper heat treatment, this will allow the required amount of Al_3ScZr particles which have a L_{12} structure to be obtained.

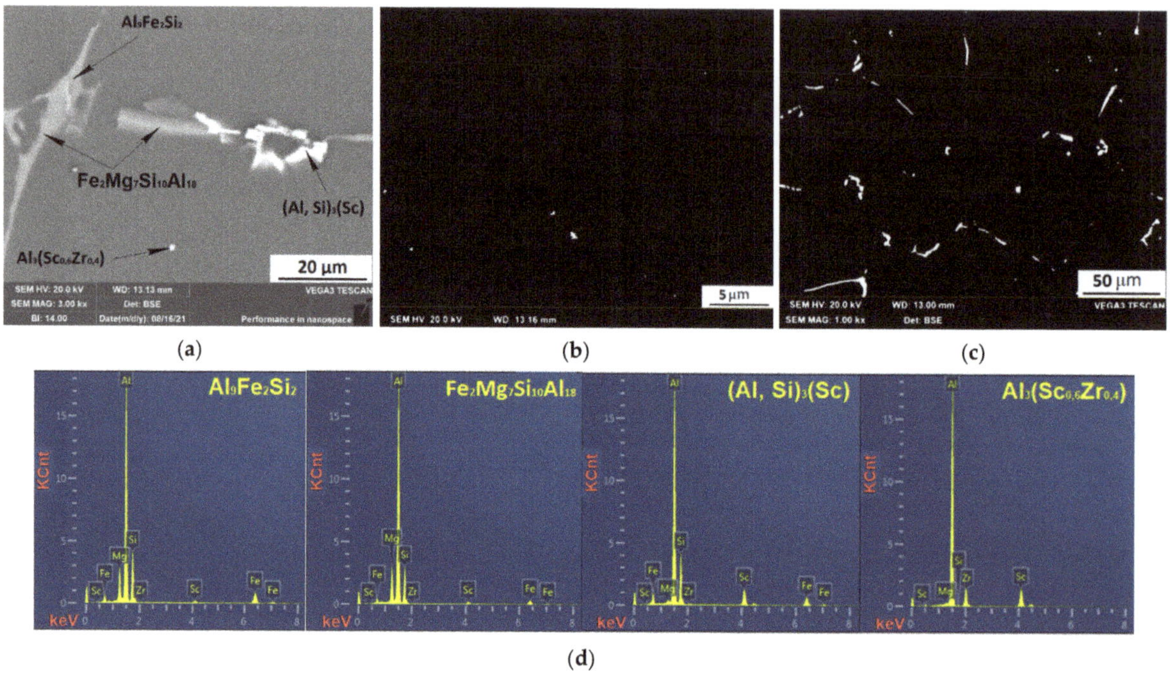

Figure 7. $Al_{0.3}Mg_1Si_{0.3}Sc_{0.15}Zr$ main types of intermetallic compounds (**a**); small dispersoids (**b**); large particles (**c**); result of the EDS investigation of chemical composition of the intermetallic compounds (**d**).

The $Al_{0.5}Mg_{1.3}Si_{0.3}Sc$ alloy contains the main types of large primary non-equilibrium intermetallic compounds $Al_9Fe_2Si_2$, $Fe_2Mg_7Si_{10}Al_{18}$, and (Al, $Si)_3Sc$. Their average size is 10.6 microns. In this case, there are spherical and needle-shaped particles, as well as large elongated eutectic particles arising along the boundaries of dendritic grains. Besides, there are many dispersoids in this alloy occupying 0.2% of the cross-sectional area, but with a rather large size.

In the $Al_{0.5}Mg_{1.3}Si_{0.3}Sc_{0.15}Zr$ alloy, large intermetallic compounds are represented by $Fe_2Mg_7Si_{10}Al_{18}$ and (Al, $Si)_3$(Sc) particles with an average diameter of 12.8 microns. There are not so many dispersoids in the alloy, however, they have a size of about 30 nm. Smaller dispersoids with an average size of 20–15 nm were detected using transmission microscopy. Considering also the fact that zirconium can practically not be observed in a supersaturated solid solution and in large intermetallic particles, it can be assumed that the main type of fine particles will be Al_3ScZr. Moreover, in this alloy Si reacts sufficiently with magnesium which makes it very promising for further research.

The results of electron microscopy are presented in:

- Table 2 (data on the size and quantity of fine particles);
- Table 3 (data on the size and quantity of large intermetallic particles);
- Table 4 (chemical composition of the main intermetallic particles) and Figures 6–9.

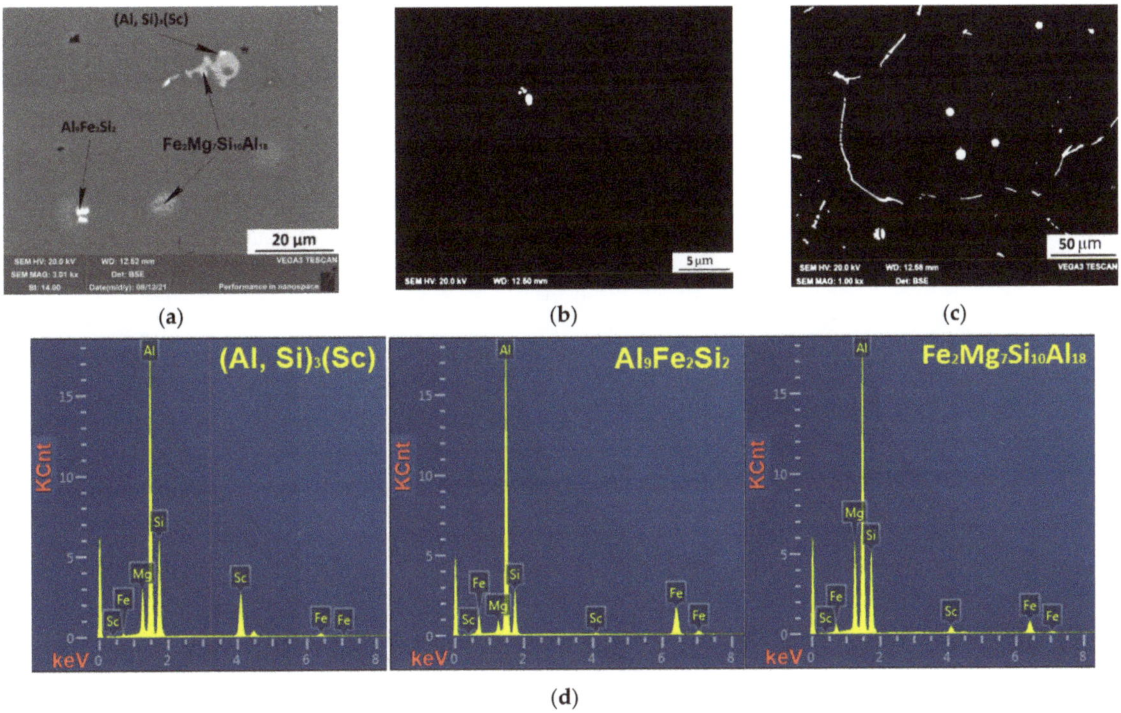

Figure 8. $Al_{0.5}Mg_{1.3}Si_{0.3}Sc$ main types of intermetallic compounds (**a**); small dispersoids (**b**); large particles (**c**); result of the EDS investigation of chemical composition of the intermetallic compounds (**d**).

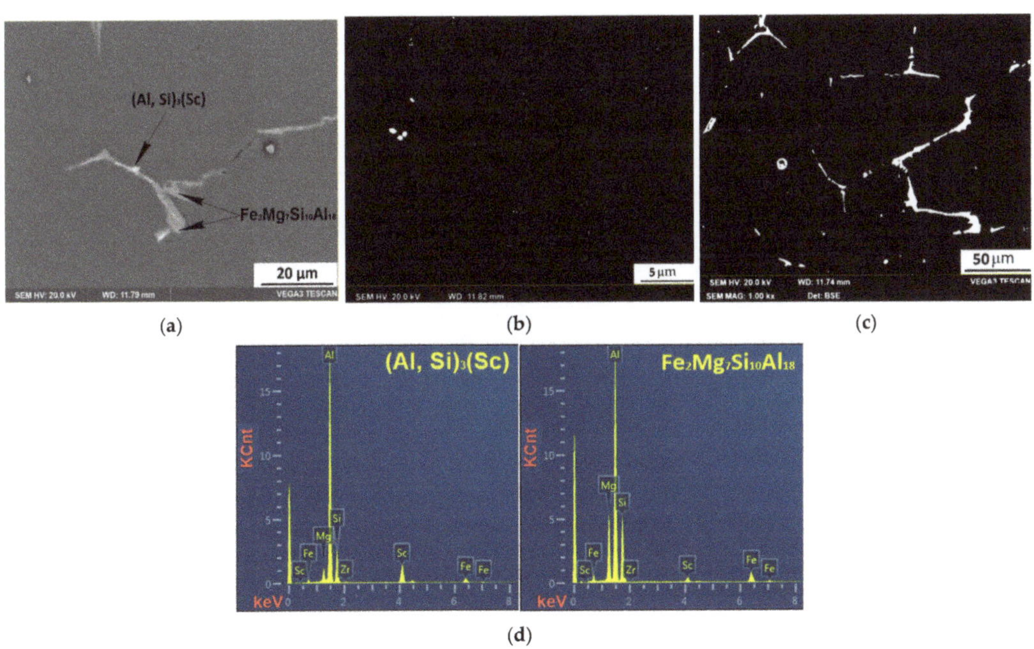

Figure 9. $Al_{0.5}Mg_{1.3}Si_{0.3}Sc_{0.15}Zr$ main types of intermetallic compounds (**a**); small dispersoids (**b**); large particles (**c**); result the EDS investigation of chemical composition of the intermetallic compounds (**d**).

Table 2. Size and amount of dispersoids in the investigated alloys.

Alloy	Area Share, %	Average Amount per 500 Microns^2	Average Size, μm
$Al_{0.3}Mg_1Si_{0.3}Sc$	0.14	21	0.05
$Al_{0.5}Mg_{1.3}Si_{0.3}Sc$	0.87	3	0.59
$Al_{0.3}Mg1Si_{0.3}Sc_{0.15}Zr$	0.2	16	0.08
$Al_{0.5}Mg_{1.3}Si_{0.3}Sc_{0.15}Zr$	0.11	21	0.03

Table 3. Size and amount of intermetallic compounds in the investigated alloys.

Alloy	Area Share, %	Average Amount per 500 Microns^2	Average Size, μm
$Al_{0.3}Mg_1Si_{0.3}Sc$	2.15	90	12.4
$Al_{0.5}Mg_{1.3}Si_{0.3}Sc$	1.5	71	10.6
$Al_{0.3}Mg_1Si_{0.3}Sc_{0.15}Zr$	1.8	78	11.6
$Al_{0.5}Mg_{1.3}Si_{0.3}Sc_{0.15}Zr$	1.7	69	12.8

Table 4. Chemical composition of intermetallic particles in the studied alloys.

Alloy	Phase	At. %					
		Al	Fe	Si	Mg	Sc	Zr
$Al_{0.3}Mg_1Si_{0.3}Sc$	$Fe_2Mg_7Si_{10}Al_{18}$	58.54	4.1	13.91	22.44		
	$Al_9Fe_2Si_2$	76.69	9.7	10.01			
	Solid solution	98.23		0.48	0.27	0.11	
$Al_{0.5}Mg_{1.3}Si_{0.3}Sc$	$Al_9Fe_2Si_2$	66.17	14.46	16.02			
	$Fe_2Mg_7Si_{10}Al_{18}$	53.99	5.35	15.43	23.6		
	$(Al, Si)_3(Sc)$	51.65	27.02	11.93			
	Solid solution	98.7		0.79	0.4	0.12	
$Al_{0.3}Mg_1Si_{0.3}Sc_{0.15}Zr$	$Fe_2Mg_7Si_{10}Al_{18}$	66.04	3.14	9.94	19.95		
	$Al_9Fe_2Si_2$	60.11	7.08	22.66			
	$(Al, Si)_3(Sc)$	61.83		24		7.06	
	$Al_3(Sc_{0.6}Zr_{0.4})$	82.88				11.44	4.61
	Solid solution	98.72		0.82	0.28	0.14	0.04
$Al_{0.5}Mg_{1.3}Si_{0.3}Sc_{0.15}Zr$	$Fe_2Mg_7Si_{10}Al_{18}$	54.08	5.3	15	24.11		
	$(Al, Si)_3(Sc)$	66.9	16.94			8.33	
	Solid solution	99.32		0.21	0.32	0.1	0.05

It must be mentioned that most of the phases listed in Tables 2–4 are non-equilibrium, as result they cannot be predicted by the diagram. The reason for this is the non-equilibrium conditions of solidification and the inevitable presence of iron even in the high frequency batch. However, the diagram on Figure 6 can be very useful for predicting phase transformations after homogenization annealing (for example, combined with quenching) when the composition is close to equilibrium.

3.5. Investigation of the Despersoid by Transmission Electron Microscopy (TEM) Method

A sufficiently large number of Al_3Sc particles is observed in the alloy $Al_{0.3}Mg_1Si_{0.3}Sc$. The size of these particles is 20 nm on average (Figure 10a,b). Particles of a given size are at the coherence loss thresholds. Therefore, some of them are completely coherent (Figure 10d), while others only partially (Figure 10d). The presented data are in good agreement with the data of scanning microscopy, according to which most of the scandium is not in a supersaturated solid solution, and at the same time, it is not observed in large particles of eutectic origin, such as $Al_{0.3}Mg_1Si_{0.3}Sc_{0.15}Zr$. In addition, it is the particles of this type that serve as the main reason for the growth of microhardness (see Figure 4a). It should also be noted that these particles contain unwanted silicon. However, considering that these particles are fully or partially coherent and considering the increase in microhardness, it is possible with a high degree of probability that they are metastable $(AlSi)_3Sc$, rather than equilibrium harmful $AlSc_2Si_2$. The appearance of such particles during solidification in the 6XXX series alloys with small additions of scandium and zirconium as a result of discontinuous precipitation has already been described in many investigations [40,41]. However, it has been described mainly in alloys that do not have such a high excess of silicon. The presence of metastable coherent and semi-coherent particles which precipitate during the cooling process after casting of the $Al_{0.3}Mg_1Si_{0.3}Sc$ alloy indicates that they can also be obtained as a result of multistage heat treatment. Thus, the cooling rate achieved in a steel chill mold without zirconium is insufficient so that $(AlSi)_3Sc$-type particles are not precipitated during cooling. It should be noted that in spite the short cooling time of the ingot, a large number of $(AlSi)_3ScZr$ dispersoids precipitate in this and other alloys investigated by TEM. This can be explained by the fact that silicon accelerates the kinetics of the precipitation of $(AlSi)_3ScZr$ [42].

A different picture of the distribution and chemical composition of fine particles is observed in the $Al_{0.5}Mg_{1.3}Si_{0.3}Sc$ alloy. It also contains the $(AlSi)_3Sc$-type particles but their number is somewhat less than in the $Al_{0.3}Mg_1Si_{0.3}Sc$ alloy (Figure 11a). However, in general, they have the same size and the same chemical composition. In addition, needle-shaped particles similar to β″ (Mg_5Si_6) appear (Figure 11b,c).

It should be noted that in spite of the rather short cooling time of the ingot, a large number of Al_3ScZr particles precipitate in this (and other alloys investigated by TEM). This can be explained by the fact that silicon accelerates the kinetics of the precipitation of these particles.

Note that emerging of β″ actually became possible due to the fact that additional magnesium appeared for the reaction with silicon. Thus, an increase in the Mg/Si ratio will contribute to obtaining an additional synergistic effect between the $(AlSi)_3Sc$ and β″ (Mg_5Si_6) particles. In general, the cooling rates after casting in a steel chill mold are also insufficient in this alloy to contain the main elements in a solid solution, and they form hardening phases $(AlSi)_3Sc$ and β″ (Mg_5Si_6).

The $Al_{0.3}Mg_1Si_{0.3}Sc_{0.15}Zr$ alloy contains a fairly large number of fine coherent or semi-coherent $(AlSi)_3ScZr$ particles with the $L1_2$-structure (Figure 12b–d). At the same time, there is a second type of elongated particles (Figure 12a,c), which can be confused with β″ (Mg_5Si_6), given the rather high content of magnesium as well as silicon in the surrounding solid solution. However, these particles are much longer than β″ (Mg_5Si_6) (from 0.5 to 1 μm). Therefore, a chemical analysis (Figure 12e) shows that they are also discontinuous precipitation of $(AlSi)_3ScZr$ particles. This fact is also confirmed by the comparison with the literature data [43], in which discontinuous Al_3Sc precipitates are practically identical in shape, size and location of relatively high-angle boundaries compared to the dispersoids observed in this work. It should be noted that, as in Ref. [44], high-angle boundaries serve as a source of discontinuous precipitation of the $(AlSi)_3ScZr$-type dispersoids. When these particles are precipitated, energy is released and the high-angle boundaries begin to move, leaving behind a large number of the $(AlSi)_3ScZr$-type dispersoids. It is noteworthy that dendritic grains, in which intermittent precipitates are found, do not contain $L1_2$-nanoparticles (Figure 12c) and vice versa. A possible explanation for this is that in discontinuous precipitation, as in recrystallization [45], there are orientations more favorable for the motion of high-angle boundaries. In grains with an intermittent release of $(AlSi)_3ScZr$, scandium and zirconium become insufficient for the appearance of more coherent particles which have $L1_2$-structure. It should be noted that, in the case of the $Al_{0.3}Mg_1Si_{0.3}Sc$ alloy, β″ (Mg_5Si_6) particles can be observed together with $(AlSi)_3Sc$ particles, in contrast to the $L1_2$ and needle shape $(AlSi)_3ScZr$ dispersoids. It can also be noted that for the discontinuous and simultaneous precipitation both needle-shaped and L1-structured dispersoids the concentration of scandium in the $Al_{0.3}Mg_1Si_{0.3}Sc$ and $Al_{0.5}Mg_{1.3}Si_{0.3}Sc$ alloys is insufficient.

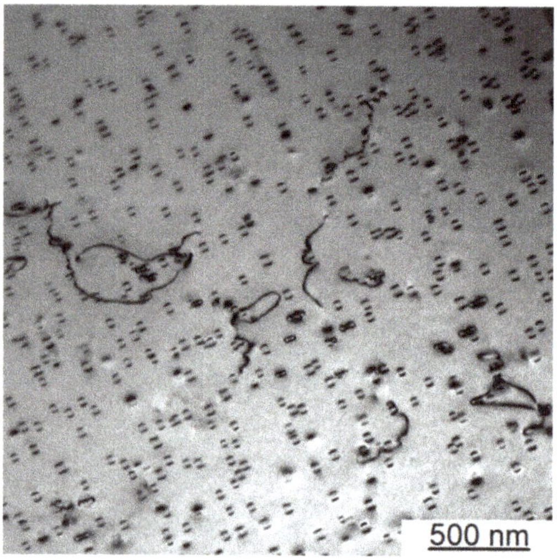

(**a**)

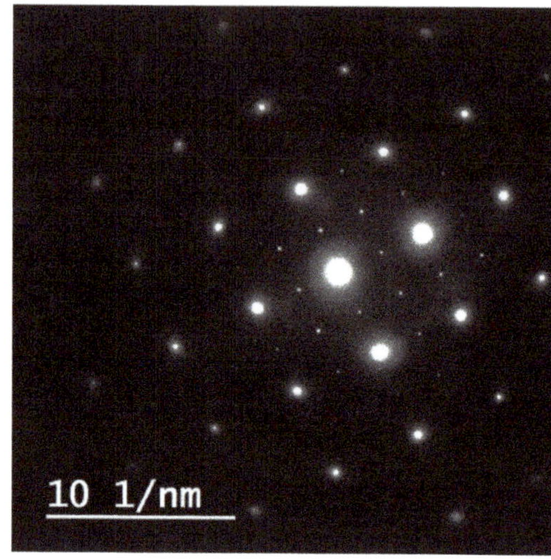

(**b**)

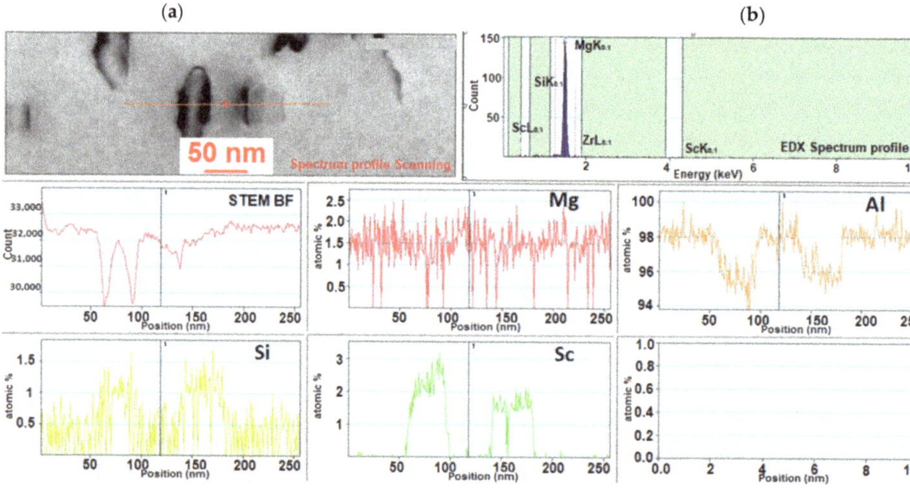

(**c**)

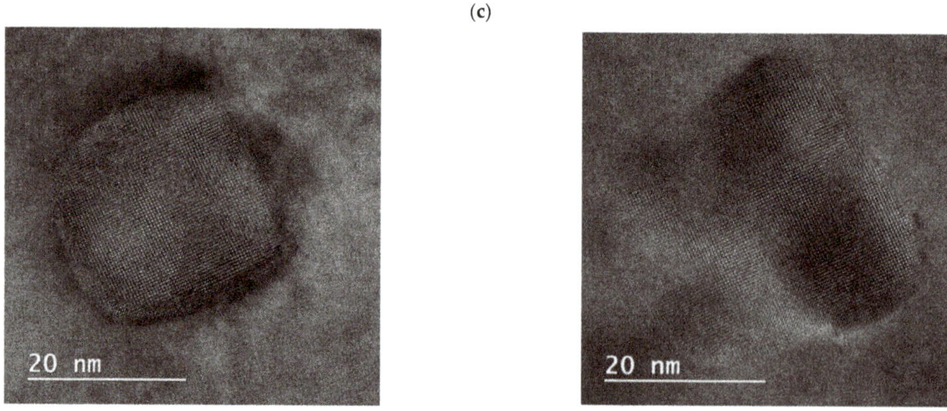

(**d**)

Figure 10. (**a**) $(AlSi)_3Sc$ coherent and semi coherent particles, (**b**) EDS profile line scan; (**c**,**d**) direct resolution of the crystal lattice with $(AlSi)_3Sc$ particles of various sizes.

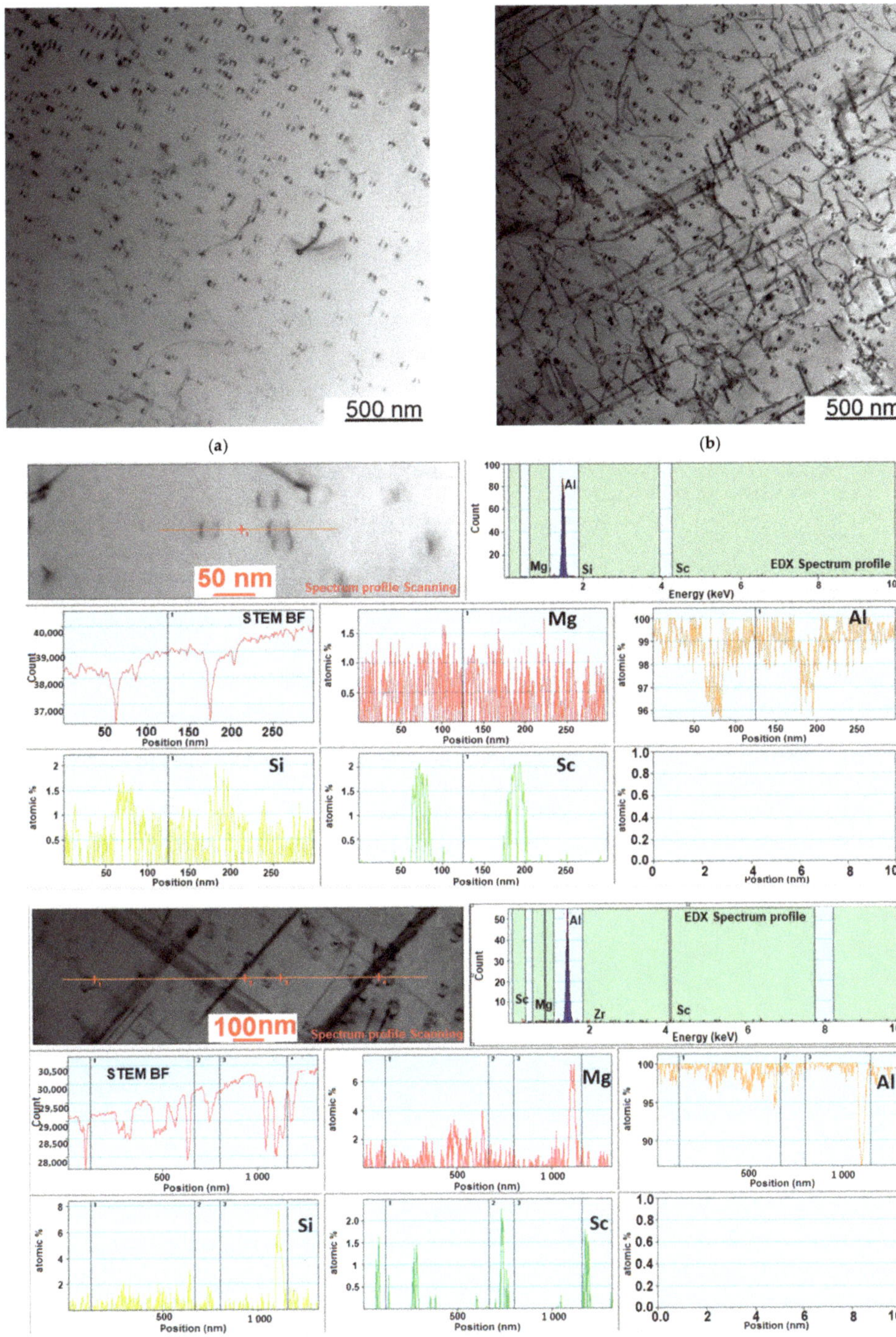

Figure 11. (**a**) Al_3Sc coherent and semi coherent particles particles; (**b**) Al_3Sc coherent and semi coherent particles + Mg_2Si (rod-like) particles (**c**) EDS profile line scan.

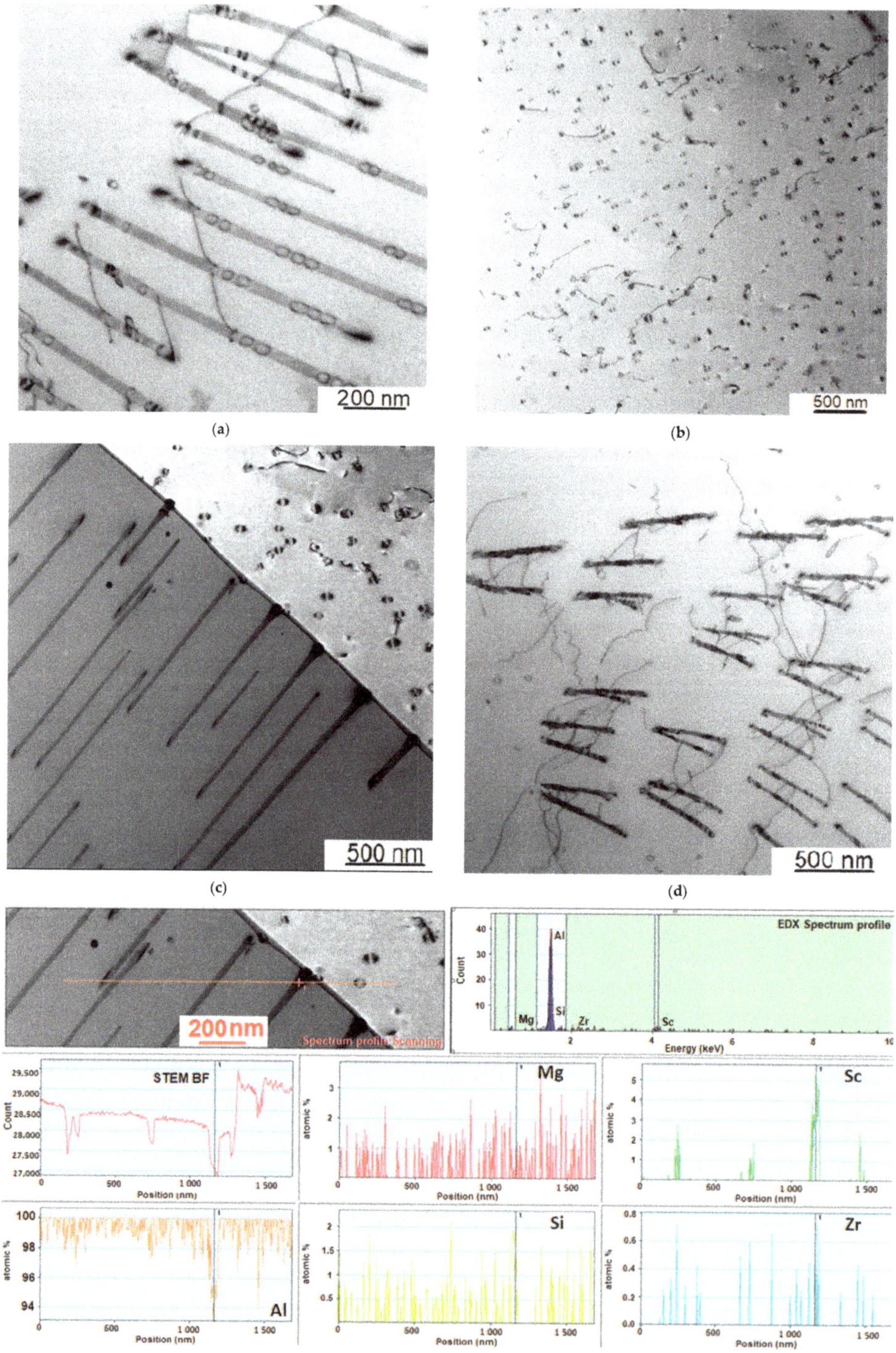

Figure 12. (**a**,**c**,**d**) Discontinuous precipitation of $(AlSi)_3ScZr$ needle-shaped particles; (**b**) $(AlSi)_3ScZr$ particles with the $L1_2$-structure, (**c**) discontinuous precipitation of $(AlSi)_3ScZr$ needle-shaped particles near high-angle boundaries. (**e**) EDS profile line scan.

Mainly coherent or semi-coherent $(AlSi)_3ScZr$-particles, which have the $L1_2$-structure (Figure 13a,b), are formed during discontinuous precipitation and are observed in the $Al_{0.5}Mg_{1.3}Si_{0.3}Sc_{0.15}Zr$ alloy. The lack of needle-shape precipitates is most likely due to the limitations of TEM associated with the small area observed. The average grain size in this alloy is 260 μm, while the survey area is 4 μm^2, at best. For the alloy $Al_{0.3}Mg_{1.3}Si_{0.3}Sc_{0.15}Zr$, where the grain is finer, it is easier to find grain boundaries around which intermittent precipitates are concentrated. In the case of $Al_{0.3}Mg_1Si_{0.3}Sc_{0.15}Zr$, the β″ (Mg_5Si_6)-phase is not detected in the alloy. Its absence can be explained by the fact that the bulk of magnesium and silicon form primary Mg_2Si-particles at this concentration of zirconium and scandium (see Figure 6).

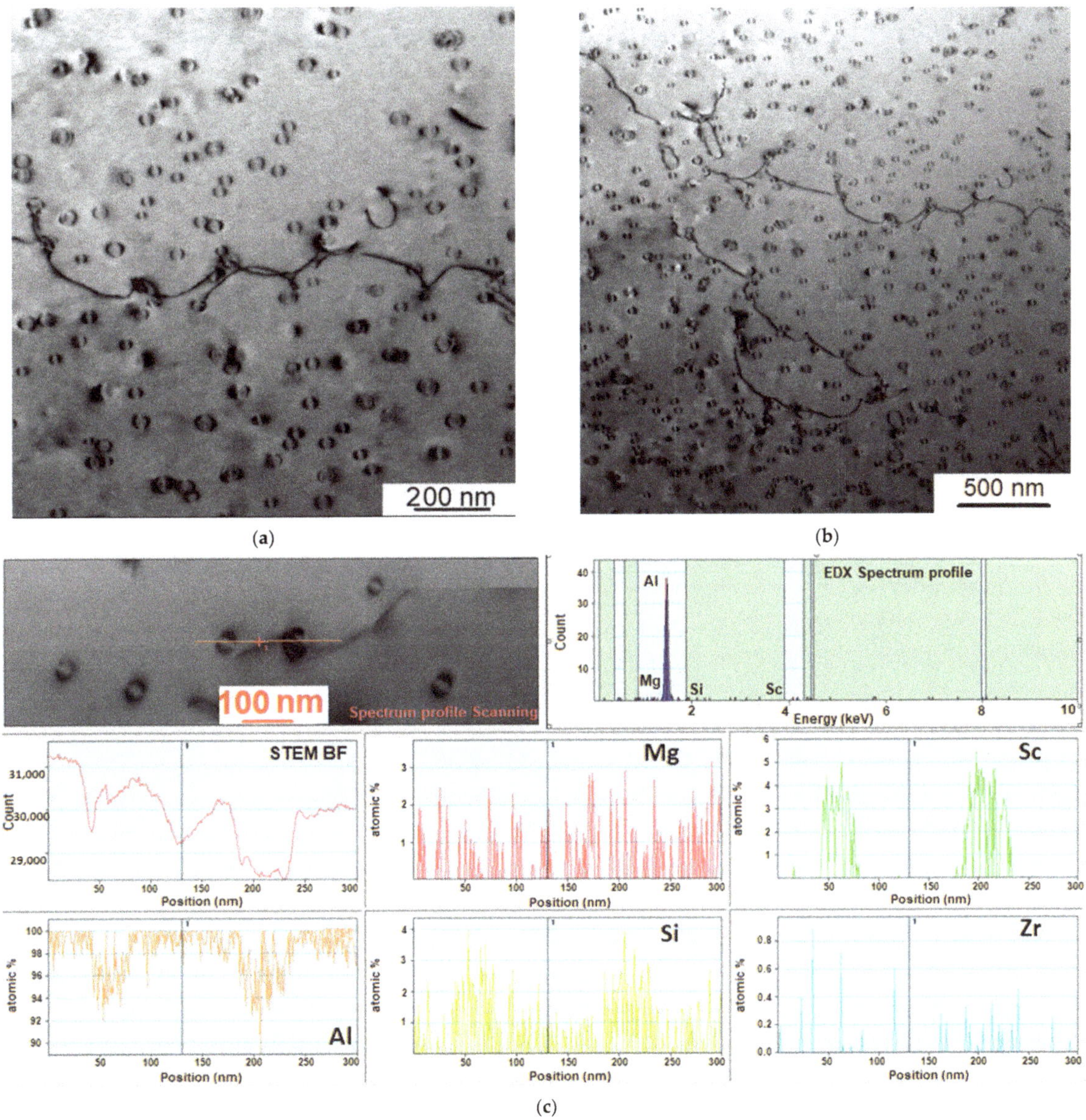

Figure 13. (**a**,**b**) $(AlSi)_3ScZr$ particles with the $L1_2$ structure, (**c**) EDS profile line scan.

4. Conclusions

Variations of Mg within 0.3 ÷ 0.5%; Si within 1 ÷ 1.3%; Sc within 0.0 ÷ 0.3%; Zr in the range of 0.0 ÷ 0.15% have a complex effect on the microstructure formed during casting of these alloys. The most significant effect on the size of the grain structure is exerted by micro-additives of Sc and Zr. In $Al_{0.3}Mg_1Si$ alloys, the complex additive $_{0.3}Sc_{0.15}Zr$ is able to refine the grain size by seven times. The refinement of the grain structure is influenced by two main factors: supercooling at the boundaries of the nucleus and the liquid phase, and the formation of Al_3Sc particles in the liquid. The latter takes place at a concentration of 0.15% zirconium and 0.24% scandium in $Al_{0.3}Mg_1Si$ alloys and 0.15% zirconium and 0.28% scandium in $Al_{0.5}Mg_{1.3}Si$ alloys. The refinement of the grain structure by supercooling is more efficient in the $Al_{0.5}Mg_{1.3}Si$ alloys, and with the primary Al_3Sc phase in the alloys of the $Al_{0.3}Mg_1Si$ type, where the scandium concentration required for the formation of this phase is reached earlier.

These alloys have a complex phase composition. After casting, it can be represented by both equilibrium and non-equilibrium phases; in addition, iron impurities have a significant effect. With regards to maximum binding of scandium to other elements, the preferable alloys are $Al_{0.5}Mg_{1.3}Si$. However, the $Al_{0.3}Mg_1Si$ alloy is better for the effective dissolution of the Mg_2Si phase, which occurs already at 450 °C. In general, compositions with combined alloying by small additions of Sc and Zr and mono additions of Zr seem to be most promising. Alloying Sc compositions with such an excess of silicon is practically unpromising.

The cooling time after solidification of the ingot is sufficient for the precipitation of fine particles. These particles are practically not tracked using scanning electron microscope SEM even at high magnification. However, they are very clearly identified using TEM, as well as by an increase in microhardness. The main type of particles is $(AlSi)_3Sc$ or $(AlSi)_3ScZr$, depending on the presence or absence of zirconium. These particles appear during discontinuous precipitates in form of the $L1_2$-structure with dimensions of 15–20 nm and are semi- and completely coherent or in needle shape form (observed only in alloys with zirconium). The formation of coherent and semi-coherent particles in these alloys opens the possibility for their further production during heat treatment. β″ (Mg_5Si_6)-particles are formed only in the alloy $Al_{0.5}Mg_{1.3}Si_{0.3}Sc$, where there is enough magnesium for their precipitation. The addition of zirconium changes the phase composition and promotes the precipitation of large Mg_2Si-type particles.

Author Contributions: Conceptualization, E.A. and S.K.; methodology, E.A., S.K. and J.H.; software, E.A. and S.K.; validation, E.A., S.K. and J.H.; formal analysis, E.A. and A.D.; investigation, E.A., M.L., V.B. and D.Z.; resources, E.A., J.H. and S.K.; data curation, E.A., J.H., S.K. and A.D.; writing—original draft preparation, E.A., J.H. and S.K.; writing—review and editing, E.A., J.H. and S.K.; visualization, E.A. and J.H.; supervision, S.K.; project administration, E.A. and S.K.; funding acquisition, S.K. All authors have read and agreed to the published version of the manuscript.

Funding: This study is funded by a grant of the Russian Science Foundation, project 21-19-00548, https://rscf.ru/project/21-19-00548/.

Institutional Review Board Statement: Not applicable.

Informed Consent Statement: Not applicable.

Data Availability Statement: The data presented in this study are available on request from the corresponding author.

Conflicts of Interest: The authors declare no conflict of interest.

References

1. Fedorov, S.; Nanocomposites, V.B. Development of Mechanical Properties of Aluminum-Silicon Alloys. *Smart Nanocompos.* **2015**, *6*, 199–202.
2. Bazhin, V.Y.; Gutema, E.M.; Savchenkov, S.A. Production Technology Features for Aluminum Matrix Alloys with a Silicon Carbide Framework. *Metallurgist* **2017**, *60*, 1267–1272. [CrossRef]

3. Deev, V.B.; Degtyar, V.A.; Kutsenko, A.I.; Selyanin, I.F.; Voitkov, A.P. Resource-Saving Technology for the Production of Cast Aluminum Alloys. *Steel Transl.* **2007**, *37*, 991–994. [CrossRef]
4. Akopyan, T.K.; Letyagin, N.V.; Belov, N.A.; Koshmin, A.N.; Gizatulin, D.S. Analysis of the Microstructure and Mechanical Properties of a New Wrought Alloy Based on the ((Al) + Al4(Ca,La)). *Eutectic. Phys. Met. Metallogr.* **2020**, *121*, 914–919. [CrossRef]
5. Sizyakov, V.M.; Bazhin, V.Y.; Vlasov, A.A. Status and Prospects for Growth of the Aluminum Industry. *Metallurgist* **2010**, *54*, 409–414. [CrossRef]
6. Belov, N.; Akopyan, T.; Korotkova, N.; Murashkin, M.; Timofeev, V.; Fortuna, A. Structure and Properties of Ca and Zr Containing Heat Resistant Wire Aluminum Alloy Manufactured by Electromagnetic Casting. *Metals* **2021**, *11*, 236. [CrossRef]
7. Hirsch, J. Aluminium in Innovative Light-Weight Car Design. *Mater. Trans.* **2011**, *52*, 818–824. [CrossRef]
8. Lamberti, M.; Escher, F. Aluminium Foil as a Food Packaging Material in Comparison with Other Materials. *Food Rev. Int.* **2007**, *23*, 407–433. [CrossRef]
9. Miller, W.S.; Zhuang, L.; Bottema, J.; Wittebrood, A.J.; De Smet, P.; Haszler, A.; Vieregge, A. Recent Development in Aluminium Alloys for the Automotive Industry. *Mater. Sci. Eng.* **2000**, *280*, 37–49. [CrossRef]
10. Rambabu, P.; Eswara Prasad, N.; Kutumbarao, V.V.; Wanhill, R.J.H. Aluminium Alloys for Aerospace Applications. In *Aerospace Materials and Material Technologies*; Prasad, N., Wanhill, R., Eds.; Indian Institute of Metals Series; Springer: Singapore, 2017; pp. 29–52. [CrossRef]
11. Pogatscher, S.; Antrekowitsch, H.; Werinos, M.; Moszner, F.; Gerstl, S.S.A.; Francis, M.F.; Curtin, W.A.; Löffler, J.F.; Uggowitze, P.J. Diffusion on demand to control precipitation aging: Application to Al-Mg-Si alloys. *Phys. Rev. Lett.* **2014**, *22*, 225701. [CrossRef]
12. Liu, C.; Zhang, X.; Tang, J.; Liu, X.; Liang, C. Effect of copper on precipitation and baking hardening behavior of Al-Mg-Si alloys. *Trans. Nonferrous Met. Soc. China* **2014**, *7*, 2289–2294. [CrossRef]
13. Kim, J.H.; Kobayashi, E.; Sato, T. Effects of Cu addition on behavior of nanoclusters during multi-step aging in Al-Mg-Si alloy. *Mater. Trans.* **2011**, *5*, 906–913. [CrossRef]
14. Hong, L.I.U.; Gang, Z.; Liu, C.M.; Liang, Z. Effects of different tempers on precipitation hardening of 6000 series aluminium alloys. *Trans. Nonferrous Met. Soc. China* **2007**, *1*, 122–127.
15. Liang, Z. Clustering and Precipitation in Al-Mg-Si Alloys. Ph.D. Thesis, Technische Universität Berlin, Berlin, Germany, 28 November 2012.
16. Werinos, M.; Antrekowitsch, H.; Ebner, T.; Prillhofer, R.; Curtin, W.A.; Uggowitzer, P.J.; Pogatscher, S. Design strategy for controlled natural aging in Al–Mg–Si alloys. *Acta Mater.* **2016**, *118*, 296–305. [CrossRef]
17. Zhen, L.; FEI, W.D.; Kang, S.B.; Kim, H.W. Precipitation behaviour of Al-Mg-Si alloys with high silicon content. *J. Mater. Sci.* **1997**, *7*, 1895–1902. [CrossRef]
18. Pogatscher, S.; Antrekowitsch, H.; Leitner, H.; Ebner, T.; Uggowitzer, P.J. Mechanisms controlling the artificial aging of Al–Mg–Si Alloys. *Acta Mater.* **2011**, *9*, 3352–3363. [CrossRef]
19. Zakharov, V.V.; Fisenko, I.A. Some Principles of Alloying of Aluminum Alloys with Scandium and Zirconium in Ingot Production of Deformed Semiproducts. *Metal Sci. Heat Treat.* **2019**, *3*, 217–221. [CrossRef]
20. Zakharov, V.V. Combined alloying of aluminum alloys with scandium and zirconium. *Metal Sci. Heat Treat.* **2014**, *3*, 281–286. [CrossRef]
21. Röyset, J.; Ryum, N. Scandium in aluminium alloys. *Int. Mater. Rev.* **2005**, *1*, 19–44. [CrossRef]
22. Shvechkov, E.I.; Filatov, Y.A.; Zakharov, V.V. Mechanical and Life Properties of Sheets from Alloys of the Al–Mg–Sc System. *Metal Sci. Heat Treat.* **2017**, *59*, 454–462. [CrossRef]
23. Aryshenskii, E.V.; Hirsch, J.; Konovalov, S.V.; Prahl, U. Specific features of microstructural evolution during hot rolling of the as-cast magnesium-rich aluminum alloys with added transition metal elements. *Metall. Mater. Trans.* **2019**, *12*, 5782–5799. [CrossRef]
24. Peng, Z.W.; Li, J.F.; Sang, F.J.; Chen, Y.L.; Zhang, X.H.; Zheng, Z.Q.; Pan, Q.L. Structures and tensile properties of Sc-containing 1445 Al-Li alloy sheet. *J. Alloys Compd.* **2018**, *747*, 471–483. [CrossRef]
25. Rokhlin, L.L.; Bochvar, N.R.; Leonova, N.P.; Sukhanov, A.V. Effect of additional doping with scandium and scandium with zirconium on strength properties of the alloys of Al–Mg 2 Si system. *Inorg. Mater.* **2016**, *15*, 1467–1471. [CrossRef]
26. Babaniaris, S.; Ramajayam, M.; Jiang, L.; Langan, T.; Dorin, T. Tailored precipitation route for the effective utilisation of Sc and Zr in an Al-Mg-Si alloy. *Materialia* **2020**, *10*, 100656. [CrossRef]
27. Zupanič, F.; Steinacher, M.; Žist, S.; Bončina, T. Microstructure and Properties of a Novel Al-Mg-Si Alloy AA 6086. *Metals* **2021**, *11*, 368. [CrossRef]
28. Dorin, T.; Ramajayam, M.; Babaniaris, S.; Jiang, L.; Langan, T.J. Precipitation sequence in Al–Mg–Si–Sc–Zr alloys during isochronal aging. *Materialia* **2019**, *8*, 100437. [CrossRef]
29. Kwon, E.P.; Do Woo, K.; Kim, S.H.; Kang, D.S.; Lee, K.J.; Jeon, J.Y. The effect of an addition of Sc and Zr on the precipitation behavior of AA6061 alloy. *Met. Mater. Int.* **2010**, *5*, 701–707. [CrossRef]
30. Babaniaris, S.; Ramajayam, M.; Jiang, L.; Langan, T.; Dorin, T. Developing an optimized homogenization process for Sc and Zr containing Al-Mg-Si alloys. *Light Met.* **2019**, *1*, 1445–1453.
31. Lityńska-Dobrzyńska, L. Precipitation of phases in Al-Mg-Si-Cu alloy with Sc and Zr additions during heat treatment. *Solid State Phenom.—Trans. Tech. Publ. Ltd.* **2007**, *130*, 163–166. [CrossRef]

32. Vlach, M.; Smola, B.; Stulíková, I.; Očenášek, V. Microstructure and mechanical properties of the AA6082 aluminium alloy with small additions of Sc and Zr. *Int. J. Mater. Res.* **2009**, *3*, 420–423. [CrossRef]
33. GOST 9450–76. *Measurements Microhardness by Diamond Instruments Indentation*; Academy of Sciences of the USSR: Moscow, Russia, 1993.
34. Thermo-Calc Software. TCAL4 Al-Based Alloy Database, Version 4. Available online: https://www.engineering-eye.com/THERMOCALC/details/db/pdf/thermo-calc/02/tcal40_extended_info.pdf (accessed on 1 May 2021).
35. Aryshenskii, E.; Hirsch, J.; Konovalov, S.; Aryshenskii, V.; Drits, A. Influence of mg content on texture development during hot plain-strain deformation of aluminum alloys. *Metals* **2021**, *11*, 865. [CrossRef]
36. Wang, F.; Qiu, D.; Liu, Z.L.; Taylor, J.A.; Easton, M.A.; Zhang, M.X. The grain refinement mechanism of cast aluminium by zirconium. *Acta Mater.* **2013**, *61*, 5636–5645. [CrossRef]
37. Yu, A.W.; Yang, C.G.; Wang, S.L.; Liu, F.C.; Zheng, Q. Effect of Sc, Zr grain refiner on the microstructure and mechanical properties of pure aluminum. *Appl. Mech. Mater.—Trans. Tech. Publ. Ltd.* **2014**, *508*, 16–21. [CrossRef]
38. Jiang, S.; Wang, R. Grain size-dependent Mg/Si ratio effect on the microstructure and mechanical/electrical properties of Al-Mg-Si-Sc alloys. *J. Mater. Sci. Technol.* **2019**, *7*, 1354–1363. [CrossRef]
39. Yong, D.; Shuhong, L.; Baiyun, H.; Chang, Y.A.; Xie, F.Y.; Ying, Y.; Chen, S.L. Thermodynamic description of the Al-Fe-Mg-Mn-Si system and investigation of microstructure and microsegregation during directional solidification of an Al-Fe-Mg-Mn-Si alloy. *Z. Für Met.* **2005**, *96*, 1351–1362.
40. Litnska-Dobrzynska, L.; Dutkiewicz, J.; Maziarz, W.; Ochin, P. Structure and properties of Al-Mg-Si alloys with Zr and Sc additions produced by melt spinning and twin rolling casting techniques. *Kov. Mater.* **2010**, *48*, 9–15. [CrossRef]
41. Dorin, T.; Ramajayam, M.; Langan, T.J. Effects of Mg, Si, and Cu on the formation of the Al_3Sc/Al_3Zr dispersoids. In Proceedings of the International Conference on Aluminium Alloys, Montréal, QC, Canada, 17–20 June 2018; Volume 16, pp. 1–11. Available online: http://www.icaa-conference.net/ICAA16/Papers/Plenary%20and%20Early%20Career/404566%20Dorin_final.pdf (accessed on 1 May 2021).
42. Booth-Morrison, C.; Mao, Z.; Diaz, M.; Dunand, D.C.; Wolverton, C.; Seidman, D.N. Role of silicon in accelerating the nucleation of Al_3(Sc,Zr) precipitates in dilute Al–Sc–Zr alloys. *Acta Mater.* **2012**, *16*, 4740–4752. [CrossRef]
43. Norman, A.F.; Prangnell, P.B.; McEwen, R.S. The solidification behaviour of dilute aluminium–scandium alloys. *Acta Mater.* **1998**, *16*, 5715–5732. [CrossRef]
44. Yashin, V.V.; Aryshensky, E.V.; Latushkin, I.A.; Stozharov, D.A. Study of kinetics of the supersaturated solid solution decomposition in alloys of the Al-Mg system with transition elements addition. *Tsvetnye Met.* **2020**, *11*, 77–84.
45. Nes, E.; Vatne, H.E. The 40 (111) Orientation Relationship in Recrystallisation/Die 40 (111) Orientierungsbeziehung bei der Rekristallisation. *Z. Für Met.* **1996**, *87*, 448–453.

metals

MDPI

Article

Hypereutectic Al-Ca-Mn-(Ni) Alloys as Natural Eutectic Composites

Evgeniya Naumova [1,*], Vitali Doroshenko [1], Mikhail Barykin [1], Tatyana Sviridova [2], Alexandra Lyasnikova [3] and Pavel Shurkin [1]

1 Department of Metal Forming, National University of Science and Technology MISiS, Leninsky Ave. 4, 119049 Moscow, Russia; v.doroshenko@mail.ru (V.D.); mr.barykin97.97@mail.ru (M.B.); pa.shurkin@gmail.com (P.S.)
2 Department of Physical Materials Science, National University of Science and Technology MISiS, Leninsky Ave. 4, 119049 Moscow, Russia; tim-17@yandex.ru
3 Department of Biotechnical and Medical Systems and Technologies, Don State Technical University, Gagarin Square 1, 344000 Rostov-on-Don, Russia; kafbma2011@yandex.ru
* Correspondence: jan73@mail.ru; Tel.: +7-9163116927

Abstract: In the present paper, Natural Metal-Matrix Composites (NMMC) based on multicomponent hypereutectic Al-Ca-(Mn)-(Ni) alloys were studied in as-cast, annealed and rolled conditions. Thermo-Calc software and microstructural observations were utilised for analysing the equilibrium and actual phase composition of the alloys including correction of the Al-Ca-Mn system liquidus projection and the solid phase distribution in the Al-Ca-Mn-Ni system. A previously unknown $Al_{10}CaMn_2$ was discovered by both electron microprobe analysis and X-ray studies. The Al-6Ca-3Mn, Al-8Ca-2Mn, Al-8Ca-2Mn-1Ni alloys with representative NMMC structure included ultrafine Ca-rich eutectic and various small-sized primary crystals were found to have excellent feasibility of rolling as compared to its hypereutectic Al-Si counterpart. What is more, Al-Ca alloys showed comparable Coefficient of Thermal Expansion values due to enormous volume fraction of Al-based eutectic and primary intermetallics. Analysis of tensile samples' fracture surfaces revealed that primary intermetallics may act either as stress raisers or malleable particles depending on their stiffness under deformation. It is shown that a compact morphology can be achieved by conventional casting without using any refining agents. Novel hypereutectic Al-Ca NMMC materials solidifying with the formation of $Al_{10}Mn_2Ca$ primary compound have the best ductility and strength. We reasonably propose these materials for high-load pistons.

Keywords: hypereutectic aluminium alloys; Al-Ca alloys; phase composition; structure; intermetallics; rolling; physical and mechanical properties

Citation: Naumova, E.; Doroshenko, V.; Barykin, M.; Sviridova, T.; Lyasnikova, A.; Shurkin, P. Hypereutectic Al-Ca-Mn-(Ni) Alloys as Natural Eutectic Composites. *Metals* **2021**, *11*, 890. https://doi.org/10.3390/met11060890

Academic Editor: Tilmann Beck

Received: 27 April 2021
Accepted: 26 May 2021
Published: 29 May 2021

1. Introduction

Currently, light materials with a given set of mechanical and physical properties are extremely in demand for manufacturing engineering parts [1–3]. Specifically, high-load pistons operated at up to 300 °C require not only specific mechanical properties (e.g., strength, hardness and ductility) but also low Coefficient of Thermal Expansion (CTE) and high thermal conductivity. Among numerous materials, hypereutectic Al-Si alloys (e.g., A390.0 and FM180 alloys), have been established due to acceptable performance in the aforementioned properties [1,4,5]. In reality, they are natural metal matrix composites (NMMC) consisting of hard (Si) crystals distributed throughout a eutectic matrix. However, they show some pivotal drawbacks such as brittleness and the need for modification operation for refining primary and eutectic (Si) phase [1,5].

In the present study, alternative Al-Ca alloys are proposed which were recently reported with excellent processability through shape casting and metal forming [6–10]. When considering hypoeutectic Al-Ca alloys, they have low density, an appropriate combination of mechanical properties and corrosion resistance [6,8]. Meanwhile, their phase composition is complicated and available thermodynamic data are relatively poor. Specifically, there is a lack in the ternary Al-Ca-X phase diagrams not to mention the quaternary systems Al-Ca-X-Y. In our previous studies, many ternary Ca-rich compounds were found including those previously unknown [7,9,11,12]. Their primary crystallisation depending on the composition matters for the design of the hypereutectic alloys. Initially, we find it expedient to develop the multicomponent phase diagrams in the regions for hypereutectic alloys is of profound scientific importance for designing novel materials.

It should be noted that a remarkably fine as-cast eutectic structure in the Al-Ca alloys can be achieved without any refining agents. Moreover, multicomponent eutectic can appear to be even thinner in comparison to a binary [(Al) + Al_4Ca] eutectic. The work [13] reported Al-Ca-Mn-Fe alloys with a set of superfine multiphase eutectics (ternary and quaternary) that provides a high-tech performance in casting and metal forming. It should be noted that the best metal forming processability was shown for Al-Ca-Mn alloys [14]. Moreover, Mn acts both as a eutectic-forming element and a solid-solution strengthening agent [1,15]. As for Al-Ca-Ni alloys, an appearance of the ternary Al_9NiCa phase together with Al_4Ca and Al_3Ni phases was found in equilibrium with (Al) solid solution [9]. Being a part of the eutectic mixture [(Al) + Al_4Ca + Al_9NiCa], the ternary compound has a submicron size and demonstrates a response to spheroidising annealing. Primary Al_9NiCa crystals exhibit a polyhedron shape that is similar to the primary (Si) phase in the hypereutectic Al-Si alloys [4]. When considering a detailed investigation of the primary crystals' shape in the multicomponent Al-Ca alloys, there has been virtually no relevant research so far. Within the range of various compounds, some may primarily solidify acquiring a compact shape without compromising the ductility but increasing both strength and the CTE value. It is known that hypereutectic Al-Si alloys are manufactured by either squeeze casting or isothermal hot stamping or hot extrusion [4]. However, massive (Si) primary crystals deteriorate the plasticity of alloys. For this reason, special approaches are being developed for refining primary crystals [1,4,5]. For ensuring processability of hypereutectic alloys through deformation, small size (50–70 μm) and uniformity in distribution of primary crystals must be maintained as well as a globular shape of eutectic-origin intermetallics as illustrated in Figure 1. Based on the preliminary results, we reasonably find hypereutectic multicomponent Al-Ca alloys very appropriate for designing NMMC materials with special physical and mechanical properties competitive to those of the Al-Si piston alloys.

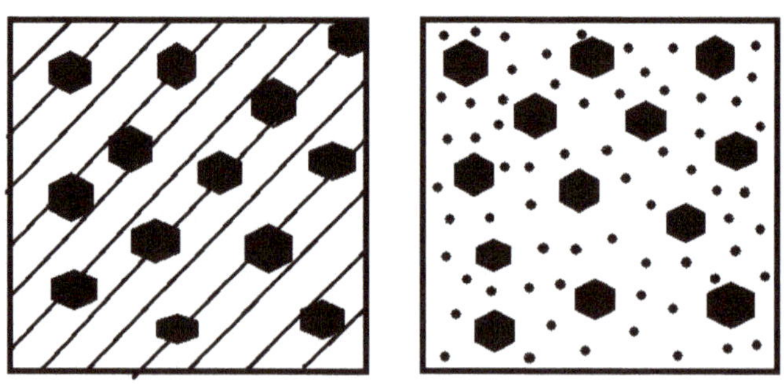

(a) (b)

Figure 1. Schematic representation of the target NMMC microstructure: (**a**) as-cast condition; (**b**) after spheroidising annealing.

The aim of this work is to explore promising hypereutectic Al-Ca alloys and study their structure and properties with Mn and Ni alloying.

2. Materials and Methods

The experimental alloys (Table 1) were prepared in an electric resistance furnace LAC (LAC, s.r.o., Židlochovice, Czech Republic) using pure aluminium (99.99% Al), calcium (99.99% Ca), silicon (99.99% Si) and Al—20% Ni and Al—20% Mn master alloys. For the sake of comparison, a commercial AlSi18Cu1Mg1Ni1 (FM180) piston alloy was prepared. The molten alloys were poured at 730–750 °C into a graphite mould (without preheating) with an inner size of 15 mm × 30 mm × 180 mm. The cooling rate under that conditions was about 10 K/s.

Table 1. Chemical composition of the experimental alloy.

Designation	Concentrations, wt. %			
	Ca	**Mn**	**Ni**	**Al**
Al-3Ca-2Mn	3	2	-	Balance
Al-6Ca-1.5Mn	6	1.5	-	Balance
Al-6Ca-2Mn	6	2	-	Balance
Al-6Ca-3Mn	6	3	-	Balance
Al-8Ca-0.5Mn	8	0.5	-	Balance
Al-8Ca-2Mn	8	2	-	Balance
Al-8Ca-3Mn	8	3	-	Balance
Al-10Ca	10	0	-	Balance
Al-10Ca-1.5Mn	10	1.5	-	Balance
Al-10Ca-2.5Mn	10	2.5	-	Balance
Al-8Ca-2Mn-1Ni	8	2	1	Balance
Al-8Ca-2Mn-2Ni	8	2	2	Balance
Al-8Ca-2Mn-3Ni	8	2	3	Balance
Al-8Ca-2Mn-4Ni	8	2	4	Balance
Al-6Ca-3Mn-2Ni	6	3	2	Balance
AlSi18Cu1Mg1Ni1 (FM180)	1Cu	0.2	1Ni + 1Mg	Balance

Heat treatment of ingots was carried out at 500 °C in a muffle electric furnace SNOL 8.2/1100 (Umega Group, AB, Utena, Lithuania) with a temperature deviation of about 3 °C. This procedure is necessary for increasing the plasticity of ingots by fragmentation and spheroidisation of eutectic intermetallics. Hot rolling of ingots was carried out at a speed of 0.2 m/s on a reversible laboratory duo-mill with a maximum rolling width of 250 mm. An initial ingot's thickness was 15 mm. We conducted rolling with a 10% reduction for each pass and an average total reduction ratio of 80%.

The microstructure was examined by scanning electron microscopy (SEM, TESCAN VEGA3, Tescan Orsay Holding, Brno, Czech Republic) equipped with an electron microprobe analysis (EMPA, Oxford Instruments plc, Abingdon, UK), and the Aztec software (Oxford Instruments plc, Abingdon, UK). Thermo-Calc software (TTAL5 database) was used to calculate the phase composition of the alloys [16]. Actual phase transformation temperatures were determined using a DSC 404 F1 Pegasus differential scanning calorimeter (Netzsch Group, Selb, Germany). X-ray analysis was carried out on a Bede D1 System (Bruker, Karlsruhe, Germany) with Cu Kα radiation (λ = 0.15406 nm) and treated by software package [17]. Electrical conductivity (EC) was determined using the eddy current method with a VE-26NP eddy structures instrument (CJSC Research institute of introscopy SPEKTR, Moscow, Russia). A high purity Al was employed for comparison purposes. Vickers hardness (HV) was determined using a Wilson Wolpert 930 N setup (Wilson Instruments, Instron Company, Norwood, MA, USA) at 50 N load and 10 s dwell time. Room temperature tensile tests for the sheet samples of 150 mm × 12 mm in dimensions were carried out using a Zwick/Z 250 setup (Zwick Roell AG, Ulm, Germany).

3. Results and Discussion

3.1. Al-Ca-Mn Alloys

Liquidus projection of the Al-Ca-Mn system (Figure 2) was calculated for substantiating the choice of the experimental hypereutectic alloys solidified under different phase fields. Experimental investigations were next conducted for revision of the isothermal boundaries. Since Al_4Ca primary crystals were reported with coarse platelet shape [7], most of the selected compositions encountering crystallisation of the primary Al_6Mn phase. Approximately 30–40 wt.% intermetallics in the structure are expected to provide essential physical and mechanical properties.

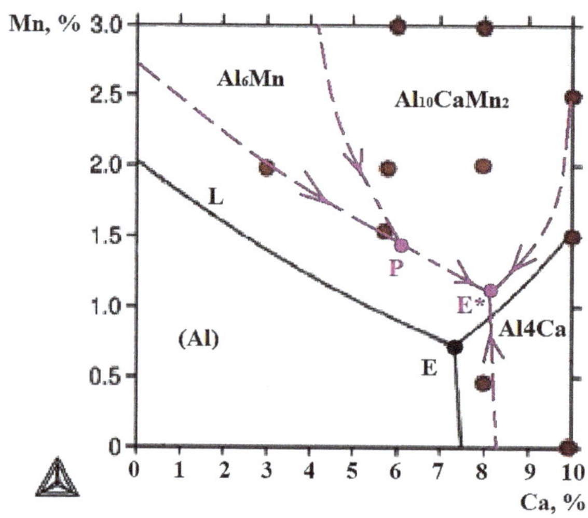

Figure 2. The predicted view of the Al-Ca-Mn phase diagram in the Al-rich region (solid line—calculation, dotted line—experiment).

Al-3Ca-2Mn, Al-6Ca-1.5Mn, Al-8Ca-0.5Mn and Al-10Ca-1.5Mn alloys are the closest to the phase-field boundaries. Microstructural observations and EMPA analysis of these alloys showed that Al-3Ca-2Mn alloy (Figure 3a) almost totally includes the colonies of the binary eutectic [(Al) + Al_4Ca]. Some insufficient Al_6Mn particles (including some calcium as solute) are also observed in the structure, while according to the equilibrium conditions their amount should have been far larger. A very fine eutectic structure in the Al-6Ca-1.5Mn alloy is decorated by equiaxed crystals of the ternary compounds (Figure 3b), namely $Al_{11}CaMn_2$, as the latter was determined according to EMPA. It should be noted that this ternary compound is not counted in the Thermo-Calc calculation but its existence indicates one more phase field in the phase diagram. Furthermore, the Al-6Ca-1.5Mn alloy displays some primary crystals of the Al_6Mn phase. The structure of the Al-8Ca-0.5Mn alloy (Figure 3c) includes eutectic [(Al) + Al_4Ca] and the primary (Al) dendrites of a small fraction. The Al-10Ca-1.5Mn alloy exhibits primary Al_4Ca crystals and a minor amount of the $Al_{11}CaMn_2$ phase. When considering the Al-10Ca-2.5Mn alloy (Figure 3e), primary Al_4Ca and $Al_{11}CaMn_2$ phases co-exist in equal fractions due to apparent phase-field boundary composition. Ultimately, primary Al_4Ca crystals and [(Al) + Al_4Ca] fine binary eutectic are observed in the structure of the Al-10Ca alloy (Figure 3f).

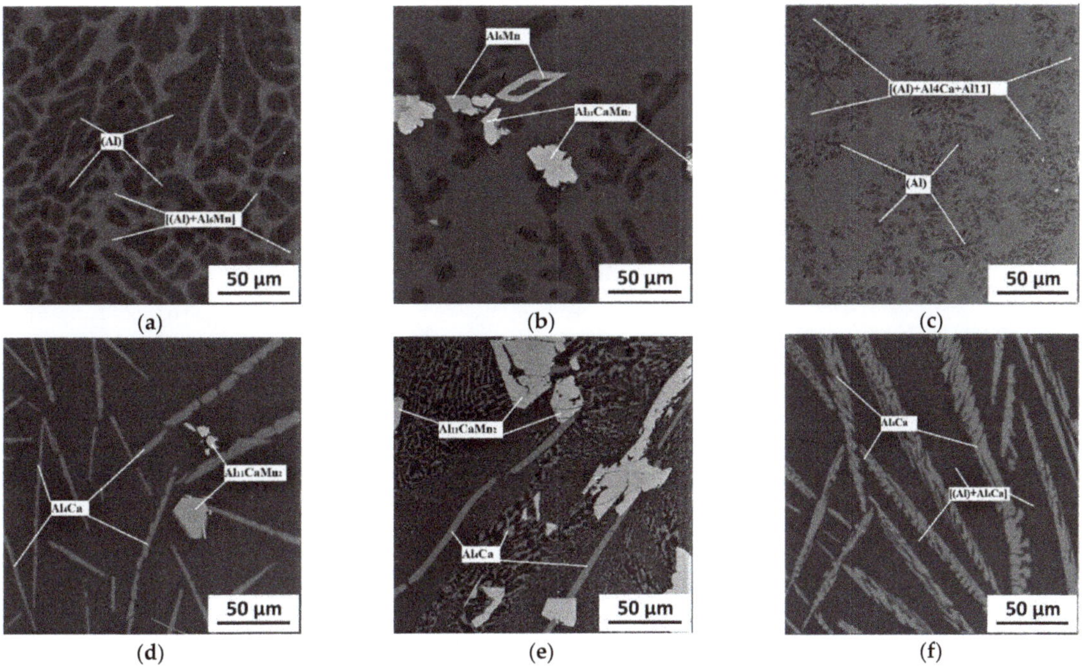

Figure 3. SEM images showing the microstructure of the experimental Al-Ca-Mn alloys in the as-cast condition: (**a**) Al-3Ca-2Mn; (**b**) Al-6Ca-1.5Mn; (**c**) Al-8Ca-0.5Mn; (**d**) Al-10Ca-1.5Mn; (**e**) Al-10Ca-2.5Mn; (**f**) Al-10Ca.

According to the aforementioned investigations, we can reasonably propose a shift of isothermal lines and extension in the (Al) existence region as it is shown in Figure 2 (dotted line). Thus, an Al-rich angle liquidus projection has been predicted. In this regard, two nonvariant reactions including peritectic L + $Al_6Mn \rightarrow (Al) + Al_{10}Mn_2Ca$ (point P) and eutectic $L \rightarrow (Al) + Al_4Ca + Al_{10}Mn_2Ca$ (point E) are expected.

The Al-6Ca-3Mn and Al-8Ca-2Mn alloys appeared to have the most favourable NMMC microstructure. Their solidification starts with the formation of the $Al_{10}CaMn_2$ phase (Figure 1a). In the Al-8Ca-2Mn alloy, a few compact angular crystals of Al_6(Mn, Fe) appeared likely due to the presence of Fe impurity. According to the calculation and structural observations, these alloys manufactured under graphite mould gravity casting (cooling rate 10 K/s) exhibit approximately 40% intermetallics (Table 2). They consist of fine eutectic (containing approximately 8%Ca and 1%Mn) and the $Al_{10}CaMn_2$ primary crystals ranged between 10 and 20 µm in linear size (Figure 4a,b). The FM180 alloy counterpart prepared under the same conditions contains less than 20% (Si) crystals of more than 40–50 µm (Figure 4c). As the eutectic structure of the Al-6Ca-3Mn and Al-8Ca-2Mn alloys is sufficiently thinner as compared to the FM180 alloy, spheroidising annealing may be utilised for improving ductility [1,10]. Some physical and mechanical properties are presented in Table 2.

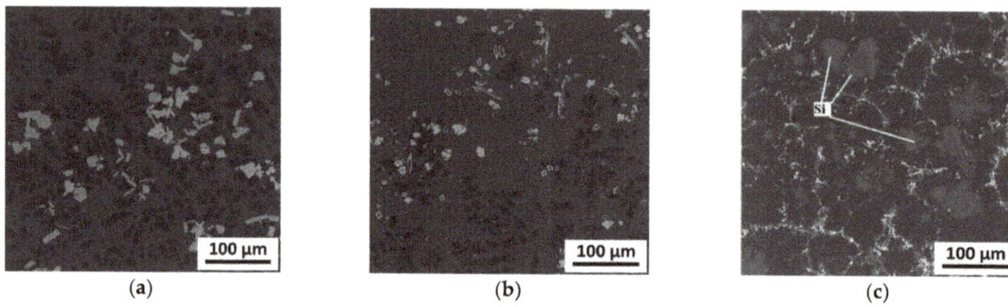

Figure 4. SEM images showing the microstructure of the experimental as-cast alloys showing a natural composite structure and FM180 alloy: (**a**) Al-6Ca-3Mn; (**b**) Al-8Ca-2Mn; (**c**) FM180.

Table 2. Calculated phase transformation temperatures, hardness and electrical conductivity of the as-cast Al-Ca-Mn alloys.

Alloy	T_L	T_S	ΔT	HV	* Ω, MSm/m
Al-6Ca-1.5Mn	665	611	54	70 ± 2	10.64 ± 0.01
Al-6Ca-2Mn	686	611	75	77 ± 3	9.45 ± 0.02
Al-6Ca-3Mn	721	611	110	72 ± 4	10.98 ± 0.01
Al-8Ca-2Mn	700	611	89	82 ± 2	9.43 ± 0.03
Al-8Ca-0.5Mn	622	611	11	63 ± 3	13.46 ± 0.01
Al-8Ca-3Mn	735	611	124	78 ± 4	9.60 ± 0.01
Al-10Ca-1.5Mn	689	611	77	93 ± 3	6.72 ± 0.12
Al-10Ca-2.5Mn	730	611	119	114 ± 10	8.37 ± 0.02

* hereinafter electrical conductivity is designated as Ω.

Instead of the $Al_{11}CaMn_2$ compound roughly estimated by EMPA, the $Al_{10}Mn_2Ca$ compound with a precise atomic composition was determined by X-ray analysis. The latter was utilised for the Al-6Ca-3Mn alloy solidified in furnace ambience as it shows a far larger fraction (12.4 vol.%) of the large ternary phase in the structure (Figure 5a). While the X-ray database does not contain the foregoing phase, the latter is known for the related systems $CaCr_2Al_{10}$ and YMn_2Al_{10} [18,19]. A crystal lattice belongs to the tP52/2 space group and has the parameters a = 1.2845 nm, c = 0.5134 nm (Figure 5b).

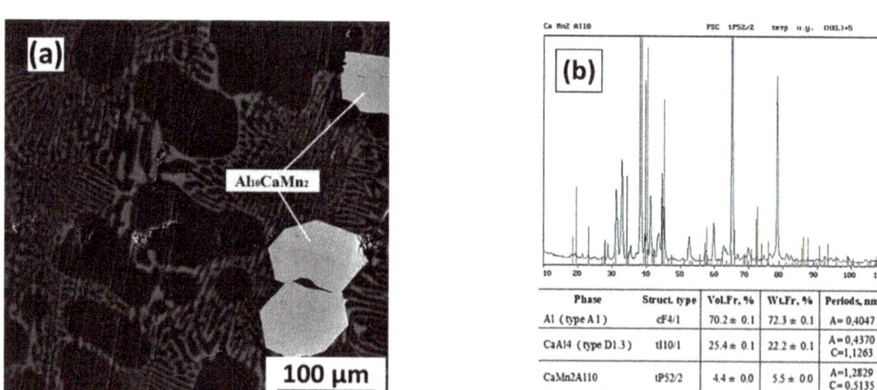

Figure 5. SEM images showing the $Al_{10}Mn_2Ca$ crystals appeared in the Al-6Ca-3Mn alloy after furnace cooling (**a**) and X-ray analysis data identified the $Al_{10}Mn_2Ca$ compound (**b**).

3.2. *Al-Ca-Mn-Ni Alloys*

In a prior study [12], Al-Ca-Ni alloys were reported with a ternary Al_9CaNi compound primary, crystals of which have a compact shape. It was assumed that nickel could improve the complex physical and mechanical properties of Al-Ca-Mn ternary alloys. Accordingly, Al-8Ca-2Mn-(1-4)Ni alloys were studied in detail (Table 1). The calculated liquidus projection of the Al-Ca-Mn-Ni system at 8% Ca is shown in Figure 6. Some physical and mechanical properties of the as-cast alloys are presented in Table 3.

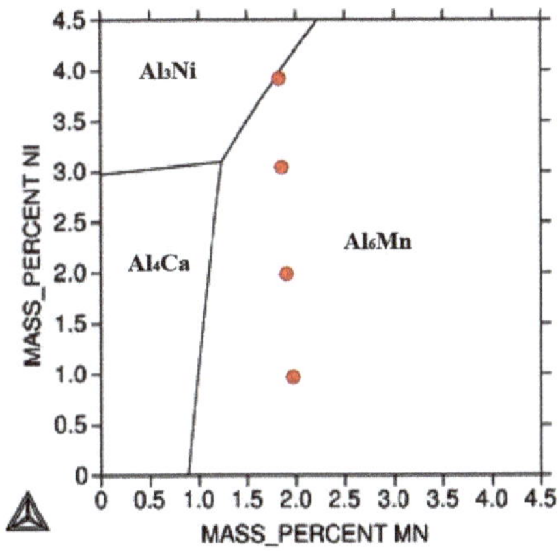

Figure 6. Calculated liquidus projection of the Al-Ca-Mn-Ni system at 8% Ca.

Table 3. Calculated phase transformation temperatures, hardness and electrical conductivity of the as-cast Al-Ca-Mn-Ni alloys.

Alloy	T_L	T_S	ΔT	HV	Ω, MSm/m
Al-8Ca-2Mn-1Ni	699	592	107	97 ± 3	9.51 ± 0.02
Al-8Ca-2Mn-2Ni	699	607	92	98 ± 3	9.18 ± 0.02
Al-8Ca-2Mn-3Ni	700	607	93	100 ± 3	9.12 ± 0.02
Al-8Ca-2Mn-4Ni	700	607	93	103 ± 3	7.00 ± 0.02
AlSi18Cu1Mg1Ni1 (FM180)	567	534	123	109 ± 3	13,7 ± 0,02

The Al-8Ca-2Mn-1Ni alloy solidified in the furnace ambience exhibits two types of primary crystals including grey-coloured Al_4Ca platelets and compact bright $Al_{10}CaMn_2$ crystals (Figure 7a). The latter contains a minor amount of nickel which likely substitutes manganese. In addition, ternary [(Al)+Al_4Ca+$Al_8Ca(Ni,Mn)_2$] and binary [(Al)+Al_4Ca] eutectics are observed in the structure. Due to the increase in the mean atomic number the Ni-rich phase is brighter in the BSE image as compared to other phases. EMPA revealed that some Mn dissolves in the Al_8CaNi_2 phase. In the furnace cooled Al-6Ca-3Mn-1Ni alloy, the primary $Al_8Ca(Ni, Mn)_2$ crystals have a compact shape and linear size of less than 10 μm (Figure 7b). Apart from the primary crystals, the presence of two binary eutectics [(Al)+Al_4Ca] and [(Al)+$Al_8Ca(Ni,Mn)_2$] is detected.

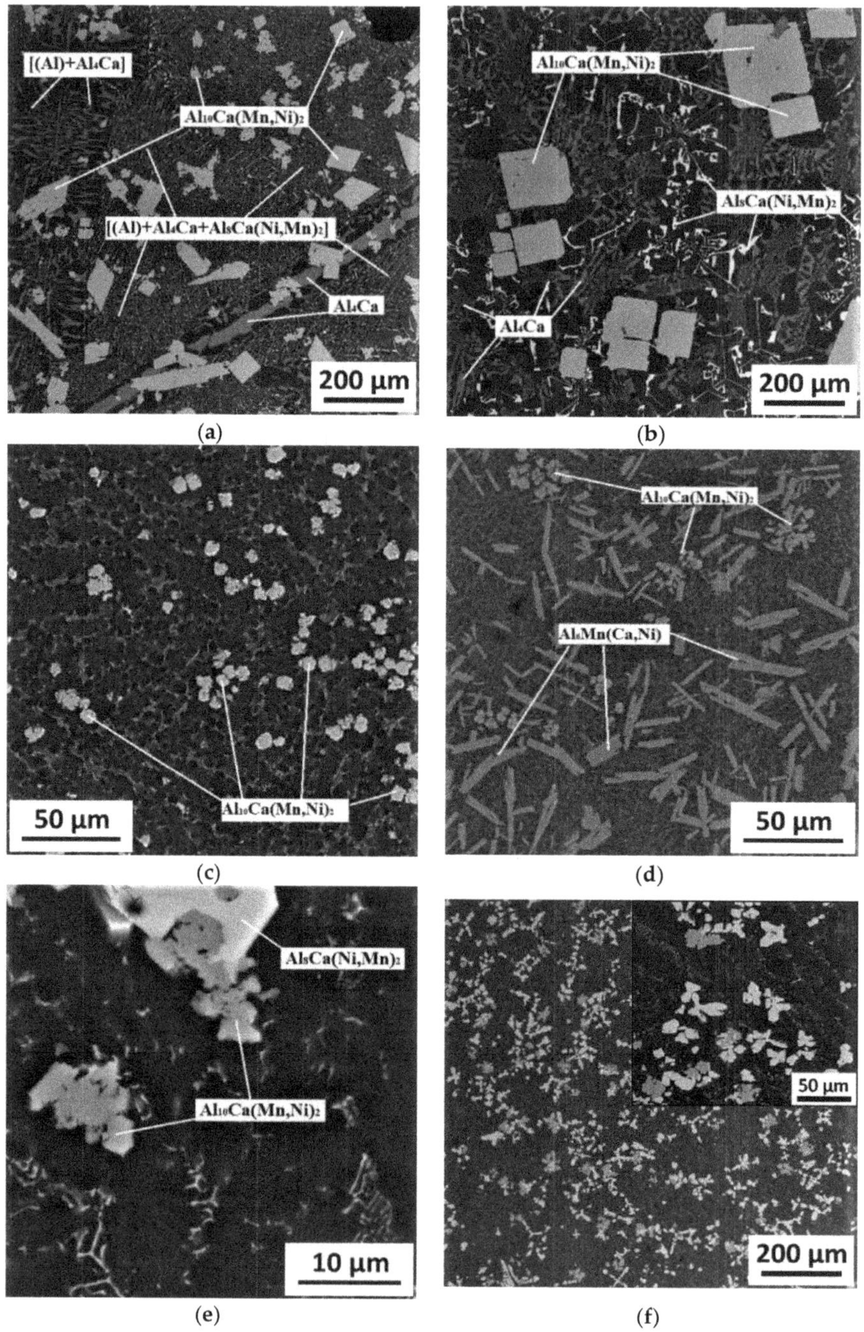

Figure 7. SEM images showing the primary crystals appeared in the as-cast Al-Ca-Mn-Ni alloys: (**a**) Al-8Ca-2Mn-1Ni (slow cooled); (**b**) Al-6Ca-3Mn-2Ni (slow cooled); (**c**) Al-8Ca-2Mn-1Ni (10 K/s); (**d**) Al-6Ca-3Mn-2Ni (10 K/s); (**e**) Al-8Ca-2Mn-2Ni (slow cooled); (**f**) Al-8Ca-2Mn-4Ni (10 K/s).

The structure of the Al-8Ca-2Mn-1Ni alloy formed at a cooling rate of 10 K/s (Figure 7c) represents a perfect natural composite and therein equiaxed small (no more than 10–15 μm) $Al_{10}Ca(Mn,Ni)_2$ ternary compound particles are evenly distributed on the superfine eutectic background. The latter contains about 8% Ca and 0.8% Mn. Similarly, the Al-6Ca-3Mn-2Ni alloy structure consists of the eutectic and two types of primary crystals established to be $Al_{10}Ca(Mn, Ni)_2$ of compact angular shape and $Al_6(Mn,Ca,Ni)$ of coarse acicular shape (Figure 7d and Table 4). It is apparent that if comparing two alloys, the Al-8Ca-2Mn-1Ni alloy has the most favourable structure as it contains fine primary crystals of less than 10–12 μm uniformly distributed in the ultrafine eutectic matrix. In the alloy Al-8Ca-2Mn-2Ni, the $Al_{10}Ca(Mn, Ni)_2$ primary crystals appeared along with bright $Al_8Ca(Ni, Mn)_2$ equiaxed crystals, having visible facets (Figure 7e). Their amount increases with Ni content. In Al-8Ca-2Mn-3Ni and Al-8Ca-2Mn-4Ni alloys manufactured at a cooling rate of 10 K/s (Figure 7f), the structure consists of dispersed eutectic and uniformly distributed compact crystals of two types including grey $Al_{10}Ca(Mn, Ni)_2$ and light $Al_8Ca(Ni, Mn)$ (Table 4).

Table 4. Composition of primary crystals according to EMPA analysis.

Alloy	Morphology	Phase	Concentration, Mass % (at.%)				
			Ca	Ni	Fe	Mn	Al
Al-6Ca-3Mn	Compact angular	$Al_{10}CaMn_2$	9.5 (7.5)	-	-	24.4 (14.2)	66.1 (78.3)
	Coarse acicular	$Al_6(Mn,Ca)$	7.8 (5.9)	-	-	15.8 (8.7)	76.4 (85.4)
Al-8Ca-2Mn	Compact angular	$Al_{10}CaMn_2$	9.7 (7.7)	-	-	22.9 (13.3)	67.4 (79)
	Coarse acicular	$Al_6(Mn,Fe)$	-	-	5.6(3.1)	18.8 (10.4)	75.6 (86.5)
Al-8Ca-2Mn-1Ni	Compact angular	$Al_{10}Ca(Mn,Ni)_2$	9.3 (7.4)	2.2 (1.2)	-	20.1 (11.6)	68.4 (79.8)
Al-6Ca-3Mn-2Ni	Compact angular	$Al_{10}Ca(Mn,Ni)$	7.8 (5.9)	0.8 (0.4)	-	18.0 (10.1)	73.4 (83.6)
	Coarse acicular	$Al_6(Mn,Ca,Ni)$	6.9 (5.1)	2.6 (1.3)	-	10.8 (5.9)	79.7 (87.7)
Al-8Ca-2Mn-(2-4)Ni	Compact angular	$Al_{10}Ca(Mn, Ni)_2$	9.4 (7.5)	3.1 (1.7)	-	20.4 (11.9)	67.1 (78.9)
	Coarse acicular	$Al_8Ca(Ni,Mn)_2$	10.5 (8.8)	26.4 (15.2)	-	4.5 (2.8)	58.6 (73.2)

For assessing processability at deformation the experimental Al-6Ca-3Mn, Al-8Ca-2Mn, Al-8Ca-2Mn-(1-4)Ni alloys were chosen as their microstructure best fits the NMMC. Before hot rolling, the ingots of all selected alloys, as well as the FM180 ingot, were annealed at a temperature of 500 °C for 3 h to improve ductility. While the FM180 alloy has been established as a casting alloy, some research reported its wrought processing. Novel Al-Ca hypereutectic alloys are expected to be uni-materials, a good choice either for casting or rolling. After annealing a substantial decrease in hardness and an increase in EC value are both due to sufficient change in eutectic particles shape.

During annealing, the fine eutectic-origin intermetallics acquire a rounded shape, the hardness of the alloys reasonably decreases and the electrical conductivity increases (Table 5). It is prominent that the FM180 alloy showed less response to spheroidisation due to the coarser eutectic structure.

Table 5. Hardness and electrical conductivity of the alloys in the annealed condition.

Alloy	HV	Ω, MSm/m
Al-6Ca-3Mn	62 ± 1	13.8 ± 0.02
Al-8Ca-2Mn	73 ± 3	12.9 ± 0.02
Al-8Ca-2Mn-1Ni	72 ± 2	13.2 ± 0.02
Al-8Ca-2Mn-2Ni	73 ± 2	11.5 ± 0.02
Al-8Ca-2Mn-3Ni	78 ± 2	10.9 ± 0.02
Al-8Ca-2Mn-4Ni	80 ± 2	10.1 ± 0.02
AlSi18Cu1Mg1Ni1 (FM180)	85 ± 3	14.5 ± 0.13

Furthermore, the Al-6Ca-3Mn, Al-8Ca-2Mn and FM180 ingots of 15 mm in thickness were rolled at 480 °C on a duo-mill 260. The rolling was carried out at a 10% reduction for each pass. In 11 passes, the Al-6Ca-3Mn and Al-8Ca-2Mn alloys reached 2.06 and 2.22 mm in thickness without any breaks and other defects (Figure 8a). The FM180 was rolled down to 3.3 mm after 10 passes and broke down during the 11th pass (Figure 8b). While the FM180 alloy is a foundry alloy, hot stamping is also applied for manufacturing pistons [20,21]. Nevertheless, the ductility of this material appeared to be insufficient for obtaining quality hot-rolled sheet products. The Al-8Ca-2Mn-(1-4)Ni alloys were rolled at 500 °C. As for the Al-8Ca-2Mn-(2-4)Ni samples, they showed transverse cracks after the first two passes. Conversely, the Al-8Ca-2Mn-1Ni alloy was rolled down to 3 mm in thickness without any visible cracks including those on the edge surfaces (Figure 8c). The primary intermetallics in all Ca-bearing alloys do not change their size and shape after deformation but become redistributed and aligned in the rolling direction. When it comes to the FM180 alloy, multiple cracks appeared in the silicon particles (Figure 9). The Al-6Ca-3Mn and Al-8Ca-2Mn hot-rolled sheets were annealed at 500 °C for 1 h and next subjected to cold rolling down to a thickness of 0.5 mm.

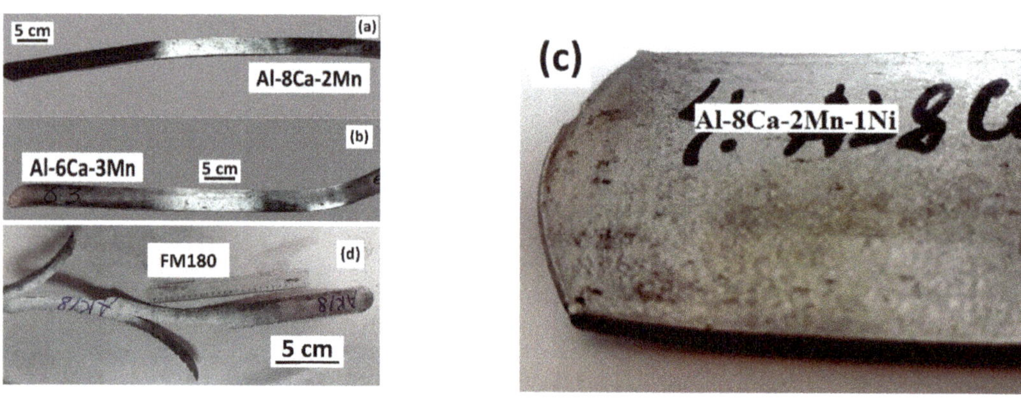

Figure 8. Hot-rolled sheets: (**a**) Al-6Ca-3Mn; (**b**) Al-8Ca-2Mn; (**c**) Al-8Ca-2Mn-1Ni; (**d**) FM180.

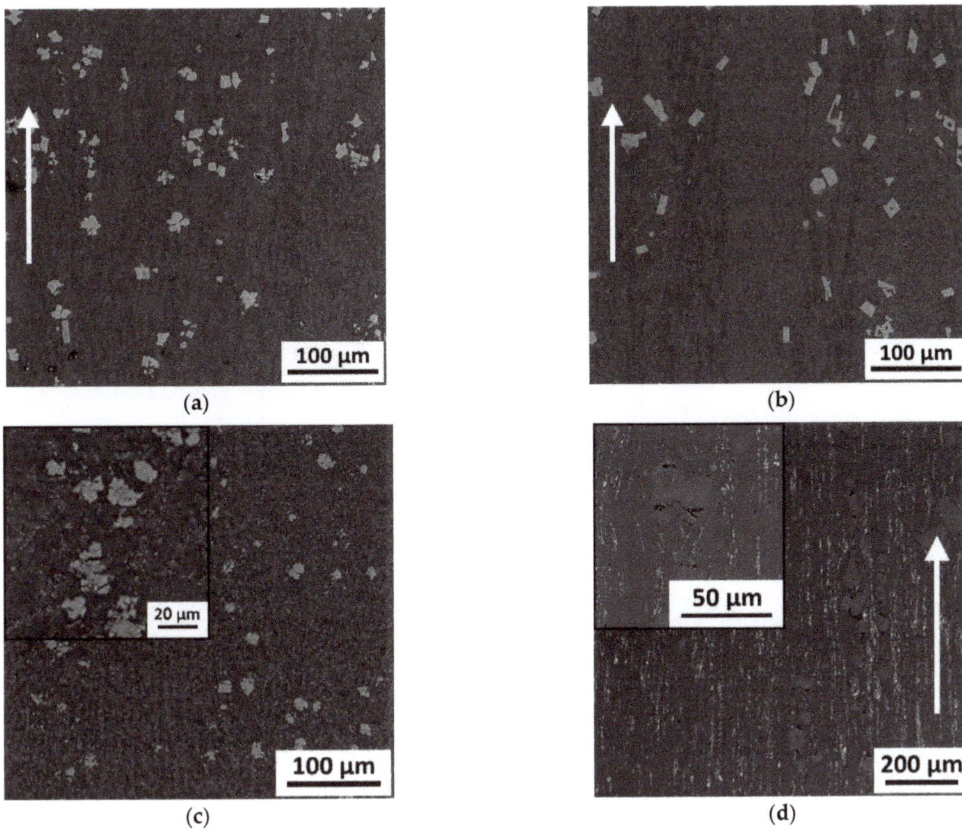

Figure 9. SEM images showing the microstructures of the hot-rolled sheets (arrows indicate the rolling direction): (**a**)—Al-6Ca-3Mn (longitudinal direction); (**b**)—Al-8Ca-2Mn (longitudinal direction); (**c**)—Al-8Ca-2Mn-1Ni (transversal direction); (**d**)—FM180 (longitudinal direction).

Then, the tensile test specimens were cut from Al-6Ca-3Mn and Al-8Ca-2Mn hot-rolled and cold-rolled sheets and a Al-8Ca-2Mn-1Ni hot-rolled sheet. Additionally, unbroken parts of the FM180 hot-rolled sheet were taken for testing. Mechanical and some physical properties of the sheets are given in Table 6. In terms of mechanical properties in the hot-rolled state, the quaternary Al-8Ca-2Mn-1Ni alloy is superior as compared to others.

Table 6. Physical and mechanical properties of the hot-rolled and cold-rolled sheets.

Alloy	Hot-Rolled Sheets					Cold-Rolled Sheets + 500,1 h, Air				
	YS, MPa	UTS, MPa	El, %	HV	Ω, MSm/m	YS, MPa	UTS, MPa	El, %	HV	Ω, MSm/m
Al-6Ca-3Mn	158	250	4.3	73	17.42	145	160	2.2	65	11.68
Al-8Ca-2Mn	162	260	3.3	82	17.10	164	166	0.25	71	15.00
Al-8Ca-2Mn-1Ni	321	428	3.2	98	15.75	-	-	-	-	-
FM180	264	315	0.7	111	18.2	-	328	0.1	131	9.52

3.3. *Coefficient of Thermal Expansion of the Ternary and Quaternary Alloys*

The alloys with the most favourable microstructure were selected for thermal expansion studying at 25–500 °C. According to [22], the CTE value of a binary hypereutectic Al—18%Ca alloy in the temperature range of 20–100 °C is 13.8 × 10^{-6} 1/°C. It is known from the literature that in the temperature range of 30–200 °C, the CTE value of the permanent mould cast hypereutectic Al alloys containing 17–19%Si is in the range $20{\cdot}10^{-6}$–22 × 10^{-6} 1/°C [23]. As is shown in Table 7, Ni alloying slightly changes CTE value in the temperature range 20–100 °C. Meanwhile, the Al-8Ca-2Mn and Al-8Ca-2Mn-1Ni alloys after 400 °C exhibit a decrease in CTE down to (19–20) × 10^{-6} 1/°C.

Table 7. The CTE values of the alloys exhibited the most favourable microstructure.

Alloy	CTE, 10^{-6} 1/°C				
	Temperature Range, °C				
	25–100	25–200	25–300	25–400	25–500
Al-6Ca-3Mn	22.72	22.76	23.24	23.31	22.18
Al-8Ca-2Mn	21.58	21.7	21.57	19.12	19.35
Al-8Ca-2Mn-1Ni	21.24	21.48	21.3	20.29	18.19
Al-8Ca-2Mn-4Ni	20.44	21.01	21.61	21.98	22.19

4. Discussion

Summing up the results, the Al-6Ca-3Mn, Al-8Ca-2Mn and Al-8Ca-2Mn-1Ni alloys may be reasonably considered as a promising basis for developing novel alloys for special applications as an alternative to the hypereutectic Al-Si alloys. Al-Ca alloys show the same fraction of primary crystals as those in the FM180 alloy (~5 vol.%) but have more favourable phase transformation temperatures. As follows from the thermal analysis (Figure 10), the Al-6Ca-3Mn and Al-8Ca-2Mn alloys have a higher solidus temperature as compared to the FM180 alloy. This feature implies the opportunity for conducting the annealing of the ingots at a higher temperature and higher thermal stability in the exploiting conditions. The results obtained by direct and differential thermal analysis appeared to be virtually similar.

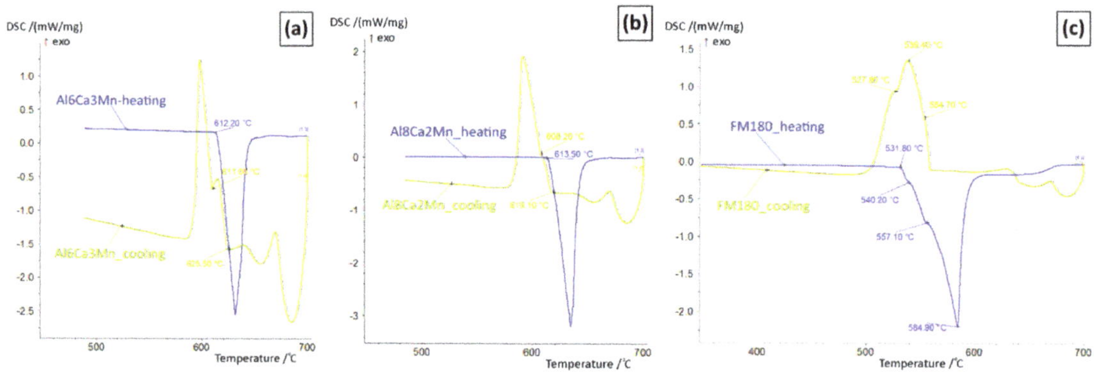

Figure 10. Results of the differential thermal analysis and direct thermal analysis on the: (**a**) Al-6Ca-3Mn; (**b**) Al-8Ca-2Mn; (**c**) FM180 alloys.

Since the Thermo-Calc software does not take into account the formation of ternary compounds in the Al-rich angle, the experimental Al-Ca-Mn-Ni liquidus projection should be different from the calculated one. Hence, in addition to the ternary $Al_{10}Ca(Mn,Ni)_2$ phase which appeared in the alloys containing more than 1%Ni, there is a ternary compound $Al_8Ca(Ni, Mn)_2$ based on the Al_8CaNi_2 phase. However, the Al_9CaNi compound formed in the Al-Ca-Ni-La alloys is indicated in [12]. In the handbook [24], the presence of both Al_8CaNi_2 and Al_9CaNi phases is alleged in the Al-rich angle of the Al-Ca-Ni ternary system. Having conducted the current study, we can propose the appearance of the Al-Ca-Mn-Ni diagram in the Al-rich angle (Figure 11). Nevertheless, the existence of two ternary compounds Al_8CaNi_2 and Al_9CaNi requires experimental verification. According to the hypothetical view of the phase diagram, there are five primary crystallisation regions: Al_4Ca+L; Al_3Ni+L; Al_6Mn+L; $Al_{10}CaMn_2$+L; Al_8CaNi_2+L. As is shown in the work [25,26], the T ($Al_{16}Mn_3Ni$) compound indicated in the diagram is present in the ternary Al-Mn-Ni system. Moreover, three invariant reactions including two peritectic L+Al_3Ni→(Al)+Al_6Mn+Al_8CaNi_2 (point P1); L+ Al_6Mn→(Al)+$Al_{10}Mn_2Ca$+Al_8CaNi_2 (point P2) and one eutectic L→(Al)+Al_4Ca+$Al_{10}Mn_2Ca$+Al_8CaNi_2 (point E) are drawn.

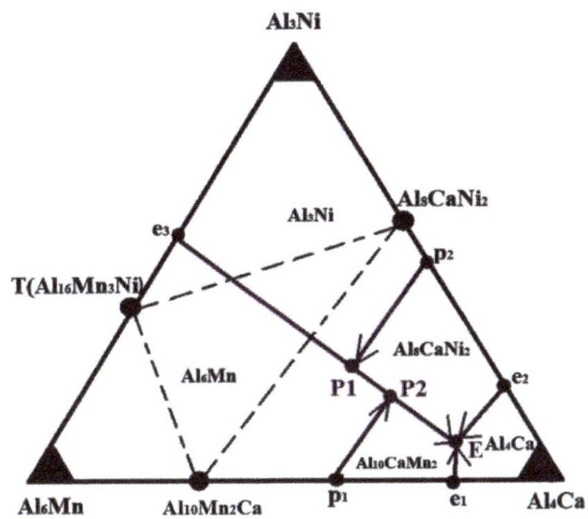

Figure 11. Hypothetical distribution of the solid phase fields in the Al-Ca-Mn-Ni system.

The Ca-bearing alloys are better in deformation than the Al-Si piston alloy due to the remarkable fineness of the eutectic structure. Poor processability of the Al-8Ca-2Mn-(2-4)Ni may be associated with a high number of the Al_8CaNi_2 phase. The latter, despite having a compact shape and small size of 40 μm, appears to act as a stress raiser causing brittle interfacial fracture (Figure 12). In contrast, the $Al_{10}Ca$ (Mn, Ni)$_2$ crystals, likely less strong, break in a fragile manner simultaneously with the matrix that ensures fewer stresses. Notwithstanding the cracks that appeared at the very beginning of the rolling, the Al-8Ca-2Mn-(2-4)Ni alloys were successful in a way similar to the FM180 alloy. That is stipulated by the superfine ductile matrix surrounding the brittle Al_8CaNi_2 crystals.

(a) (b)

Figure 12. Fracture surfaces of the hot-rolled samples: (**a**) Al-8Ca-2Mn-1Ni; (**b**) Al-8Ca-2Mn-2Ni.

Thus, we can reasonably conclude that, despite the promising natural composite structure of the Al-8Ca-2Mn-(2-4)Ni alloys, they are not feasible in hot deformation due to the presence of the high fraction of brittle compound. Therefore, the stiffness of the primary crystals should be comparable to that of the matrix (surrounding eutectic structure) that may be noted as a requirement to the structure of the new Al-Ca hypereutectic natural composites. With a noticeable difference in strength between matrix and primary crystals, even if the latter has a small size and uniform distribution within the bulk, they appear to become places where brittle deformation occurs. Summing up, the $Al_{10}Mn_2Ca$ crystals mostly meet the foregoing requirements.

5. Conclusions

(1) The composition and morphology of the primary crystals that appeared in the hypereutectic Al-(3-10)Ca-(0-3)Mn-(0-4)Ni alloys were studied using experimental and computational methods including the construction of the liquidus projections. As a result, the hypothetic Al-Ca-Mn иAl-Ca-Mn-Ni phase diagrams were proposed in an Al-rich angle.

(2) An appearance of the $Al_{10}CaMn_2$ compound, unknown in the X-ray databases, was established in the Al-Ca-Mn system. It was found that the crystal lattice belongs to the tP52/2 space group and has the parameters a = 1.2845 nm and c = 0.5134 nm. The new compound was observed with a relatively compact shape and medium size of 20 µm.

(3) The ternary Al-6Ca-3Mn and Al-8Ca-2Mn alloys were found to have high processability at hot and cold rolling. A crack-free hot-rolled sheet was manufactured from the quaternary Al-8Ca-2Mn-1Ni alloy. High plasticity at the rolling of the hypereutectic Al-Ca alloys was substantiated.

(4) A comparison of the physical and mechanical properties along with processability of the Al-6Ca-3Mn, Al-8Ca-2Mn and Al-8Ca-2Mn-1Ni experimental alloys and the commercial FM180 piston alloy with FM180 grade silumin was performed. The Al-8Ca-2Mn-1Ni was found to be an effective basis for the development of new-generation natural metal matrix composite alternatives to the Al-Si counterparts. What is more, experimental alloy exhibit microstructure characterised both by small-sized primary crystals and fine eutectic structure. In this respect, no special modification operation during melting and casting is required.

(5) It is shown that several features of the primary crystals must be maintained for achieving the appropriate performance of the Al-Ca hypereutectic alloys. While some primary crystals with equiaxed shape and uniform distribution cannot act as a

structural component in the new alloys, the $Al_{10}Mn_2Ca$ compound was revealed to be most appropriate.

Author Contributions: Conceptualisation, E.N.; investigation, V.D. and T.S.; methodology, V.D., M.B. and A.L.; resources, V.D., M.B., T.S. and A.L.; supervision, E.N.; writing—original draft, E.N.; writing—review and editing, E.N. and P.S. All authors have read and agreed to the published version of the manuscript.

Funding: This research was funded by the Russian Science Foundation (project no. 20–19–00746).

Institutional Review Board Statement: Not applicable.

Informed Consent Statement: Not applicable.

Data Availability Statement: Data are available from the corresponding author on reasonable request.

Acknowledgments: The authors are grateful to Aleksandr Koshmin, a postgraduate student of the Department of Metal Forming, NUST MISiS, for experimental alloys rolling.

Conflicts of Interest: The authors declare no conflict of interest.

References

1. Glazoff, M.; Khvan, A.; Zolotorevsky, V.; Belov, N.; Dinsdale, A. *Casting Aluminum Alloys: Their Physical and Mechanical Metallurgy*, 2nd ed.; Elsevier: London, UK, 2018; p. 608. [CrossRef]
2. Gloria, A.; Montanari, R.; Richetta, M.; Varone, A. Alloys for Aeronautic Applications: State of the Art and Perspectives. *Metals* **2019**, *9*, 662. [CrossRef]
3. Graf, A. Aluminum alloys for lightweight automotive structures. In *Materials, Design and Manufacturing for Lightweight Vehicles*, 2nd ed.; Woodhead Publishing in Materials: Cambridge, UK, 2021; pp. 97–123. [CrossRef]
4. Jorstad, J.; Apelian, D. Hypereutectic Al-Si Alloys: Practical Casting Considerations. *Int. J. Metalcast.* **2009**, *3*, 13–36. [CrossRef]
5. Zhang, H.-H.; Duan, H.; Shao, G.; Xu, L. Microstructure and mechanical properties of hypereutectic Al-Si alloy modified with Cu-P. *Rare Met.* **2008**, *27*, 59–63. [CrossRef]
6. Naumova, E.A. Use of Calcium in Alloys: From Modifying to Alloying. *Russ. J. Non-Ferr. Met.* **2018**, *59*, 284–298. [CrossRef]
7. Belov, N.A.; Naumova, E.A.; Akopyan, T.K.; Doroshenko, V.V. Phase Diagram of the Al-Ca-Fe-Si System and Its Application for the Design of Aluminum Matrix Composites. *JOM* **2018**, *70*, 2710–2715. [CrossRef]
8. Belov, N.A.; Naumova, E.A.; Akopyan, T.K.; Doroshenko, V.V. Design of Multicomponent Aluminium Alloy Containing 2 wt.% Ca and 0.1 wt.% Sc for Wrought and Cast Products. *JALCOM* **2018**, *762*, 528–536. [CrossRef]
9. Belov, N.A.; Naumova, E.A.; Bazlova, T.A.; Doroshenko, V.V. Phase Composition and Hardening of Castable Al-Ca-Ni–Sc Alloys Containing 0.3% Sc. *Met. Sci. Heat Treat.* **2017**, *59*, 76–81. [CrossRef]
10. Naumova, E.A.; Akopyan, T.K.; Letyagin, N.V.; Vasina, M.A. Investigation of the structure and properties of eutectic alloys of the Al-Ca-Ni system containing REM. *Non-Ferr. Met.* **2018**, *2*, 24–29. [CrossRef]
11. Rogachev, S.O.; Naumova, E.A.; Vasileva, E.S.; Magurina, M.Y.; Sundeev, R.V.; Veligzhanind, A.A. Structure and mechanical properties of Al-Ca alloys processed by severe plastic deformation. *Mater. Sci. Eng. A* **2019**, *767*, 13841. [CrossRef]
12. Akopyan, T.K.; Belov, N.A.; Naumova, E.A.; Letyagin, N.V.; Sviridova, T.A. Al-matrix composite based on the Al-Ca-Ni-La system additionally reinforced by the L12 type nanoparticles. *Trans. Nonferrous Met. Soc. China* **2020**, *30*, 850–862. [CrossRef]
13. Belov, N.A.; Naumova, E.A.; Doroshenko, V.V.; Bazlova, T.A. Effect of manganese and iron on the phase composition and microstructure of aluminum-calcium alloys. *Tsvetnye Met.* **2017**, *8*, 66–71. [CrossRef]
14. Rogachev, S.O.; Naumova, E.A.; Sundeev, R.V.; Tabachkova, N.Y. Structural and phase transformations in a new eutectic Al-Ca-Mn-Fe-Zr-Sc alloy induced by high pressure torsion. *Mater. Lett.* **2019**, *243*, 161–164. [CrossRef]
15. Muggerud, A.M.F.; Mørtsell, E.A.; Li, Y.; Holmestad, R. Dispersoid strengthening in AA3xxx alloys with varying Mn and Si content during annealing at low temperatures. *Mater. Sci. Eng. A* **2013**, *567*, 21–28. [CrossRef]
16. Thermo-Calc Software TTAL5 Al-Alloys. Available online: www.thermocalc.com (accessed on 20 April 2021).
17. Shelekhov, E.V.; Sviridova, T.A. Programs for X-ray analysis of polycrystals. *Met. Sci. Heat Treat.* **2000**, *42*, 309–313. [CrossRef]
18. Yaniv, G.; Vidal, D.; Fuks, D.; Meshi, L. Bonding and Stability of Ternary Structures in the CeT2Al20 (T=Ta, W, Re) and YRe2Al20 Alloys. *Metals* **2020**, *10*, 422. [CrossRef]
19. Janssen, Y.; Angst, M.; Dennis, K.W.; McCallum, R.W.; Canfield, P.C. Differential thermal analysis and solution growth of intermetallic compounds. *J. Cryst. Growth* **2005**, *285*, 670–680. [CrossRef]
20. Zhu, Q.; Rassili, A.; Midson, S.P.; Hu, X.G. Thixoforming of Hypereutectic AlSi12Cu2NiMg Automotive Pistons. *Solid State Phenom.* **2019**, *285*, 446–452. [CrossRef]
21. Prudnikov, A.N. Deformable heatproof transeutectic silumin for pistons. *Steel Transl.* **2009**, *39*, 456. [CrossRef]
22. Naumova, E.A.; Petrov, M.A.; Stepanov, B.A.; Vasilieva, E.S. Stamping with torsion of the Al—Ca alloy workpiece with high concentration of Al_4Ca. *Tsvetnye Met.* **2019**, *1*, 66–71. [CrossRef]

23. Hatch, J.E. *Aluminum: Properties and Physical Metallurgy*; American Society for Metals: Cleveland, OH, USA, 1984; pp. 320–350.
24. Petzow, G.; Effenberg, G. *Ternary Alloys: A Comprehensive Compendium of Evaluated Constitutional Data and Phase Diagrams*; Wiley-VCH: Berlin, Germany; Weinheim, Germany, 1990; Volume 3, p. 647.
25. Belov, N.A.; Aksenov, A.A.; Eskin, D.G. *Multicomponent Phase Diagrams: Applications for Commercial Aluminum Alloys*; Elsevier: Amsterdam, The Netherlands, 2005; pp. 223–256. [CrossRef]
26. Mondolfo, L.F. *Aluminium Alloys: Structure and Properties*; Butterworths: London, UK, 1976; pp. 806–841. [CrossRef]

metals

MDPI

Article

Influence of Mg Content on Texture Development during Hot Plain-Strain Deformation of Aluminum Alloys

Evgenii Aryshenskii [1], Jurgen Hirsch [1,2], Sergey Konovalov [1,*], Vladimir Aryshenskii [3] and Alexander Drits [3]

[1] Department of Metals Technology and Aviation Materials, Samara National Research University, Moskovskoye Shosse 34, 443086 Samara, Russia; ar-evgenii@yandex.ru (E.A.); juergen.hirsch@hydro.com (J.H.)
[2] Research and Development, Hydro Aluminium Rolled Products GmbH, D-53117 Bonn, Germany
[3] JSC Arconic SMZ, Alma-Atinskaya 33/34, 443051 Samara, Russia; vladimir.aryshensky@arconic.com (V.A.); Alexander.Drits@arconic.com (A.D.)
* Correspondence: ksv@ssau.ru; Tel.: +7-84-6267-4640

Abstract: The study addresses the effect of magnesium and other alloying elements on rolling "β-fiber" texture formation during hot deformation of aluminum alloys. For the study, flat cast ingots from three aluminum alloys with variable magnesium content were deformed in a Gleeble testing unit with different parameters of thermomechanical treatment. Immediately after completion of deformation, the samples were quenched using an automatic cooling system and the microstructure and crystalline texture was analyzed by optical microscopy and X-ray analysis. The analysis demonstrated that an increase in alloying components, magnesium in particular, leads to an increase in brass-type texture and a decrease in S and copper-type texture. The reason was that the simulation of the deformation texture development revealed a great contribution of impurity atoms rather than the decrease in stacking fault energy.

Keywords: aluminum alloys; deformation; magnesium content; texture; simulation

Citation: Aryshenskii, E.; Hirsch, J.; Konovalov, S.; Aryshenskii, V.; Drits, A. Influence of Mg Content on Texture Development during Hot Plain-Strain Deformation of Aluminum Alloys. *Metals* **2021**, *11*, 865. https://doi.org/10.3390/met11060865

Academic Editor: Frank Czerwinski

Received: 26 April 2021
Accepted: 21 May 2021
Published: 25 May 2021

1. Introduction

The improvement of casting and thermomechanical treatment practices is one of the most demanding aspect of modern material science [1–6]. Combination of high plasticity, fairly low density and reasonable strength makes aluminum alloys indispensable in multiple present-day industries [7–10]. Still, one of the main problems associated with the processing of aluminum alloys is the anisotropy of their physical and mechanical properties. Anisotropy is caused by the crystallographic texture formed during thermomechanical processing [11]. Significant reduction of anisotropy can be achieved by a careful selection of specific texture components, e.g., balancing the cube texture, giving high plasticity at 0°/90° to the rolling direction with the "β-fiber" texture of plane strain deformation (i.e., rolling) causing high plasticity at 45° [11]. The β-fiber itself is composed of three texture components (see Table 1 below), with the central S component giving maximum plasticity at 45° to the rolling direction.

Table 1. Euler angles of ideal texture components including "β-fiber" [12].

Component	Miller Indices {hkl}⟨uvw⟩	Euler Angles		
		φ_1	Φ	φ_2
Cube	{001}⟨100⟩	0°	0°	0°/90°
Cube RD	{013}⟨100⟩	0°	22°	0°/90
Cube ND	{001}⟨310⟩	22°	0°	0°/90°
Goss	{011}⟨100⟩	0°	45°	0°/90°
Bs	{011}⟨211⟩	35°	45°	0°/90°
Cu	{112}⟨111⟩	90°	30°	45°
S	{123}⟨634⟩	60°	35°	65°

The effect of the remaining two components on anisotropy has a more complicated nature. It can be illustrated by the earing profile, i.e., the non-uniform height and thickness of the wall after deep drawing of a cup from a cylindrical workpiece. Peaks are formed in the areas with higher plasticity along the rolling direction, while valleys form in the areas with lower plasticity. The authors of [13] showed that the earing profile and, as a result, anisotropy for each of the three β-fiber components is individual. Therefore, it is obvious that efficient anisotropy reduction will be possible by an individual texture component management, rather than controlling the overall volume of deformation texture.

Besides adequate control of the deformation process and the resulting distribution of β-fiber components, it is the management of the corresponding recrystallization processes that defines the quality of aluminum semi-products, based on final grain size and orientations and resulting mechanical properties. Here, the rolling texture component "S" has a high influence on the cube recrystallization texture formation via oriented growth due to its preferred 40° orientation relationship along the crystallographic direction <111> relative to the cube texture [14]. Therefore, the generation of an exact S texture component quantity is critical.

One of the factors influencing the distribution of texture components within the β-fiber is the stacking fault energy. For example, pure copper with medium stacking fault energy level develops as pronounced "β-fiber" rolling texture—named "copper texture" [12]. The main texture in low stacking fault energy FCC metals such as α-brass formed during plain-strain (e.g., rolling) deformation is also called "brass texture" [12,15,16].

Some of the recent studies [17] established a more continuous transition path from Cu-type texture to Bs-type texture and an orientation distribution "balance center" is identified, depending on SFE. High SFE values cause its shift towards the copper-type texture, while low values shift it towards the brass-type texture. The effect of a high value in SFE on deformation strengthening is caused by stimulating transverse and double transverse dislocations to glide, enabling dislocations to bypass obstacles. Such an effect reduces the intensity of deformation strengthening (strain hardening). During the simulation of plastic deformation by shear along slip planes, such a drop in strength intensity means a drop in critical shear stress as a function of shear strain between slip planes. In Reference [17], it is suggested that the intensification of strain hardening results in the activation of a larger number of slip planes for the same type of crystal deformation. Some slip systems may be favorably oriented, while others are less favorable for glide initiation, a behavior that is numerically expressed by the Taylor factor.

If there is no strain hardening, the glide pattern would follow the most favorably oriented slip systems with minimum shear stress action. In this case, the crystal would reach the theoretically stable position at relatively low strain. However, due to deformation strengthening, slip on the systems, which are favorably oriented with respect to the applied external force, can be less energetically favorable compared to slip on other slip systems with less strain hardening. A larger number of slip systems is activated and some orientation dispersion advances at the same strain level, resulting in less pronounced copper-type texture and relatively larger fraction of α-fiber textures, including the brass-type texture, which is attributed to both the α- and β-fiber. As mentioned in several studies [15–17], the evolution of the deformation texture starts with the development of an α-fiber texture, which consists of multiple texture components in the Euler angle space between the Goss texture and Bs texture, followed by the development of the β-fiber texture as deformation progresses.

As far as aluminum is concerned, there are various options for the distribution of texture components. This strongly depends on the content of alloy elements, specifically magnesium. In industrially pure aluminum alloy Al 99.5 (AA1050) the β-fiber intensity is higher on the copper side {112}<111> (φ_2 = 45°) and shifts at higher strains towards the S texture {123}<634> (φ_2 = 65°), whereas the brass component {110}<112> (φ_2 = 0°/90°) develops relatively low [18]. On the contrary, the Al-Mg alloy AA5182, having high (up to 5%) magnesium content, reveals a shift of the texture towards the brass component [19]. This texture pattern change is mainly associated with the effect of magnesium to maximally decrease the stacking fault energy. In pure aluminum, it reaches 174 mJ/m^2, and in the alloy with 6.28% magnesium content, it reaches 151.6 mJ/m^2 [20].

Thus, magnesium content increase has a strong effect on texture composition during plane hot deformation and subsequent recrystallization (due to oriented growth mechanism influence). Therefore, accounting for magnesium effect on texture evolution during aluminum alloys' thermomechanical treatment presents one of the key factors of the product anisotropy reduction. It should be noted that the listed above papers on aluminum just state the significant effect of magnesium on texture composition emerging during deformation. However, there are no detailed studies of quantitative evaluation of magnesium effect on β-fiber texture distribution. At the same time, there are studies providing detailed quantitative evaluation of Zn effect on deformation texture distribution in silver [21] and copper [22] alloys. The authors of [23] evaluated the Mn effect on deformation texture composition. Likewise, there are no studies accounting for magnesium effect during deformation textures' modeling. However, similar studies were conducted for copper alloys with Mn additions [24]. Adequate texture composition control during materials' thermodynamic processing is not possible without proper understanding of the quantified magnesium effect on aluminum alloys' deformation texture composition as well as specific features of this factor modeling.

The objective of this study is to investigate the effect of magnesium content on the development of the key β-fiber texture components during hot deformation. In addition, effective methods of this factor accounting during deformation texture modeling shall be demonstrated.

2. Experiment Description

To study the effect of magnesium content on texture formation during thermomechanical treatment, plain-strain deformation tests were performed on three representative aluminum alloys: Al95, AlMg1 and AlMg4. The exact alloy compositions are presented in Table 2 below.

Table 2. Chemical composition of the studied alloys.

Alloy	Element										
	Fe	Si	Mn	Cr	Ti	Al	Cu	Be	Mg	Zn	Impurities
Al95	Below 0.03	Below 0.03	-	-	Below 0.02	99.95	Below 0.015	-	-	-	-
Al-Mg1	Up to 0.7	Up to 0.3	Up to 0.2	Up to 0.1	-	97.15–99.5	Up to 0.2	-	1	Up to 0.2	others, each 0.05; total 0.15
Al-Mg4	Up to 0.5	Up to 0.4	0.2–0.7	0.05–0.25	Up to 0.15	93.15–96.23	Up to 0.1	0.0002–0.005	4.4	Up to 0.25	others, each 0.05; total 0.15

Al95 has no magnesium, Al-Mg1 has about 1%, and Al-Mg4 has 4.4% magnesium content. The samples were cast and homogenized at:

- 600–615 °C during 6 h for Al95;
- 595–610 °C during 6 h for Al-Mg1;

- 485–500 °C during 8 h for Al-Mg4.

Prior to deformation, the dimensions of the initial samples were as follows: length a = 30 mm, width b = 35 mm, thickness h = 10 mm.

Table 3 lists plain-strain deformation temperatures and rates for the different alloys Al-95, Al-Mg1 and Al-Mg4. The degree of rolling strain was 80% for all modes.

Table 3. Plain-strain deformation temperature and rate modes for the studied alloys.

Alloy	Mode ID	Plain-Strain Deformation Mode	
		Temperature, °C	Strain Rate, s^{-1}
Al95	91	300	1
	92	450	1
	93	300	10
	94	450	10
AlMg1	11	300	1
	12	450	1
	13	300	10
	14	450	10
AlMg4	41	300	1
	42	450	1
	43	300	10
	44	450	10

Figure 1 illustrates the test device for the deformation which was close to plain strain. TC are thermocouples—their locations in the sample and in the tool are indicated with arrows. Trapezoidal hammers are located at right angles to the rolling direction (see Figure 1). The hammer head part's width (w) is 10 mm.

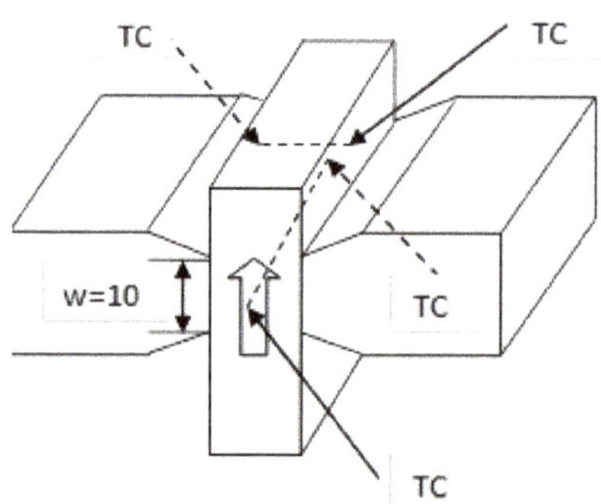

Figure 1. Test device for the deformation by trapezoidal hammers (TC are thermocouples).

The samples were heated in low vacuum up to deformation temperatures, listed in Table 3, at 3 °C/s heating rate. Slight compression (<5 kN) during heating ensured uniform contact between hammers and samples along the entire contact area. After the preset temperature was reached, followed by 3 min isothermal holding time, the sample was compressed to the defined strain degree with the preset strain rate. One second after deformation completion, the sample was cooled to room temperature by water flushing. Specimens were cut out from the workpiece middle section and prepared for grain structure and texture examination by optical microscopy and X-ray texture analysis.

The microstructure was examined using optical microscope Axiovert 40 MAT (Carl Zeiss AG, Oberkochen, Germany). Microsections of specimens were prepared by cutting out, mechanical grinding, polishing and electrolytic polishing in fluoroboric acid electrolyte, having the following composition: boric acid—11 g; fluorhydric acid—30 mL; distilled water—2200 mL.

Deformed samples cut out from the middle sections were used for texture determination by direct pole figure measurement with the survey plane perpendicular to the direction of compression, exerted by the hammers. The textures were measured by four incomplete pole figures {111}, {200}, {220} and {311}, obtained by the "diffraction" method, using X-ray diffractometer DRON-7 (Bourevestnik JSC, Saint-Petersburg, Russia) in $Co_{k\alpha}$-radiation. The incline angle range α was 0° to 70° and the rotation angle range β was 0° to 360° with 5° increments each. The intensity drop in the pole figure periphery due to defocusing effect was compensated by correction factors, calculated by accounting for the pole figures' X-ray photography conditions. The orientation distribution function (*ODF*) was calculated from the measured pole figures, presented as a superposition of a large number (2000) of standard distributions with identical low scattering. The same orientation distribution function (*ODF*) was used to calculate complete pole figures and inverse pole figures for three mutually perpendicular sample directions: normal to the rolling plane, rolling direction and transverse direction.

In addition, the chemical composition of the aluminum matrix and large intermetallic particles after treatment according to modes 91, 11 and 41 shown in Table 3 was investigated by electron microscopy. The JEOL 6390 A scanning electron microscope (SEM) (JEOL Ltd., Tokyo, Japan) was used for this purpose. The sample preparation technique consisted of mechanical grinding, polishing and electropolishing. Electropolishing of the specimen was carried out at 85–110 °C and the voltage of 10–30 V in electrolyte of the following composition: 500 mL of H_3PO_4, 300 mL of H_2SO_4, 50 g of CrO_3 and 50 mL of H_2O.

As mentioned earlier, rolled FCC lattice metals develop both copper (Cu) and brass (Bs)-type orientations. Rolling textures of high SFE metals and alloys show pronounced Cu orientations, while low SFE materials form strong Bs textures [24].

The deformation texture simulation model, described in [25], was applied for modeling the texture evolution during hot deformation. The incorporated modifications enabled to consider the effect of magnesium content on the evolution of the major β-fiber texture components due to the decrease in stacking fault energy. As stated in the Introduction, the evolution of texture intensity depends on the nature of deformation strengthening, thus requiring some modifications, introduced in the model in reference [25]. The model in reference [25] accounts only for the ratio, relating critical shear stress and shear strain rate, and is written as:

$$\gamma^{\alpha} = \gamma_0^{\alpha}\left(\frac{\tau_r^{\alpha}}{\tau_c^{\alpha}}\right)^m, \quad (1)$$

where γ^{α} is actual shear strain rate, $\gamma_0^{\propto}$ is basis shear strain rate value, $\tau_r^{\propto}$ is actual critical shear stress value, τ_c^{α}—basis critical shear stress value.

This study offers the following ratio for τ_r^{α} calculation:

$$\tau_r^{\propto} = \tau_c^{\propto}\varepsilon_{ac}^{\alpha\ n}, \quad (2)$$

where $\varepsilon_{ac}^{\alpha}$ is the parameter, characterizing deformation strengthening, (measure of crystalline structure imperfections (dislocations) accumulated in α-type slip system).

$$n = n_0\left(\ln(\dot{\varepsilon}) + \frac{Q}{RT}\right) \quad (3)$$

where $\dot{\varepsilon}$ is macroscopic sample strain rate, n_0 is basis n factor value within 0.3–0.5 range, Q is plastic deformation activation energy, taken as 140,000 J/m^3 for Al-95, 165,000 J/m^3 for AlMg1, 175,000 J/m^3 for AlMg4, R is absolute gas constant.

The parameter $\varepsilon_{ac}^{\alpha}$ is of generalized nature, accounting for various strengthening components: an SFE-specific ability of dislocations to bypass obstacles, dislocations' damping in solid solution, dislocations' retardation by forest dislocations and impurity atmosphere.

The parameter $\varepsilon_{ac}^{\alpha}$ is calculated using shear strain increments $d\varepsilon^{\alpha} = \gamma^{\alpha} dt$, where dt is a time increment:

$$\begin{bmatrix} d\varepsilon_{ac}^{1} \\ d\varepsilon_{ac}^{2} \\ d\varepsilon_{ac}^{3} \\ \cdots \\ \cdots \\ d\varepsilon_{ac}^{10} \\ d\varepsilon_{ac}^{11} \\ d\varepsilon_{ac}^{12} \end{bmatrix} = \begin{Vmatrix} q & p & p & r & r & \cdots & r & r & r & r & r \\ p & q & p & r & r & \cdots & r & r & r & r & r \\ p & p & q & r & r & \cdots & r & r & r & r & r \\ \cdots & \cdots & \cdots & \cdots & \cdots & \cdots & \cdots & \cdots & \cdots & \cdots & \cdots \\ \cdots & \cdots & \cdots & \cdots & \cdots & \cdots & \cdots & \cdots & \cdots & \cdots & \cdots \\ r & r & r & r & r & \cdots & r & r & q & p & p \\ r & r & r & r & r & \cdots & r & r & p & q & p \\ r & r & r & r & r & \cdots & r & r & p & p & q \end{Vmatrix} \begin{bmatrix} d\varepsilon^{1} \\ d\varepsilon^{2} \\ d\varepsilon^{3} \\ \cdots \\ \cdots \\ d\varepsilon^{10} \\ d\varepsilon^{11} \\ d\varepsilon^{12} \end{bmatrix} \quad (4)$$

Considering 12 slip systems, vector $d\varepsilon^{\alpha}$ $\alpha = 1.12$ was converted to vector $d\varepsilon_{ac}^{\alpha}$ $\alpha = 1.12$ using ratio (3). In ratio (3), q is the factor accounting for an increase in dislocation density in this particular slip system, dislocations' damping and dislocations' retardation by impurities atmosphere. p is the factor accounting for slip system dislocation interactions in intersecting planes. This factor is responsible for dislocations' retardation due to the intersection with other dislocations and the formation of Lomer-Cottrell jogs.

3. Results and Discussion

The as-cast Al-95 has a coarse dendritic columnar structure with elongated grains up to 2 mm with 300 µm thickness, as shown in Figure 2a. However, more equiaxed grains with up to 1 mm size are located in the center. In the area of the deformed sample adjacent to the die, a retained dendritic structure zone is observed, which is due to locally impeded deformation, obstructing structure workover, as shown in Figure 2b [26]. It should be noted, however, that both columnar and equiaxed grains align in the horizontal direction and are in line with their plastic strain. At the same time, severely directionally elongated grains are observed in the center. With an increase in the deformation temperature to 450 °C in the alloy Al-95, the presence of fine newly recrystallized grains, with the size of several tens of microns are observed as shown in Figure 2c. The higher the deformation temperature, the greater the number of such formations. This can be explained by the inability to generate the microstructure instantly, since it takes several seconds to develop it after the onset of deformation. Due to the high temperature of the sample and therefore, high migration rate of large-angle boundaries [23], this is sufficient for the growth of the most desired texture components. In AlMg1 alloy, the dendritic structure has much smaller sizes and ranges, from 600 µm to 800 µm on average; besides, it is far more equiaxed, as shown in Figure 2d. Such conditions can be explained by a high content of impurities, maximizing undercooling between the liquid and the surface of nuclei, formed during crystallization, thus facilitating recrystallization [27]. During upsetting, an impeded deformation zone is also observed along the workpiece boundaries with elongated grains in the center, as shown in Figure 2e. Additionally, effective methods of this factor accounting during deformation texture modeling shall be demonstrated.

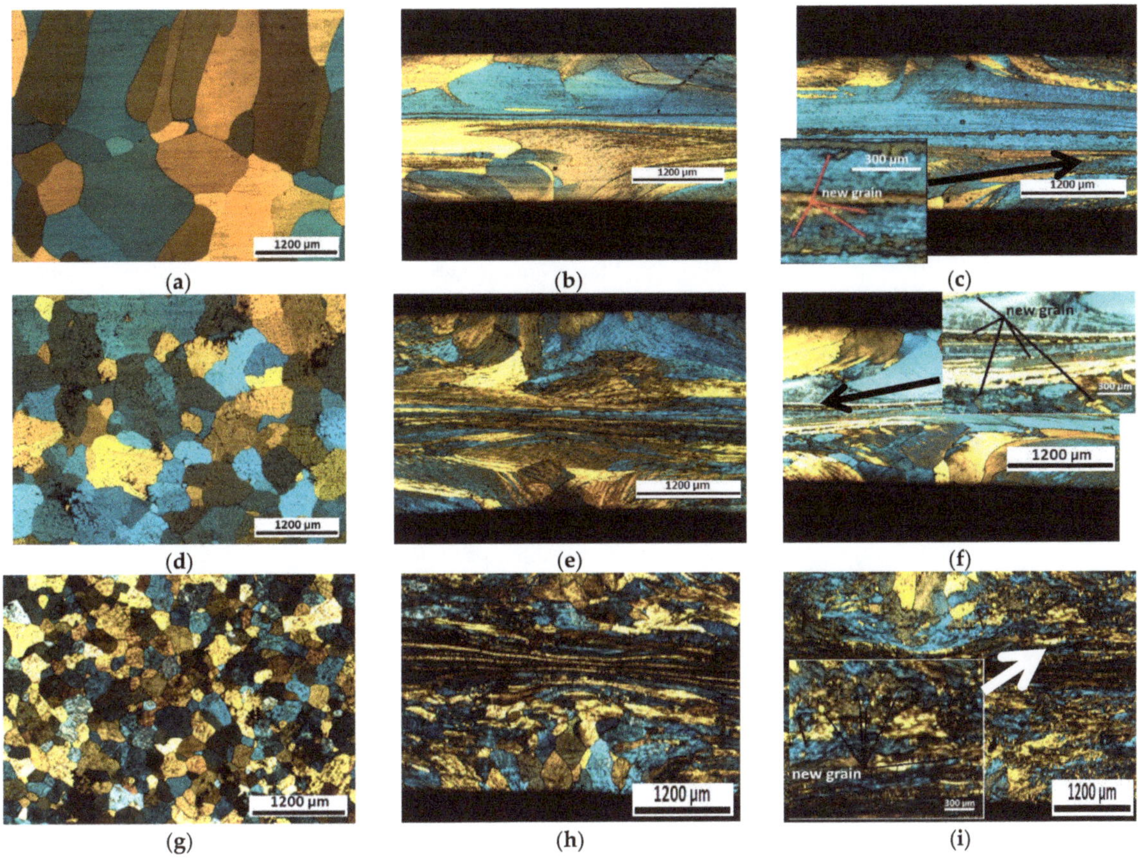

Figure 2. Grain structure of three alloys, 80% hot deformed in plain-strain condition at different strain rates and temperatures: (**a–c**) Al-95 alloy: (**a**) as-cast state, (**b**) 80%, strain rate 1 s^{-1}, 300 °C (**c**) 80%, strain rate 1 s^{-1}, 450 °C; (**d–f**) AlMg1 alloy: (**d**) as-cast state, (**e**) 80%, strain rate 1 s^{-1}, 300 °C, (**f**) 80% strain rate 1 s^{-1}, 450 °C; (**g–i**) AlMg4 alloy: (**g**) as-cast state, (**h**) 80%, strain rate 1 s^{-1}, 450 °C, (**i**) 80%, strain rate 10 s^{-1}, 450 °C.

With Al-Mg1 temperature increase, as shown in Figure 2f, much more new grains' nuclei are observed, which is associated with aluminum alloys' recrystallization kinetics boost, caused by magnesium content increase [28].

Compared to the Al-Mg1 alloy, Al-Mg4 has a finer as-cast structure with average grain size ranging from 200 µm to 300 µm, as shown in Figure 2h, which is explained by even higher content of Mg as well as the presence of Mn, Cr, Ti and other elements. After thermomechanical treatment using a Gleeble unit, an impeded deformation zone is also observed, and elongated grains occur in the center. It should be noted that none of the alloys change the pattern of grain structure distribution over the deformation center thickness under the variation of temperature-rate parameters of the deformation, as shown in Figure 4h. In all cases, an impeded deformation zone with retained dendritic structures and elongated grains are observed in the workpiece center. At 450 °C, significantly larger amount of recrystallization nuclei is observed in this alloy, compared to A-95 and Al-Mg4, which is illustrated in Figure 2i.

Electron microscopy results demonstrated a negligible amount of intermetallic particles in Al-95 alloy, as shown in Figure 3a. Aluminum matrix chemical analyses revealed that it is almost completely composed of aluminum with a small fraction of silicon (within margin of error). This is explained by high purity of this aluminum alloy. Therefore, it is rather difficult to detect the locations with other dissolved elements using chemical analysis by electron microscopy method. The results of the analysis enable the conclusion that the presence of such elements does not produce notable effect on texture evolution. Al-Mg1, as shown in Figure 3b, has a specific for such alloys' $ALFe_6$-type particles, and others contained in this alloy's transition elements, e.g., Cu, dissolve in them. The latter attract Mg, which can be traced in such particles. However, main Mg content is observed in aluminum matrix, while other elements' content lies within determination error margin. Thus, it may be concluded that it is magnesium which plays the key role in strengthening, hence texture evolution. AlFe6 particles with magnesium traces are observed in Al-Mg_4, as shown in Figure 3c. Magnesium in similar $Al(FeSi)_6$ particles was also observed in magnesium-rich 5454 alloy, which was explained by examination method imperfections (i.e., aluminum matrix portion trapping in addition to particles for analysis). In addition, Al3Mg_2 particles are observed, they partially withdraw magnesium from solid solution and its content is slightly over 3%, while the content of other impurity elements in solid solution is within measurement error margin. Thus, it is not the other elements, but magnesium will produce the major effect on crystalline texture development during hot deformation.

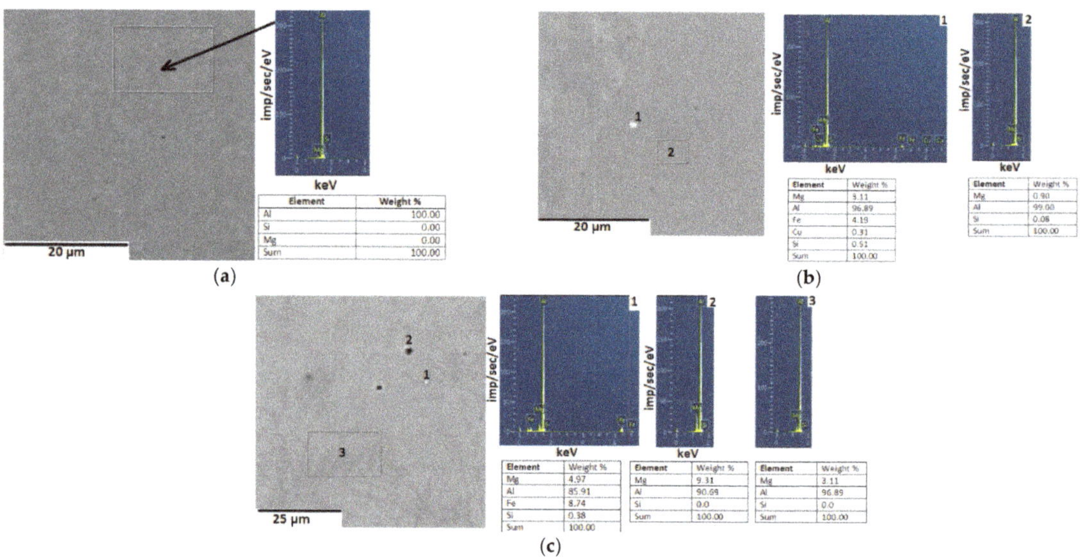

Figure 3. Electron microscopy results for the following alloys: (**a**) Al-95; (**b**) Al-Mg1; (**c**) Al-Mg4.

It may be caused by Al_3Mg_2-type particles forming due to nonequilibrium crystallization conditions and magnesium concentration exceeding 1 (it is proven by electron microscopy results). The higher the magnesium content is, the less effect will its further build-up have on its increasing content in the solid solution (hence strengthening), it will be more represented by Al3Mg_2-, Mg_2Si-type intermetallic particles. Besides, it may be assumed that until 1% concentration is reached, texture formation behavior changes due to Cu and S abundance decrease and Bs increase, followed by equalization of slipping along all planes. However, fine studies, deploying scanning microscopy methods, are required for this aspect of the investigation.

In Figure 4 the location of the specific for aluminum alloys texture components are shown in the Euler space (for $\varphi_2 = 90°$, $\varphi_2 = 45°$ and $\varphi_2 = 45°$ sections). The values of the Euler angles for Figure 4 is taken from Table 1.

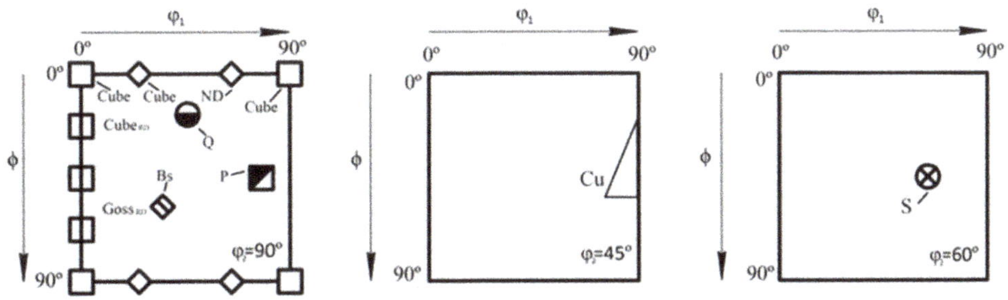

Figure 4. Ideal texture orientations on *ODF*.

The deformation of commercially pure aluminum with a deformation rate of 1 s^{-1} and at 300 °C, results in the development of a strong pronounced Cu and S deformation texture (Figure 5a). This perfectly matches the microstructure analysis, namely the presence of severely elongated (i.e., deformed) grains. Besides, a moderately pronounced brass deformation texture, as well as a small number of Goss textured grains are also present. The latter can develop both during deformation and recrystallization. Deformation with 1 s^{-1} rate and at 450 °C (Figure 5b) results in a moderately pronounced Cu deformation texture, with the Bs component being totally absent. This is associated with the temperature rise causing a reduction in strengthening (strain hardening) in all main slip planes specific to face-centered cubic lattice metals [29]. At a strain rate increase to 10 s^{-1} at 300 °C (Figure 5c), the Goss texture is better pronounced, the volume of Cu and S also increases, while the Bs texture is greatly diminished at such thermomechanical treatment modes. This is explained by the fact that strengthening increases with strain rate, resulting in a maximal deployment of slip systems, specific to Cu and S texture components. At a rise in temperature to 450 °C at 10 s^{-1} (Figure 5d), the strain rate decreases the volume of Cu and S and, in some way, boosts the brass texture (Bs) volume, while Goss disappears completely.

Such a pattern is explained by a decrease in strain hardening with temperature rise, while the brass-type texture-associated slip systems once again become favorable. In addition, a small volume of cube texture appears, as the result of a small number of fine recrystallized grains. The general feature, specific for this alloy, consists in the deployment of slip systems, associated with Cu and S texture components and—to a lesser degree—with the brass texture component Bs. It should be noted that commercially pure aluminum features a high stacking fault energy, which brings it closer to copper with respect to texture formation.

During AMg1 hot deformation at 1 s^{-1} strain rate and 300 °C temperature (Figure 6a), all three β-fiber texture specific components, i.e., Cu, S and Bs, are observed. At the same time, unlike the previous alloy, it shows a strongly pronounced Bs texture component, while the other two rolling texture components are less pronounced. A temperature rise causes some dissipation of the Bs texture volume, while the amount of Cu and S slightly increases (Figure 6b). A strain rate increase (Figure 6c) to 10 s^{-1} causes an increasing in volume of S and especially Cu rolling texture components. At 10 s^{-1} strain rate and 450 °C (Figure 6d), no clearly defined deformation textures are observed, and all three rolling texture components are very weak.

Such conditions can be explained by a stronger effect of temperature on strength in such alloys, causing far more rapid strength decrease with temperature rise, compared to Al-95 alloy. In addition, due to the high temperature and high strain rate, a large number of recrystallized grains is also present. Some of them will contribute to cube texture, and others to a random texture constituent. Thus, the majority of elements, occurring in this alloy, will stimulate the development of multiple second-phase particles with the tendency of forming new recrystallized nuclei by particle-stimulated nucleation (PSN), having a mostly random texture constituent [30,31].

A strain rate increase leads to an increase in the Zener-Hollomon value, hence an increase of energy, required for the recrystallization process. Therefore, a rate increase to 450 °C temperature will facilitate the dispersion of textures' components, rather than their strengthening, as might be expected. At the same time, the grain boundary mobility in this alloy at 300 °C temperature is not sufficient for significant grain growth during the metal cooling period. Thus, regardless of the high Zener-Hollomon parameter value in all thermomechanical treatment modes, recrystallization at 300 °C temperature does not have significant effect on the texture's composition. It should be pointed out, unlike commercially pure aluminum, that mainly brass texture-associated slip systems are deployed in Al-Mg1 alloy. Strengthening, stimulated by the strain rate increase, starts deploying slip systems, associated with the Cu and S components. Strengthening related to temperature, strain rate and many impurity elements, especially, magnesium, will have a different influence on the activation of various slip systems.

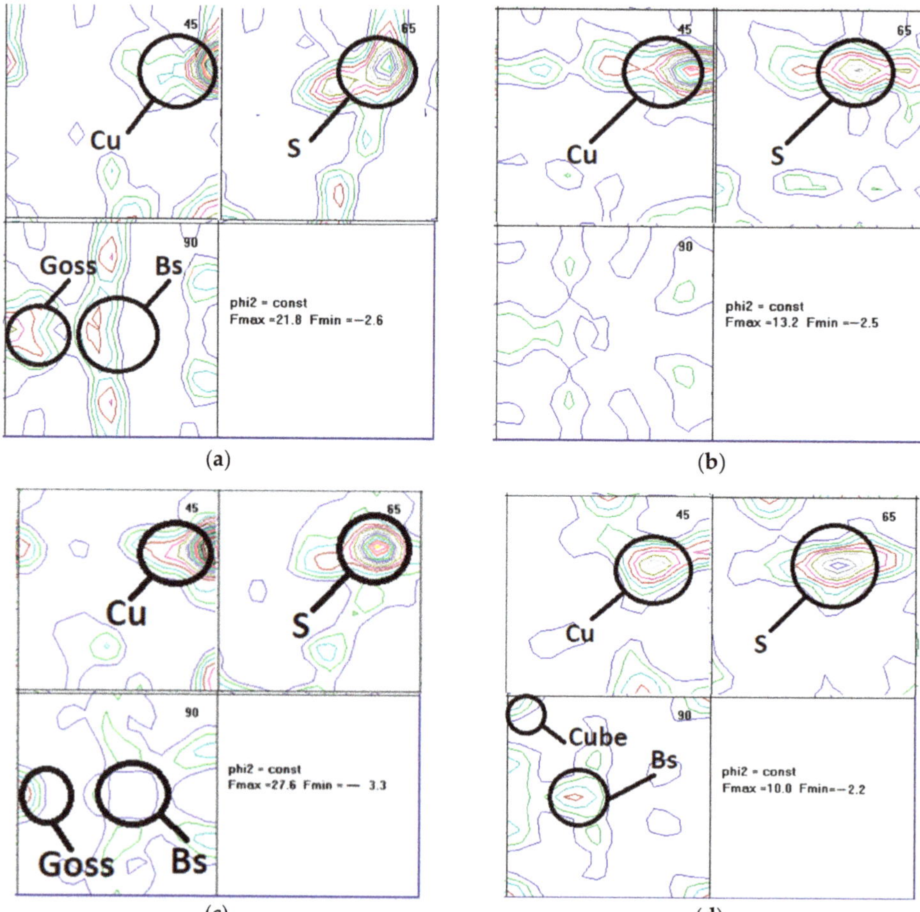

Figure 5. *ODF*, aluminum 99.95% (for composition refer to Table 1) compressed under plain strain to 80% engineering strain (**a**) strain rate 1 s^{-1}, temperature 300 °C, (**b**) strain rate 1 s^{-1}, temperature 450 °C, (**c**) strain rate 10 s^{-1}, temperature 300 °C (**d**) strain rate 10 s^{-1}, temperature 450 °C.

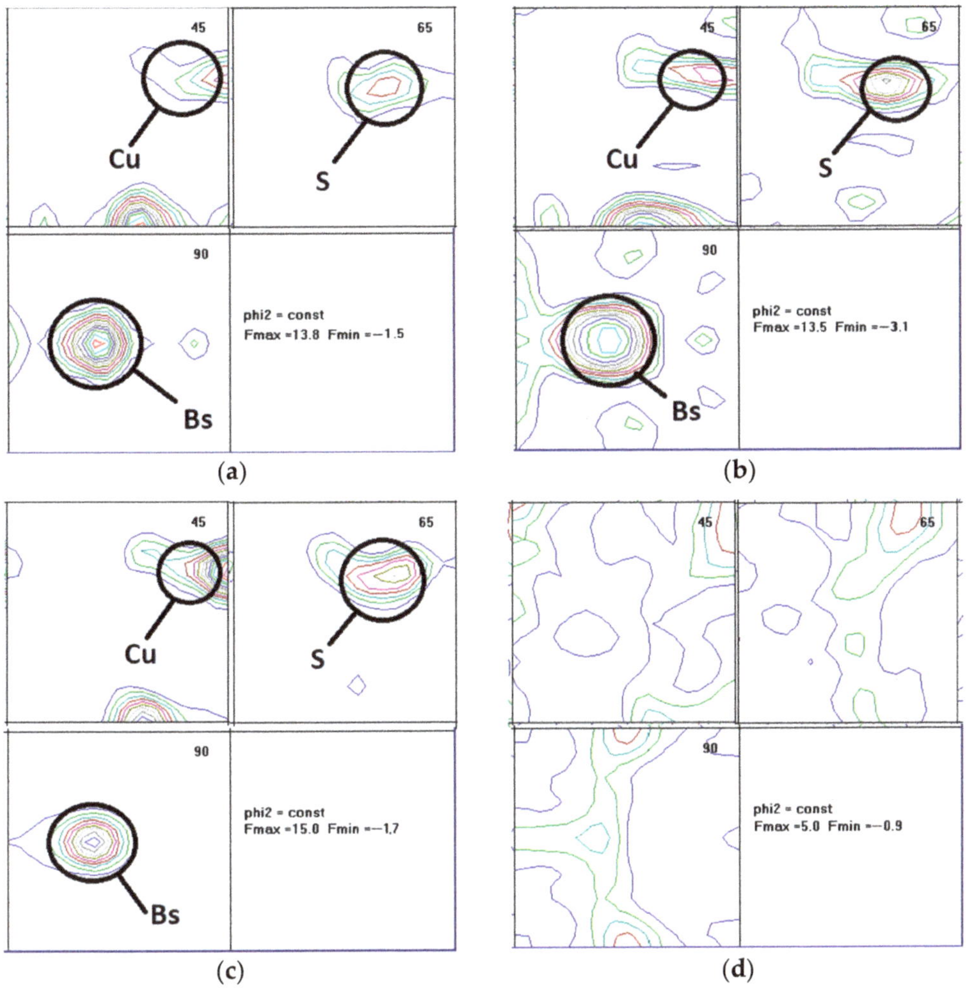

Figure 6. *ODF*, Al + 1% Mg (composition, see Table 2) compressed under plain strain to 80% engineering strain (**a**) strain rate 1 s^{-1}, temperature 300 °C, (**b**) strain rate 1 s^{-1}, temperature 450 °C, (**c**) strain rate 10 s^{-1}, temperature 300 °C (**d**) strain rate 10 s^{-1}, temperature 450 °C.

At 1 s^{-1} strain rate and 300 °C, three well-formed major β-fiber components are observed in Al-Mg4 alloy (Figure 7a). As the temperature rises to 450 °C (Figure 7b), the volume of all three deformation texture components slightly decreases, which is explained by strain hardening. As the strain rate rises to 10 s^{-1} (Figure 7c), the volume of the three major rolling texture (β-fiber) components also increases. At 10 s^{-1} strain rate and 450 °C (Figure 7d), the total amount of β-fiber texture decreases compared to the mode, illustrated in Figure 3c. This can be also attributed to the initiation of the recrystallization process. Note that all deformation texture components are less pronounced in this alloy, compared to Al-Mg1. This is explained by the high Mg content of Al-Mg4, which is close to AA5182 alloy. The latter alloy fully recrystallizes within 10 s under these temperature-rate strain modes, but in all major cases, recrystallization duration varies from 50 to 100 s^{-1}. Due to a larger amount of newly formed 10–30 μm grains, observed in Al-Mg4 (Figure 4i), the deformation texture is weaker. At the same time, like in Al-Mg1, this alloy has a clearly defined Bs texture.

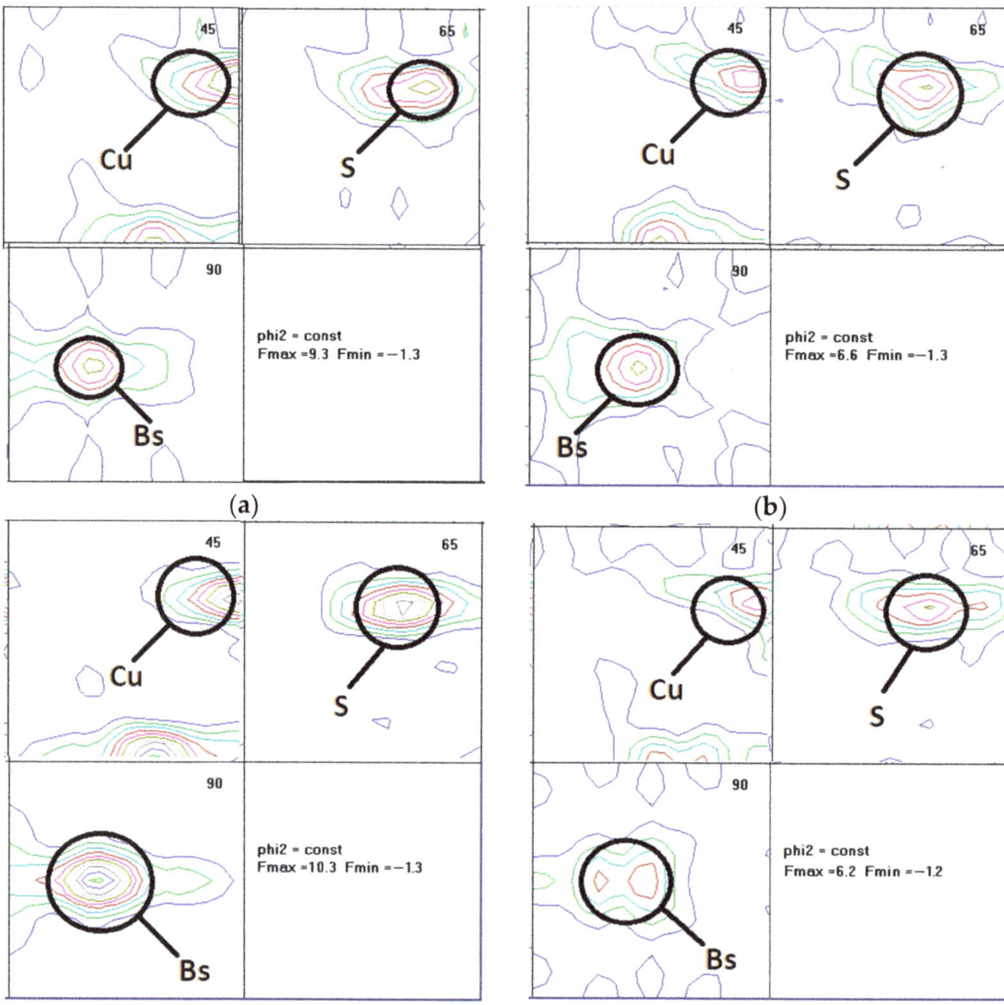

Figure 7. *ODF*, Al + 4% Mg (composition, see Table 3) compressed under plain strain to 80% engineering strain, material initially as cast, strain rate (**a**,**b**) 1 s^{-1}; (**c**,**d**) 10 s^{-1}; temperatures (**a**,**c**) 300 °C; (**b**,**d**) 450 °C.

For clarity, the various β-fiber textures' distribution within 45° < φ_2 < 90° interval is illustrated in Figure 8.

It is obvious, from Figure 8a, that the copper-type texture and similar components are abundant in Al-95 under all processing modes. Their intensity gradually declines to 60°, i.e., to the S texture. It is followed by a sharp decline in texture intensity until, finally, the same values at φ_2 = 90° are reached, which is also characteristic for the Bs component. In the Al-Mg1 alloy, the texture intensity declines within 45° < φ_2 < 60° interval, i.e., around the C and S components; at the same time, brass-type orientation intensity builds up. As mentioned earlier, this is due to strain hardening, which impairs the advantage of gliding on the slip systems, attributed to copper- and S-type textures. At the same time, the building up of the brass-type textures is observed, as gliding on its systems makes such strengthening level more favorable.

All texture intensities decline in the Al-Mg4 alloy, which is explained not only by better recrystallization capability of this alloy, but by strengthening reaching the values, which disable gliding in slip systems, associated both with S and Cu components, preferring the brass-type textures related slip systems. It should be stressed that increase in strain hardening is caused not only by an increase in *SFE*, but also by an increase in magnesium impurity atoms, dissolved in solid solution [32]. It should also be considered that the *SFE* of pure aluminum (annealed) is equal to 174 mJ/m^2, while the *SFE* of the alloy with 6.28% magnesium content is 151.6 mJ/m^2. In other words, even deformed alloys with extremely high magnesium content maintain a high level of *SFE*, compared to other materials [20]. Considering the data, presented by this source, it may be assumed that Al-95 alloy (with account for its high purity) energy will be equal to 177 mJ/m^2, Al-Mg1—168.0 mJ/m^2, Al-Mg4—154.7 mJ/m^2, thus *SFE* difference for two extreme alloys will be slightly above 22 mJ/m^2.

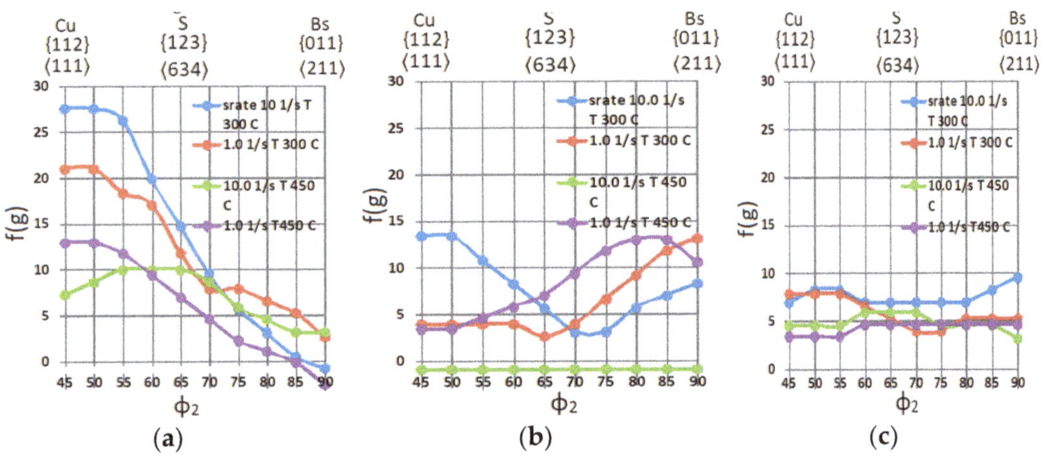

Figure 8. Texture components' intensities along the β-fiber: (**a**) Aluminum Al-99.5 (AA1050); (**b**) Al-Mg1; (**c**) Al-Mg4.

Therefore, opposite to copper and copper-based alloys, *SFE* cannot be considered as a prevailing factor in shifting towards Bs-type textures when referred to aluminum alloys.

The calculations with variable values: $q = 0.25$–1.5, $p = 0.1$–0.5 and $r = 0.25$–0.7 demonstrate a greater effect of impurity atoms. The calculations show the best agreement for Al-95 alloy at $q = 0.32$, $p = 0.19$, $r = 0{,}32$; for Al-Mg1 alloy at $q = 0.51$, $p = 0.24$, $r = 0.43$; for Al-Mg4 at $q = 1.1$, $p = 0.36$, $r = 0.54$, thus, the impurity atoms-related factor q has a greater effect than the *SFE*-related factors p and r.

Figure 9 shows the results of *ODF* simulations of thermomechanical treatment for the three studied alloys at strain rate of 10 s^{-1} and 300 °C. For Al95 alloy, as shown in Figure 3a, the calculated and the observed *ODF* demonstrate a high convergence for the S and Cu components. At the same time, the Bs texture components are slightly overestimated, while Goss texture values are underestimated. The AlMg1 alloy has less pronounced Cu and S textures, while a minor Bs texture fraction is observed.

Note that in the model, unlike in the actual process, all deformation texture components are less clearly developed. Al-Mg4 has three sufficiently clearly defined components with a dominating cube texture, which accurately agrees with the observed pattern.

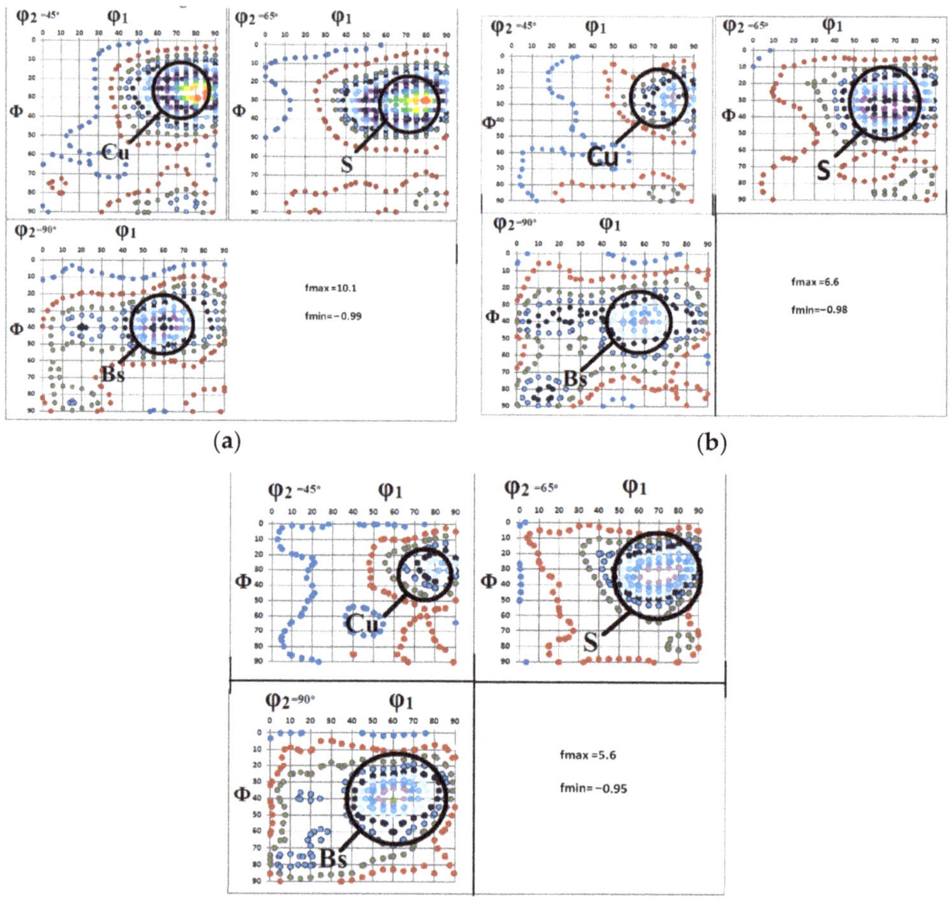

Figure 9. Example of calculated *ODF*: (**a**) Aluminum 1050; (**b**) AlMg1; (**c**) AlMg4, strain rate 10 s^{-1}, temperature 300 °C.

For the calculations, $n_0 = 0.46$ was assumed and τ_c^α to be the same in all slip systems. For aluminum, it was taken as 1, based on the data in ref. [33], for the Al 1% Mg alloy τ_c^α = 2.5 and for 4% magnesium alloy τ_c^α = 4.0. The results are shown in Figure 10.

As can be seen in Figure 8, the calculation matches the conditions, under which the data in Figure 7 were obtained; in general, the calculation adequately reflects the transformation of texture components in case of modifications in alloy composition and deformation parameters. However, it underestimates the intensity values of the copper and S-type texture components as well as of the brass-type texture; some inaccuracies manifest themselves clearly at 300 °C temperature. For Al-Mg1, inaccuracies consist in a slightly overestimated brass texture and underestimated copper texture. The best match is observed in Al-Mg4 alloy, emphasizing a greater effect of magnesium atoms and other components, dissolved in the aluminum matrix, on strain hardening (crystalline strengthening), compared to an increase in *SFE*. To further improve the calculation accuracy requires a further in-depth investigation of physical effects on matrix components (1), various dislocation movements and crystalline strengthening due to their interactions with various obstacles, represented by dislocation forest and Lomer-Cottrell jogs.

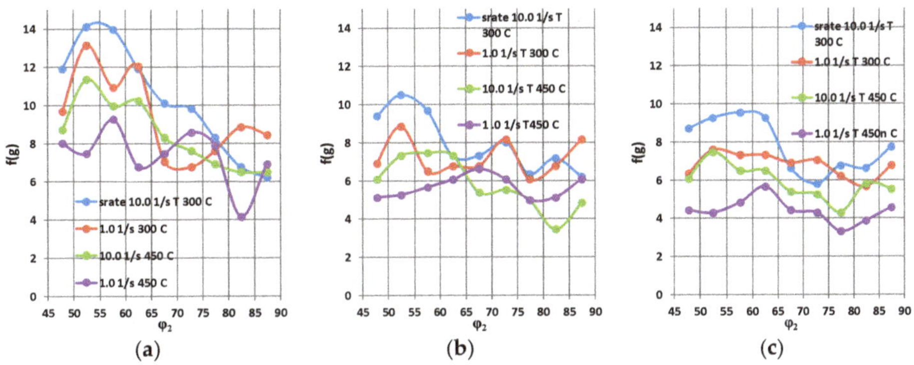

Figure 10. Model-based calculation of texture components' re-distribution with increasing deformation strengthening: (**a**) Aluminum Al 99.5 (AA1050); (**b**) Al-Mg1; (**c**) Al-Mg4.

Magnesium content effect on the main texture components during different thermo-mechanical treatment modes is illustrated in Figure 11. At 300 °C, temperature and strain rate of 10 s^{-1} Cu and S textures abundance drops from 27 f(g) and 15 f(g) to 7 f(g) and 7.3 f(g) respectively with magnesium content increase. With increasing magnesium content, Bs component abundance grows from 0.7 f(g) to 9.5 f(g). At lower temperature, all orientations' abundance decreases, which is related to recrystallization effect and strengthening decrease. However, overall nature of magnesium effect on major texture components stays the same: with magnesium increase from 0% to 4% S and Bs abundance drops from 13 f(g) and 7 f(g) to 3.4 f(g) and 4.6 f(g) respectively. It shall be pointed out that the sharpest Cu and S orientations decrease and Bs abundance increase are observed with magnesium content ranging from 0% to 1%. It is followed by further texture components' abundance change; however, it has steadier behavior. The main cause could be incomplete magnesium solubility in aluminum matrix due to nonequilibrium crystallization conditions and magnesium content over 1%; in this case, Al_3Mg_2-type intermetallic compounds will be formed. The latter will be preserved during homogenization, though they will become smaller in size. Besides, it may be assumed that until 1% concentration is reached, texture formation behavior changes due to Cu and S abundance decreases and Bs increase, followed by equalization of slipping along all planes. However, fine studies, deploying scanning microscopy methods, are required for this aspect of the investigation.

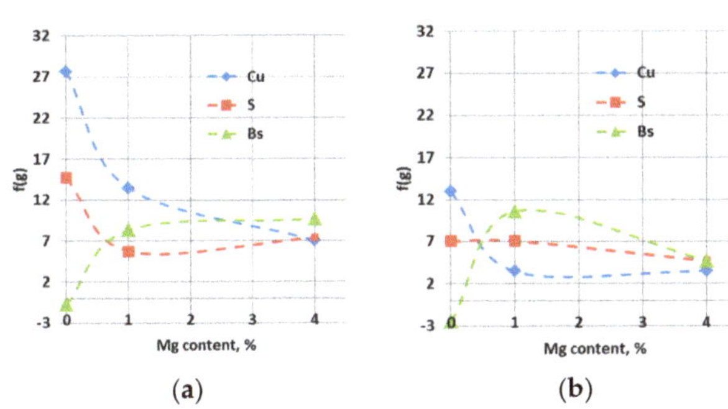

Figure 11. Magnesium content effect on the main β-fiber texture components' abundance: (**a**) 10 s^{-1}, temperatures 300 °C; (**b**) 10 s^{-1}, temperatures 450 °C.

4. Conclusions

The study showed that an increase in the content of magnesium and other alloying elements has a significant effect on the development of rolling texture (β-fiber) components during hot deformation. In commercially pure aluminum, mainly copper and S texture-related slip planes are deployed. As the magnesium content increases, the intensities of these components deplete, and a pronounced brass texture appears. There are two explanations for this fact: Firstly, the stacking fault energy (*SFE*) decreases. This factor undoubtedly contributes to the stain hardening, since the SFE decreases for magnesium-rich alloys by 10–15%. The simulation demonstrates the greater effect of impurity atoms on crystallographic strengthening, i.e., strain hardening. To further improve the model accuracy requires in-depth analysis of strengthening matrix components' relationship with physical phenomena, emerging during dislocations' movements and interactions with various obstacles.

Author Contributions: Conceptualization, E.A. and V.A.; methodology E.A., V.A. and J.H.; software E.A. and S.K.; formal analysis E.A., A.D. and J.H..; investigation E.A. and S.K.; resources E.A., A.D., and S.K.; data curation E.A., A.D. and S.K.; writing—original draft preparation E.A. and J.H.; writing—review and editing, E.A., J.H., A.D. and S.K.; visualization E.A. and J.H. supervision S.K. and V.A.; project administration E.A., V.A. and S.K.; funding acquisition E.A.; All authors have read and agreed to the published version of the manuscript.

Funding: This study is funded by a grant of the Russian Science Foundation, project number 18-79-10099.

Data Availability Statement: The data presented in this study are available on request from the corresponding author.

Conflicts of Interest: The authors declare no conflict of interest.

References

1. Fedorov, S.; Nanocomposites, V.B. Development of Mechanical Properties of Aluminum-Silicon Alloys. *Smart Nanocompos.* **2015**, *6*, 199–202.
2. Bazhin, V.Y.; Gutema, E.M.; Savchenkov, S.A. Production Technology Features for Aluminum Matrix Alloys with a Silicon Carbide Framework. *Metallurgist* **2017**, *60*, 1267–1272. [CrossRef]
3. Deev, V.B.; Degtyar, V.A.; Kutsenko, A.I.; Selyanin, I.F.; Voitkov, A.P. Resource-Saving Technology for the Production of Cast Aluminum Alloys. *Steel Transl.* **2007**, *37*, 991–994. [CrossRef]
4. Akopyan, T.K.; Letyagin, N.V.; Belov, N.A.; Koshmin, A.N.; Gizatulin, D.S. Analysis of the Microstructure and Mechanical Properties of a New Wrought Alloy Based on the ((Al) + Al4(Ca,La)) Eutectic. *Phys. Met. Metallogr.* **2020**, *121*, 914–919. [CrossRef]
5. Sizyakov, V.M.; Bazhin, V.Y.; Vlasov, A.A. Status and Prospects for Growth of the Aluminum Industry. *Metallurgist* **2010**, *54*, 409–414. [CrossRef]
6. Belov, N.; Akopyan, T.; Korotkova, N.; Murashkin, M.; Timofeev, V.; Fortuna, A. Structure and Properties of Ca and Zr Containing Heat Resistant Wire Aluminum Alloy Manufactured by Electromagnetic Casting. *Metals* **2021**, *11*, 236. [CrossRef]
7. Hirsch, J. Aluminium in Innovative Light-Weight Car Design. *Mater. Trans.* **2011**, *52*, 818–824. [CrossRef]
8. Lamberti, M.; Escher, F. Aluminium Foil as a Food Packaging Material in Comparison with Other Materials. *Food Rev. Int.* **2007**, *23*, 407–433. [CrossRef]
9. Miller, W.S.; Zhuang, L.; Bottema, J.; Wittebrood, A.J.; De Smet, P.; Haszler, A.; Vieregge, A. Recent Development in Aluminium Alloys for the Automotive Industry. *Mater. Sci. Eng. A* **2000**, *280*, 37–49. [CrossRef]
10. Rambabu, P.; Eswara Prasad, N.; Kutumbarao, V.V.; Wanhill, R.J.H. Aluminium Alloys for Aerospace Applications. *Aerosp. Mater. Mater. Technol.* **2017**, 29–52. [CrossRef]
11. Engler, O. Control of Texture and Earing in Aluminium Alloy AA 3105 Sheet for Packaging Applications. *Mater. Sci. Eng. A* **2012**, *538*, 69–80. [CrossRef]
12. Engler, O.; Hirsch, J.; Lücke, K. Texture development in Al-1.8 wt% Cu depending on the precipitation state-II. Recrystallization textures. *Acta Metall. Mater.* **1995**, *43*, 121–138. [CrossRef]
13. Engler, O.; Hirsch, J. Polycrystal-Plasticity Simulation of Six and Eight Ears in Deep-Drawn Aluminum Cups. *Mater. Sci. Eng. A* **2007**, *452*, 640–651. [CrossRef]
14. Engler, O. Simulation of the Recrystallization Textures of Al-Alloys on the Basis of Nucleation and Growth Probability of the Various Textures Components. *Textures Microstruct.* **1997**, *28*, 197–209. [CrossRef]
15. Hirsch, J.; Lücke, K. Mechanism of Deformation and Development of Rolling Textures in Polycrystalline f.c.c. Metals-II. Simulation and Interpretation of Experiments on the Basis of Taylor-Type Theories. *Acta Metall.* **1988**, *36*, 2883–2904. [CrossRef]
16. Hirsch, J. Walztexturentwicklung in Kubisch-Flächenzentrierten Metallen. Ph.D. Thesis, Rheinisch-Westfälische Technische Hochschule Aachen, Aachen, Deutschland, December 1984. [CrossRef]
17. Miraglia, M.; Miraglia, M.; Dawson, P.; Leffers, T. On the Influence of Mechanical Environment on the Emergence of Brass Textures in FCC Metals. *Acta Mater.* **2007**, *55*, 799–812. [CrossRef]
18. Evgenii, A.; Kawalla, R.; Vladimir, A. Investigation of Texture and Structure Evolution during Hot Rolling of 1070, 3104 and 8011 Aluminum Alloys in Continuous Mill. *La Metall. Ital.* **2017**, *109*, 11–21.

19. Aryshenskii, E.V.; Aryshenskii, V.Y.; Grechnikova, A.F.; Beglov, E.D. Evolution of Texture and Microstructure in the Production of Sheets and Ribbons from Aluminum Alloy 5182 in Modern Rolling Facilities. *Met. Sci. Heat Treat.* **2014**, *56*, 347–352. [CrossRef]
20. Zhao, D.; Løvvik, O.M.; Marthinsen, K.; Li, Y. Impurity Effect of Mg on the Generalized Planar Fault Energy of Al. *J. Mater. Sci.* **2016**, *51*, 6552–6558. [CrossRef]
21. Wang, J.; Zheng, J.H.; Sekine, K. Rolling Texture Evolution of Ag and Ag–Zn Alloys Having Very Low Stacking Fault Energies. In Proceedings of the ICOTOM 11, Xi'an, China, 16–20 September 1996; Liang, Z., Zuo, L., Chu, Y., Eds.; International Academic Publishers: Beijing, China, 1996; pp. 362–367.
22. Kallend, J.S.; Davies, G.J. The development of texture in copper and copper–zinc alloys. *Texture* **1972**, *1*, 51–69. [CrossRef]
23. Engler, O.; Pithan, C.; Lücke, K. Rolling texture development in Cu–Mn alloys. *Mater. Sci. Forum* **1994**, *157–162*, 679–683. [CrossRef]
24. Crumbach, M.; Muckopadhyay, P.; Aretz, H.; Pomana, G.; Wagner, P.; Gottstein, G. A Taylor Type Deformation Texture Model Considering Grain Interaction and Material Properties. Part II–Experimental Validation and Coupling to FEM. In Proceedings of the First Joint International Conference on Recrystallisation and Grain Growth, Aachen, Germany, 27–31 August 2001; Gottstein, G., Molodov, D.A., Eds.; Springer: Berlin/Heidelberg, Germany, 2001; pp. 1061–1068.
25. Evgenii, A.; Erkin, B.; Hirsch, J.; Vladimir, A.; Segrey, K. Development of the New Fast Approach for Calculation of Texture Evolution during Hot Deformation of Aluminum Alloys. *Procedia Manuf.* **2019**, *37*, 492–499. [CrossRef]
26. Aryshenskii, E.; Hirsch, J.; Yashin, V.; Sergei, K.; Kawalla, R. Influence of Local Inhomogeneity of Thermomechanical Treatment Conditions on Microstructure Evolution in Aluminum Alloys. *J. Mater. Eng. Perform.* **2018**, *27*, 6780–6799. [CrossRef]
27. Vatne, H.; Furu, T.; Ørsund, R. Modelling Recrystallization after Hot Deformation of Aluminium. *Acta Mater.* **1996**, *44*, 4463–4473. [CrossRef]
28. Wells, M.A.; Samarasekera, I.V.; Brimacombe, J.K.; Hawbolt, E.B.; Lloyd, D.J. Modeling the microstructural changes during hot tandem rolling of AA5XXX aluminum alloys: Part I. Microstructural evolution. *Metall. Mater. Trans. B* **1998**, *29*, 611–620. [CrossRef]
29. Mathur, K.; Dawson, P. On Modeling the Development of Crystallographic Texture in Bulk Forming Processes. *Int. J. Plast.* **1989**, *5*, 67–94. [CrossRef]
30. Humphreys, F.J. The Nucleation of Recrystallization at Second Phase Particles in Deformed Aluminium. *Acta Metall.* **1977**, *25*, 1323–1344. [CrossRef]
31. Chan, H.M.; Humphreys, F.J. Effect of Particle Stimulated Nucleation on Orientation of Recrystallized Grains. *Met. Sci.* **1984**, *18*, 527–529. [CrossRef]
32. Morris, J.G. Dynamic Strain Aging in Aluminum Alloys. *Mater. Sci. Eng.* **1974**, *13*, 101–108. [CrossRef]
33. Totten, G.E.; MacKenzie, D.S. *Handbook of Aluminum*; CRC Press: New York, NY, USA, 2003; Volume 2.

metals

MDPI

Article

Effect of Rolling Temperature and Subsequence Ageing on the Mechanical Properties and Microstructure Evolution of an Al-Cu-Li Alloy

Lin Wang [1,2], Charlie Kong [3], Puneet Tandon [4], Alexander Pesin [5,*], Denis Pustovoytov [5] and Hailiang Yu [1,2,*]

1 State Key Laboratory of High Performance Complex Manufacturing, College of Mechanical and Electrical Engineering, Central South University, Changsha 410083, China; 153711078@csu.edu.cn
2 Light Alloys Research Institute, Central South University, Changsha 410083, China
3 Mark Wainwright Analytical Centre, University of New South Wales, Sydney, NSW 2052, Australia; c.kong@unsw.edu.au
4 deLOGIC Lab, PDPM Indian Institute of Information Technology, Design and Manufacturing, Jabalpur 482005, Madhya Pradesh, India; ptandon@iiitdmj.ac.in
5 Laboratory of Mechanics of Gradient Nanomaterials, Nosov Magnitogorsk State Technical University, 455000 Magnitogorsk, Russia; pustovoitov_den@mail.ru
* Correspondence: pesin@bk.ru (A.P.); yuhailiang@csu.edu.cn (H.Y.)

Abstract: The mechanical properties and microstructure evolution of an Al-Cu-Li alloy sheet processed via hot rolling (HR) (at 400 °C and 500 °C) or cryorolling (CR) (at −100 °C and −190 °C) and subsequence aging at 160 °C for 10 h were investigated. Before aging, the highest ultimate tensile strength of 502 MPa was achieved when the sheets were cryorolled at −190 °C, while the better ultimate tensile strength of 476 MPa and the best elongation rate of 11.1% was achieved simultaneously when the sheets were cryorolled at −100 °C. The refined grains and numerous uniform deformation-induced dislocations microstructures were responsible for the improved strength and enhanced ductility of the cryorolled sheets compared to that of the alloy processed by hot rolling with a low dislocation density zone (LDDZ) and high dislocation density zone (HDDZ). After aging at 160 °C for 10 h, the ultimate tensile strength further improved resulted from the greater precipitation strengthening, and the increased precipitates provided greater resistance to dislocations movement resulting in the increased ductility although the dislocation density decreased. The uniform dislocation microstructures in the cryorolled sheets provide numerous nucleation sites for the precipitates, leading to higher strength after aging.

Keywords: Al-Cu-Li alloy; mechanical property; cryorolling; aging

Citation: Wang, L.; Kong, C.; Tandon, P.; Pesin, A.; Pustovoytov, D.; Yu, H. Effect of Rolling Temperature and Subsequence Ageing on the Mechanical Properties and Microstructure Evolution of an Al-Cu-Li Alloy. *Metals* **2021**, *11*, 853. https://doi.org/10.3390/met11060853

Academic Editor: Elisabetta Gariboldi

Received: 26 April 2021
Accepted: 20 May 2021
Published: 22 May 2021

1. Introduction

Aluminum-copper-lithium (Al-Cu-Li) alloys have been widely used as structural materials for aerospace and aircraft vehicles [1–3]. A comprehensive review about the development of the third generation Al-Li alloy was reported by Rioja et al. [4] and quantitative calculations of the Al-Cu-Li alloy during the deformation process were operated [5]. The alloy is a typical precipitation strengthening alloy resulting from the T_1 (Al_2CuLi) or θ' (Al_2Cu) phases [6]. Zhou et al. studied the effects of creep aging parameters on mechanical properties and found that the yield strength and the ultimate tensile strength increased to 493 MPa and 540 Mpa, respectively, after being aged at 163 °C for 25 h [7]. With the increase of the Cu element, proper heat treatment can further improve the strength via increasing precipitation strengthening [8–10].

In the past 30 years, people have developed several severe plastic deformation (SPD) techniques to fabricate ultrafine-grained (UFG) materials, including high-pressure torsion [11], equal channel angular pressing [12], accumulative roll bonding (ARB) [13], and asymmetric cryorolling (ACR) [14]. During these techniques, rolling has been widely used to fabricate UFG metallic sheets. Yu et al. [15] reported that the properties of the copper sheets subjected to ACR were better than that of the copper sheets subjected to asymmetric rolling because of the restriction of dislocation movement. Magalhães et al. investigated the mechanical properties and the microstructure evolution of AA6061 processed by conventional rolling, CR, asymmetric rolling, ACR before and after post-deformation heat treatments, respectively. They found that uniform elongation of the materials annealed after ACR increased, while the alloys' texture intensity reduced resulting from recovery and recrystallization phenomena [16]. During CR, refined grains and accumulated dislocations that resulted from the increased strain hardening were reasonable for the enhanced mechanical properties [17].

The deformation-induced dislocation microstructures could act as nucleation sites for precipitation, which can further improve the mechanical properties [18]. The excellent mechanical properties, the yield strength of 485 MPa, the ultimate tensile strength of 540 MPa, and the ductility of 11%, were achieved for an Al–Cu alloy through CR and subsequence aging [19]. Deng et al. reported that the mechanical properties of an Al−Cu−Li alloy could be improved significantly via CR followed by aging at 160 °C for 32 h [20]. Li et al. proposed that the temperature corresponding to peak hardness is negatively related to rolling reduction and is positively related to rolling temperature [21]. However, there are few reports about the mechanical properties and microstructure evolution of the Al−Cu−Li alloys processed by CR under different temperatures. When the rolling temperature was set as liquid nitrogen temperature, grains could be refined uniformly and the dislocation network formed. If the rolling temperature was set as a middle temperature during liquid nitrogen temperature and room temperature, the reports about the microstructure evolution were rare.

In this study, the mechanical properties and microstructure evolution of Al-Cu-Li alloy sheets fabricated by HR (at 400 °C and 500 °C) or CR (at −100 °C and −190 °C) and subsequence aging at 160 °C for 10 h were investigated.

2. Materials and Methods

The original materials were Al–Cu–Li alloy sheets with an initial size of 60 mm × 100 mm × 2 mm self-melted in the laboratory. The chemical composition listed in Table 1 has been confirmed by the Inductive Coupled Plasma-Optical Emission Spectroscopy (ICP-OES) test. The processing schedule was shown in Figure 1. The size of the tensile sample was shown in the lower-left corner of the figure. The alloy sheets were placed in the furnace at 500 °C for 1 h for homogenization. Rolling experiments were carried out on a multifunction 4-high rolling mill with a maximum rolling width of 300 mm and a work roll diameter of 80 mm. The rolling speed was set as 2 m/min. For each pass during CR or HR, the rolling reduction of each pass was 16% and the final thickness of the alloy sheet after six passes was about 0.7 mm. After HR or CR at different temperatures, the samples were air-cooled to room temperature. The rolling experiments were operated in a "dry friction" condition and the work rolls surfaces were smooth after being polished. Before each pass of rolling at 400 °C or 500 °C (HR 400 °C or HR 500 °C), the sheets were placed in the furnace for 3 min. Before each CR pass at −100 °C or −190 °C (CR −100 °C or CR −190 °C), the sheets were placed in the cryogenic furnace for >8 min. After rolling, the rolled sheets were aged at 160 °C for 10 h in an aging furnace.

Table 1. Chemical composition of Al-Cu-Li alloy.

Composition (Wt%)	Cu	Mg	Zn	Li	Ag	Mn	Zr	Al
Al-Cu-Li alloy	3.72	0.58	0.46	0.96	0.30	0.58	0.31	91.6

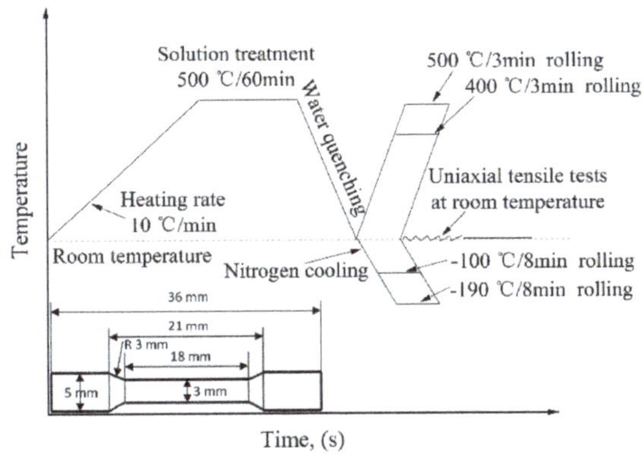

Figure 1. Processing schedule.

The tensile tests were carried out on an AGS-5/10 kN testing machine produced by Shimadzu Company (Suzhou, China) at a strain rate of 0.03 s^{-1} at room temperature. Each tensile test was repeated three times for the mean values to ensure accuracy. The fracture surface was examined utilizing the scanning electron microscopy (SEM) equipped on the platform of a Phenom Prox Desktop and the accelerating voltage was set as 15 kV. The morphology of grains after processing was studied using the optical microscope (the platform of VHX−5000) after the samples were etched with Keller's reagent for about 16 s on the normal plane. A Philips CM200 transmission electron microscope, with a field-emission gun operating at 200 kV, was used to examine the cross-section microstructures in the rolling direction. A Thermo Fisher Helios G4 PFIB was used to prepare the transmission electron microscopy (TEM) specimens via the in situ lift-out technique. The range from 30° to 110° was detected by X-ray on a Bruker D8 Advance diffractometer. The step size was set as 0.02° and the Cu target was utilized in the diffraction test.

3. Results and Discussion

Figure 2 shows the mechanical properties of the Al–Cu–Li alloy sheets processed by rolling at different temperatures and subsequence aging at 160 °C for 10 h. The ultimate tensile strength and the ductility of the alloy after rolling and aging were summarized in Figure 2b and Figure 2c,d respectively. After solution treatment at 500 °C for 1 h, the ultimate tensile strength of 270 MPa and the elongation rate of 11.0% was obtained. When the Al–Cu–Li alloy was hot rolled at 400 °C, the lowest ultimate tensile strength of 356 MPa and the lowest elongation of 8.1% was achieved. The ultimate tensile strength increased to 449 MPa for HR 500 °C, to 476 MPa for CR −100 °C and to 502 MPa for CR −190 °C. The ductility increased to 9.5% for CR −190 °C, to 10.7% for HR 500 °C, to 11.1% for CR −100 °C. The yield strength of the sheet processed by HR 400 °C increased when the rolling temperature increased to 500 °C and it further increased with the decreasing of the rolling temperatures. To elucidate the effects of rolling temperature on strain rate, the mechanism of dislocation movement and interaction was considered. More slip systems were easily activated during CR and the slip plane of screw dislocation changed to other planes which the Burgers vector b of the dislocation is common to. One possible origin of the screw dislocations pair was a thermal cross-slip of screw dislocations at different stress levels and the termination of the dipoles were edge dislocations [22]. The limit of dislocations multiplication increased when the alloy was subjected to CR in the temperature range. Additionally, the ultrafine grains with high dense dislocations deformed during CR facilitate the nucleation and growth of precipitates resulting from the lower activation energy barrier and increased free energy [23].

After aging at 160 °C for 10 h, the ultimate tensile strength and the ductility increased and the ultimate tensile strength sequence was similar to that before aging. The highest ultimate tensile strength of 549 MPa and the elongation rate of 13.9% was achieved simultaneously when the sheets were processed by CR at −190 °C and subsequence aging for 10 h.

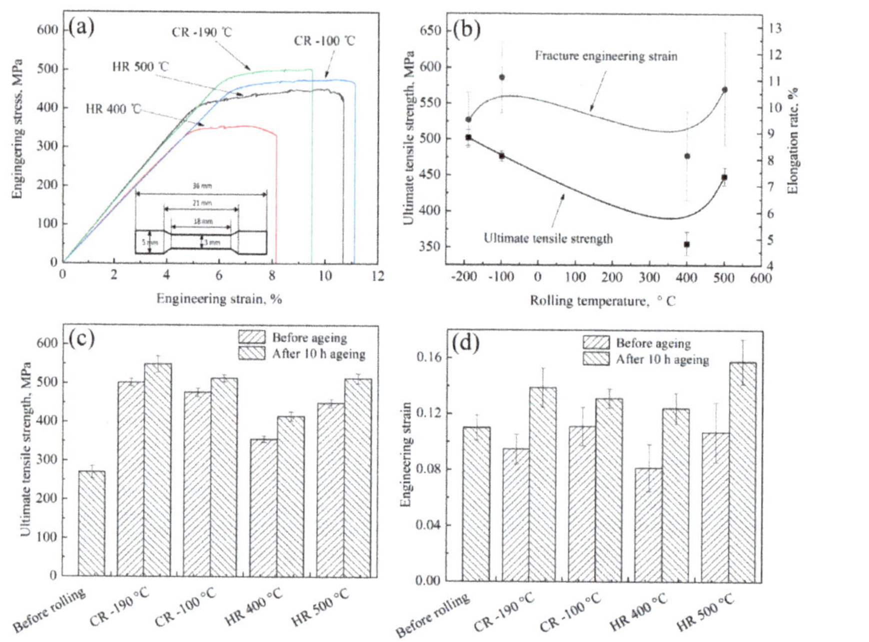

Figure 2. Mechanical properties of Al-Cu-Li alloy sheets after rolling and aging. (**a**) Engineering stress vs engineering strain curve of the Al-Cu-Li alloy sheets subjected to different processes; (**b**) Ultimate tensile stress and elongation vs rolling temperature; (**c**) Ultimate tensile stress vs rolling temperature after 10 h aging at 160 °C; (**d**) Engineering strain vs rolling temperature after aging at 160 °C for 10 h.

Figure 3 shows the hardness's of the alloy sheets processed at different conditions. After solution treatment, the hardness of 118 HV was obtained. After rolling, the hardness increased due to the refined grains, increased dislocations, and precipitations. The highest hardness value of 148 HV was achieved when the alloy sheet was processed by CR at −190 °C. After aging at 160 °C for 10 h, it increased to 158 HV. The variation trend of hardness was similar to that of ultimate tensile strength.

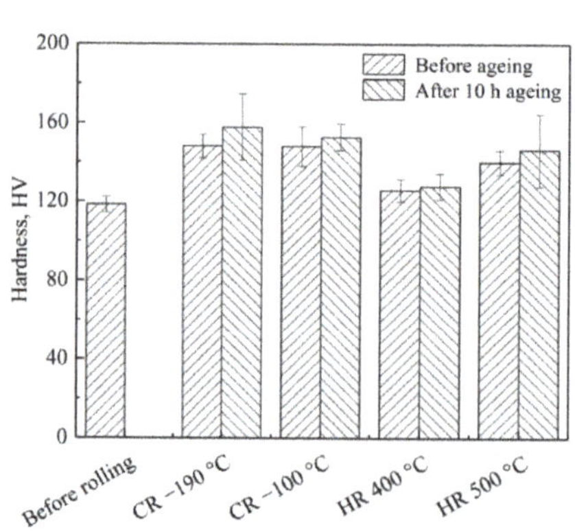

Figure 3. Hardness of Al-Cu-Li alloy sheets after rolling and aging.

Figure 4 shows optical microscopies of the Al-Cu-Li alloy by different processes. The grain size of the alloy after solution treatment was relatively large. After rolling, grains were refined to an elongated shape. With the decrease of rolling temperature, the mean grain size decreased obviously. After aging at 160 °C for 10 h, the grain size slightly increased and the sequence was similar to that before aging.

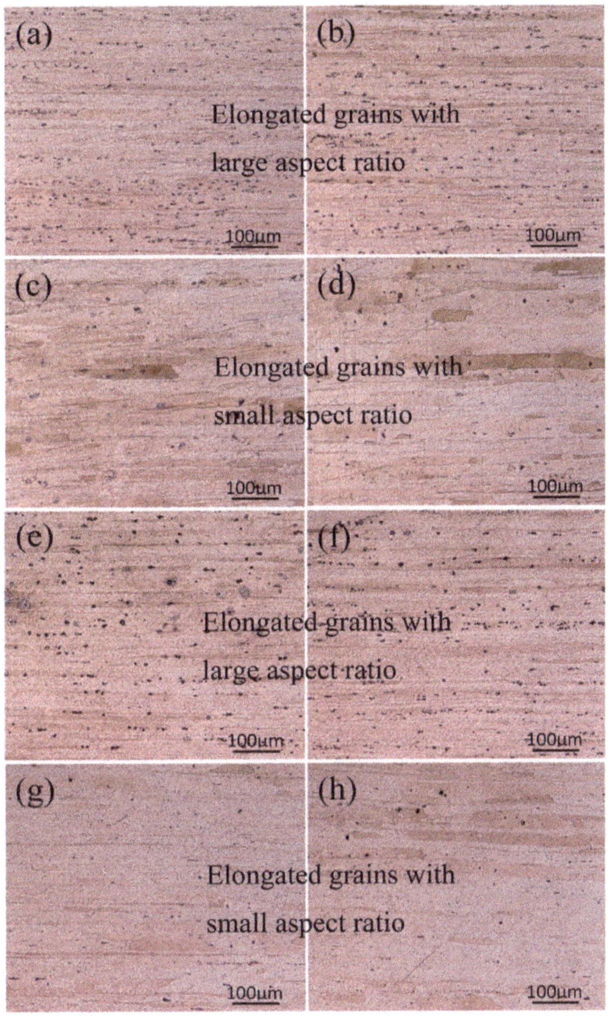

Figure 4. Optical microscopies of Al-Cu-Li alloys subjected to CR (at (**a**) −190 °C and (**b**) −100 °C) and HR (at (**c**) 400 °C and (**d**) 500 °C), as well as CR at (**e**) −190 °C, and (**f**) −100 °C, HR at (**g**) 400 °C, (**h**) 500 °C and subsequent aging at 160 °C for 10 h.

Figure 5 shows the TEM images of microstructures of sheets rolled at different temperatures. After rolling, numerous dislocation lines were formed and significantly entangled into dislocation clusters and into dislocation walls along grain boundaries and subgrain boundaries, which were marked by yellow lines and red lines. With the decrease of rolling temperature, more dislocations progressively accumulated and the dislocation debris, with a size of 80 nm, gradually moved to the dislocations cluster and dislocation walls entangled with them. Another great difference between the alloys processed by HR and CR was that the dislocation distribution in the cryorolled sample was relatively uniform while LDDZ and HDDZ were obviously observed and they were distributed alternately, which have been marked in Figure 5c,d. When the low deformation strain was applied, vacancies and dislocations gradually formed and dislocations accumulated and entangled, forming dislocations clusters and dislocations walls with the increase of applied strain. As the hindering of the dislocation movement and the suppressing of the dynamic recovery, more dislocation networks were formed in the cryorolled samples. As the deformation strain further increased, the stress fields caused by dislocations started interacting with each other, leading to the relaxation and rearrangement of the dislocation network. The dislocation network gradually transformed into low angled dislocations cells, which were relatively thick and consisted of lesser geometrical necessary dislocation density. The dislocation cell boundaries prefer changing to subgrain boundaries leading to the formation of ultrafine subgrain microstructure in the alloy. Hence, a subgrain structure was observed in alloy processed by HR 500 °C and the microstructure free of dislocations was due to recovery phenomena at the relatively high temperature. HDDZ was observed because the dislocation cell boundaries were relatively stable compared to another dislocation network. Moreover, the second-phase particles have a certain pinning effect on the dislocation movement. A similar nonuniform dislocation structure was obtained when ferritic steel was pre-deformed at 25 °C by Tanaka et al. [24]. In addition to observing dislocations, the secondary-phase particles appeared beside the dislocation walls or dislocations clusters after HR, as shown in Figure 5c,d. With the increase of roll temperature, the fraction of the secondary-phase particles increased due to the larger nucleation energy.

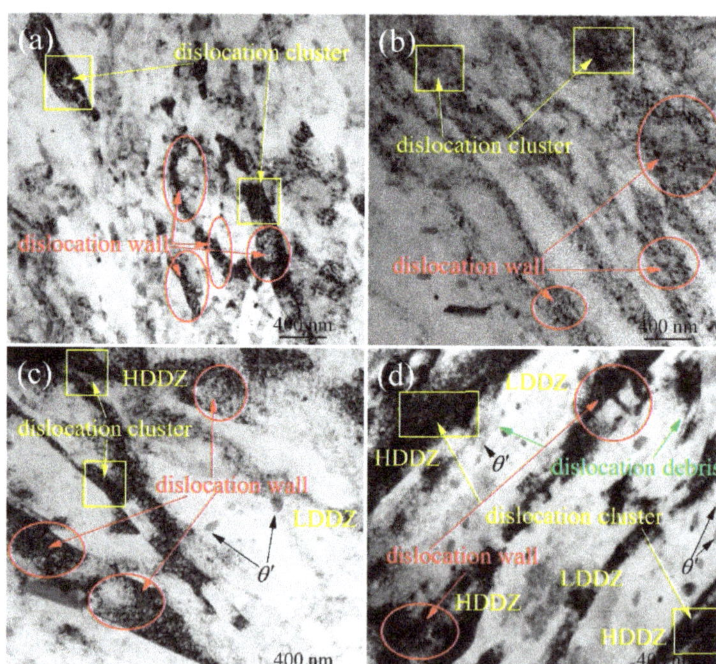

Figure 5. TEM images of Al-Cu-Li alloy subjected to CR at (**a**) −190 °C, and (**b**) −100 °C, HR at (**c**) 400 °C, (**d**) 500 °C.

Figure 6 shows the TEM images on a scale of 50 nm of the alloy subjected to CR −190 °C and HR 500 °C. With the reduction of 65%, numerous stack faults were formed in grains or at grain boundaries. When the alloy was processed by CR, the deformation-induced stack faults arranged along more than five directions while the two directions, which have marked by black dotted lines, were observed in the alloy processed by HR 500 °C.

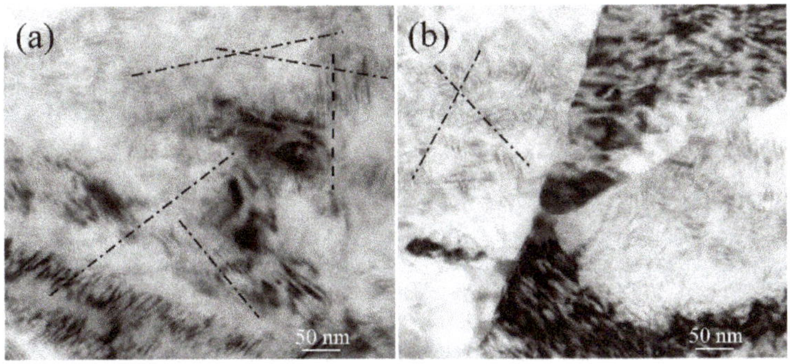

Figure 6. TEM images on a scale of 50 nm of Al-Cu-Li alloy subjected to (**a**) CR at −190 °C, and (**b**) HR at 500 °C.

Figure 7 shows TEM images of the samples after aging at 160 °C for 10 h. θ′ was the main strengthening phase of the alloy. With the increase of the aging time, the secondary-phase particles gradually nucleated and grew while the dislocations decreased through annihilation. When the alloys were processed by CR −190 °C, the higher dislocation density in the alloy provided more nucleation sites for the precipitates leading to higher precipitation strengthening. While the rolling temperature increased from 400 °C to 500 °C, the diffusion energy of the solute atoms increased making it easy to precipitation. The number and area fraction of the secondary-phase particles in the samples rolled by CR or HR changes significantly.

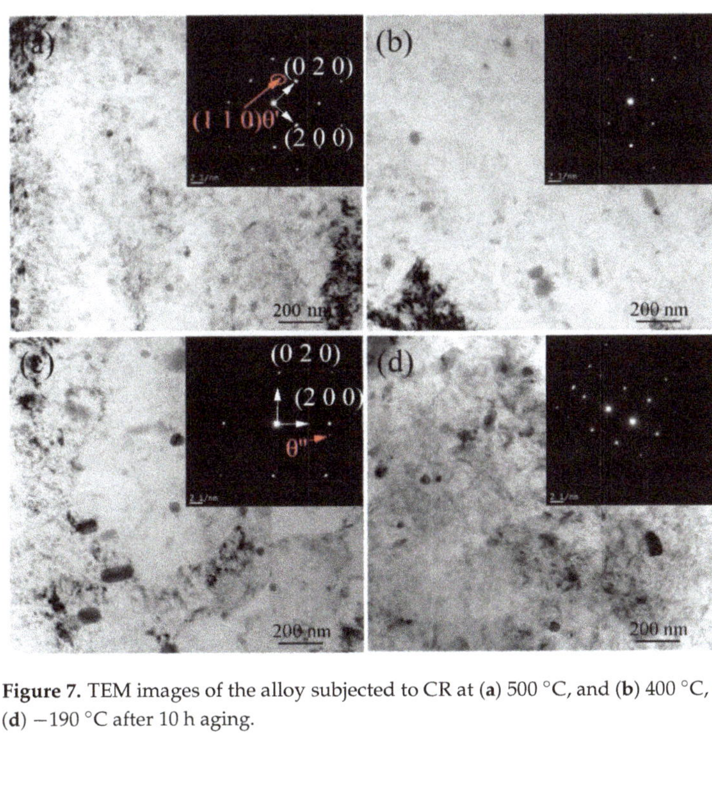

Figure 7. TEM images of the alloy subjected to CR at (**a**) 500 °C, and (**b**) 400 °C, HR at (**c**) −100 °C, (**d**) −190 °C after 10 h aging.

Figure 8 shows XRD patterns of the alloy processed by rolling at different temperatures and subsequence aging. The Williamson–Hall method was utilized to calculate the dislocation density [25]. The dislocation density and microstrain are shown in Figure 9. The highest dislocation density of 1.86×10^{14} m^{-2} was achieved when the alloy was processed by CR −190 °C. With the increase of the rolling temperature, dislocation density decreased sharply. After aging at 160 °C, dislocation density slightly decreased compared to the rolled samples at the corresponding temperature.

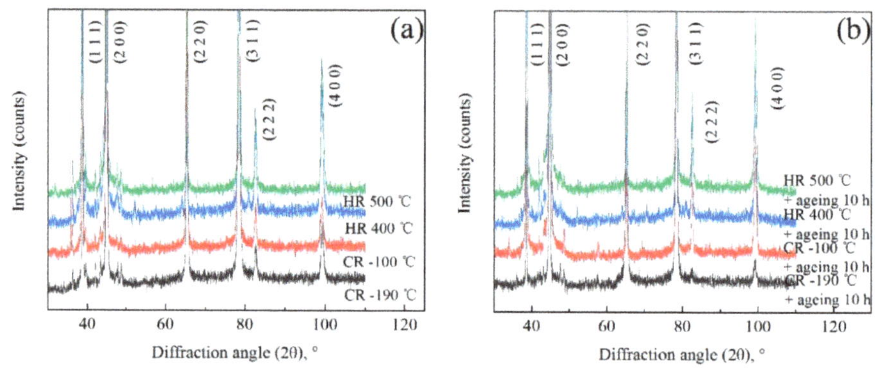

Figure 8. XRD patterns of the alloy subjected to (**a**) rolling (**b**) and subsequence aging at 160 °C for 10 h.

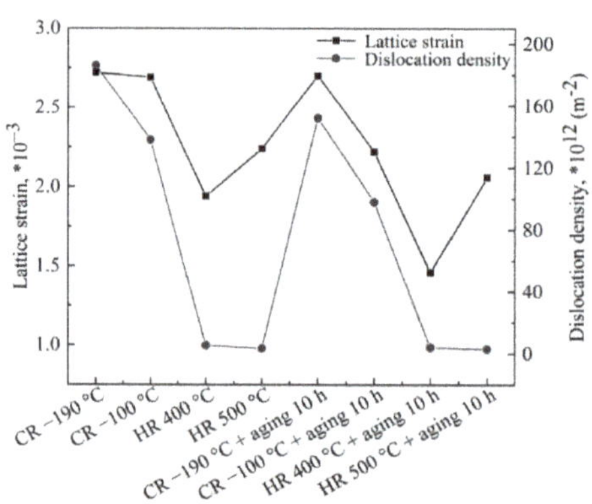

Figure 9. Lattice microstrain and dislocation density of the alloy processed under different conditions.

Figure 10 shows the fracture morphology of the Al–Cu–Li alloy sheets after rolling. The fracture mode is a mixture of ductile fracture and brittle fracture. The Al–Cu–Li alloy subjected to CR at −100 °C and HR at 500 °C show more dimples on the fracture which have been marked by yellow lines, and their ductility is good. However, for the Al–Cu–Li alloy sheets subjected to HR at 400 °C, there are few shallow dimples. Moreover, there also are a few flat areas on the fracture surface that have been marked by red lines, and the ductility of the Al–Cu–Li alloy is the lowest.

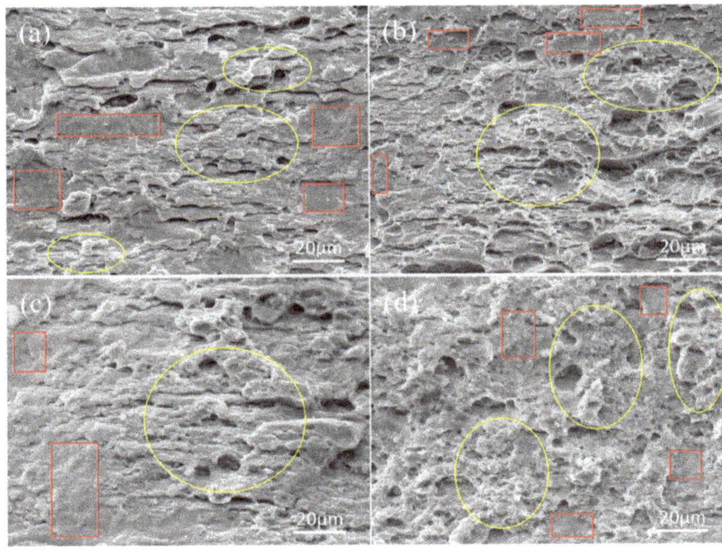

Figure 10. Fracture morphologies of the sheets subjected to CR at (**a**) −190 °C and (**b**) −100 °C; HR at (**c**) 400 °C, (**d**) 500 °C.

Figure 11 shows fracture morphologies of the alloy rolled at different temperatures and subsequence aging at 160 °C for 10 h. A comparison between Figures 10 and 11 shows that the dimples turned deeper and increased, which indicated the increase of the ductility of the sheets.

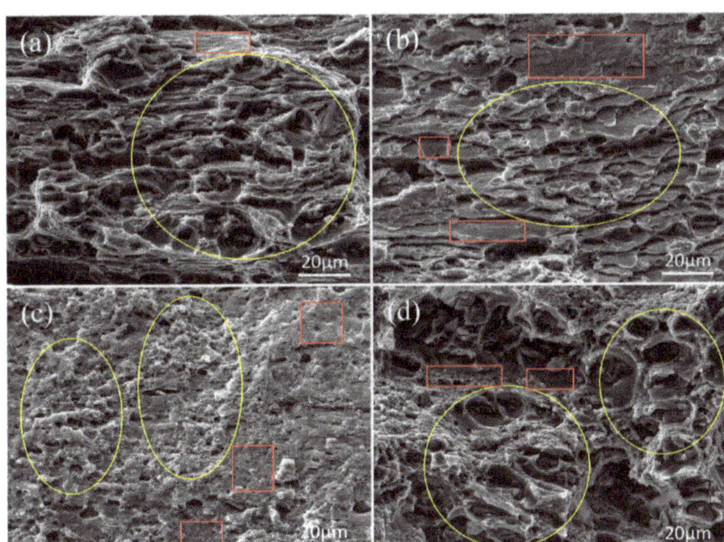

Figure 11. Fracture morphologies of the sheets subjected to CR at (**a**) −190 °C, and (**b**) −100 °C, HR at (**c**) 400 °C, (**d**) 500 °C after 10 h aging.

From the above results, it is obvious that the rolling temperature plays a significant role in the fracture surface and mechanical properties of the Al-Cu-Li alloy sheets. It can be seen from Figure 2 that the differences of the ultimate tensile strength of 146.5 MPa are generated between the sheets processed by SR 400 °C and SR −190 °C. The grain sizes were refined with the decreasing rolling temperature. The formation of small sized sub-grains appears during CR, which can be the reason for the increase of the ultimate tensile strength according to the Hall–Peach relationship [26]. In the processing of HR 500 °C, there are some clusters and a few amounts of precipitation, which can lead to better ultimate tensile strength compared with sheets processed by HR 400 °C. When the sheets were processed by CR, the driving energy of the dislocation movement was relatively low and dislocation movement and dynamic recovery were suppressed. Hence, new slip systems were active resulting in the appearance of more directions of stack fault arrangement compared with HR. When the rolling temperature increased from 400 °C to 500 °C, the size of the second-phase particles decreased and the number increased due to the higher solubility and a larger nucleation drive energy, which is shown in Figure 4, except the slight increase of the grain size. The fine second phase can hinder the movement of dislocations during tensile tests, resulting in enhanced strength and better ductility. Similar results have been obtained by Pathak et al. on the bulk ultrafine grain Al 2014 alloy. The percentage elongation of the alloy processed by CR + WR 170 °C (9.5%) was higher than that of the alloy processed by CR + WR 100 °C (4.5%) due to the formation of fine metastable spherical phase Al_2Cu and solid strengthening [27]. Interestingly, the CR processed sheets present better ductility than the sheets processed at 400 °C, resulting from the restrain of dynamic recovery and dislocation movement in the cryogenic temperature. Similar results have been reported in an Al-Mg alloy [28], high manganese steel [29], and CuZr alloy [30]. After aging at 160 °C for 10 h, the secondary-phase particle θ' and θ'' nucleated and grew, which has great strengthening effects. When the sheets were processed by CR at −190 °C and subsequence aging at 160 °C, the strengthening effects from refined grains, accumulated dislocations, precipitation were the biggest, leading to the highest strength.

4. Conclusions

The mechanical properties and microstructure evolution of an Al-Cu-Li alloy processed by CR and HR were investigated. The conclusions can be drawn as follows:

(1) CR can significantly improve the tensile strength and ductility of Al-Cu-Li alloy sheets, the best ultimate tensile strength of 502 MPa was achieved for the specimens subjected to CR at −190 °C, and the best ductility of 11.1% was obtained for the specimens subjected to CR at −100 °C before aging. The highest ultimate tensile strength of 549 MPa and the elongation rate of 13.9% was achieved simultaneously when the sheets were processed by CR at −190 °C and subsequence aging for 10 h.

(2) The mechanical properties of Al-Cu-Li alloy sheets are dependent on the grain sizes, precipitations and clusters, the dislocation microstructure, and the back stress distribution during rolling. Grains in cryorolled samples were seriously refined compared to that in the alloy processed by HR. Numerous deformation-induced dislocation networks were observed. HDDZ and LDDZ occurred in the hot-rolled samples, while it distributed uniformly with a more arranged direction in the cryorolled samples. After aging at 160 °C for 10 h, grains size slightly increased compared to that in the alloy rolled at the corresponding temperature. The deformation-induced dislocations can facilitate the nucleation and growth of the main strengthening phases θ', which resulted in enhanced strength.

Author Contributions: Conceptualization, H.Y. and A.P.; methodology, L.W.; investigation, L.W., C.K., P.T., D.P.; resources, H.Y.; writing—original draft preparation, H.Y.; writing—review and editing, all authors; supervision, H.Y.; project administration, H.Y. and A.P.; funding acquisition, H.Y., and A.P. All authors have read and agreed to the published version of the manuscript.

Funding: This research was funded by the National Key Research and Development Program (Grant No.: 2019YFB2006500), National Natural Science Foundation of China (Grant number: 5167040503), Innovation Driven Program of Central South University (Grant No.: 2019CX006), the Science and Technology Innovation Program of Hunan Province (Grant No.: 2020RC2002), the Research Fund of the Key Laboratory of High Performance Complex Manufacturing at Central South University. A.P.

and D.P. appreciate the financial support of the research which was carried out within the framework of the implementation of the Resolution of the Government of the Russian Federation of April 9, 2010, No. 220 (Contract No. 075-15-2019-869 from May 12, 2019) and a grant from the Russian Science Foundation (project No. 20-69-46042 of 20.05.2020).

Institutional Review Board Statement: Not applicable.

Informed Consent Statement: Not applicable.

Data Availability Statement: The data presented in this study are available on request from the corresponding author.

Conflicts of Interest: The authors declare no conflict of interest.

References

1. Triolo, A.; Lin, J.S.; Triolo, R. Early and late stages of demixing of a commercial Al-Li alloy. *J. Mater. Sci.* **2002**, *37*, 1207–1213. [CrossRef]
2. Nazarov, S.A.; Ganiev, I.N.; Ganieva, N.I.; Calliari, I. Microstructure and mechanical properties of Al + 6%Li alloy containing rare-earth metals. Vestnik Magnitogorskogo Gosudarstvennogo Tekhnicheskogo Universiteta im. G.I. Nosova. *Vestn. Nosov Magnitogorsk State Tech. Univ.* **2017**, *15*, 63–68. [CrossRef]
3. Nazarov, S.A.; Ganiev, I.N.; Norova, M.T.; Ganieva, N.I.; Calliari, I. Potentiodynamic study of the Al + 6% Li alloy doped with yttrium in nacl solution. Vestnik Magnitogorskogo Gosudarstvennogo Tekhnicheskogo Universiteta im. G.I. *Vestn. Nosov Magnitogorsk State Tech. Univ.* **2016**, *14*, 95–100. [CrossRef]
4. Rioja, R.J.; Liu, J. The evolution of Al-Li base products for aerospace and space applications. *Metall. Mater. Trans. A* **2012**, *43*, 3325–3337. [CrossRef]
5. Li, H.; Liu, X.; Sun, Q.; Ye, L.; Zhang, X. Superplastic Deformation Mechanisms in Fine-Grained 2050 Al-Cu-Li Alloys. *Materials* **2020**, *13*, 2705. [CrossRef]
6. Peng, Z.; Li, J.; Sang, F.; Chen, Y.; Zhang, X.; Zheng, Z.; Pan, Q. Structures and tensile properties of Sc-containing 1445 Al−Li alloy sheet. *J. Alloys Compd.* **2018**, *747*, 471–483. [CrossRef]
7. Zhou, C.; Zhan, L.; Shen, R.; Zhao, X.; Yu, H.; Huang, M.; Li, H.; Yang, Y.; Hu, L.; Liu, D.; et al. Creep behavior and mechanical properties of Al-Li-S4 alloy at different aging temperatures. *J. Cent. South Univ.* **2020**, *27*, 1168–1175. [CrossRef]
8. Li, J.F.; Liu, P.L.; Chen, Y.L.; Zhang, X.H.; Zheng, Z.Q. Microstructure and mechanical properties of Mg, Ag and Zn multi-microalloyed Al-(3.2-3.8)Cu-(1.0-1.4)Li alloys. *Trans. Nonferrous Met. Soc. China* **2015**, *25*, 2103–2112. [CrossRef]
9. Gao, Z.; Chen, J.H.; Duan, S.Y.; Yang, X.B.; Wu, C.L. Complex Precipitation Sequences of Al-Cu-Li-(Mg) Alloys Characterized in Relation to Thermal Ageing Processes. *Acta Metall. Sin.* **2016**, *29*, 94–103. [CrossRef]
10. Gao, C.; Luan, Y.; Yu, J.C.; Ma, Y. Effect of thermo-mechanical treatment process on microstructure and mechanical properties of 2A97 Al-Li alloy. *Trans. Nonferrous Met. Soc. China* **2014**, *24*, 2196–2202. [CrossRef]
11. Sun, Y.; Aindow, M.; Hebert, R.J.; Langdon, T.G.; Lavernia, E.J. High-pressure torsion-induced phase transformations and grain refinement in Al/Ti composites. *J. Mater. Sci.* **2017**, *52*, 12170–12184. [CrossRef]
12. Liu, F.; Yuan, H.; Goel, S.; Liu, Y.; Wang, J.T. Bulk nanolaminated nickel: Preparation, microstructure, mechanical property, and thermal stability. *Metall. Mater. Trans. A* **2018**, *49*, 576–594. [CrossRef]
13. Wang, H.; Su, L.H.; Yu, H.L.; Lu, C.; Tieu, K.; Liu, Y.; Zhang, J. A new finite element model for multi-cycle accumulative roll bonding process and experiment verification. *Mater. Sci. Eng. A* **2018**, *726*, 93–101. [CrossRef]
14. Yu, H.L.; Du, Q.L.; Godbole, A.; Lu, C.; Kong, C. Improvement in Strength and Ductility of Asymmetric-Cryorolled Copper Sheets under Low-Temperature Annealing. *Metall. Mater. Trans. A* **2018**, *49*, 4398–4403. [CrossRef]
15. Yu, H.L.; Lu, C.; Tieu, K.; Li, H.J.; Godbole, A.; Kong, C.; Zhao, X. Simultaneous grain growth and grain refinement in bulk ultrafine-grained copper under tensile deformation at room temperature. *Metall. Mater. Trans. A* **2016**, *47*, 3785–3789. [CrossRef]
16. Magalhães, D.C.C.; Kliauga, A.M.; Ferrante, M.; Sordi, V.L. Asymmetric cryorolling of AA6061 Al alloy: Strain distribution, texture and age hardening behavior. *Mater. Sci. Eng. A* **2018**, *736*, 53–60. [CrossRef]
17. Gholinia, A.; Humphreys, F.; Prangnell, P. Processing to ultrafine grain structures by conventional routes. *Mater. Sci. Technol.* **2000**, *16*, 1251–1255. [CrossRef]
18. Han, Y.; Shi, J.; Xu, L.; Cao, W.Q.; Dong, H. Effect of hot rolling temperature on grain size and precipitation hardening in a Ti-microalloyed low-carbon martensitic steel. *Mater. Sci. Eng. A* **2012**, *553*, 192–199. [CrossRef]
19. Shanmugasundaram, T.; Murty, B.S.; Sarma, V.S. Development of ultrafine grained high strength Al-Cu alloy by cryorolling. *Scr. Mater.* **2006**, *54*, 2013–2017. [CrossRef]
20. Deng, Y.; Huang, G.; Cao, L.; Wu, X.; Huang, L.; Xia, M.; Liu, Q. Improvement of strength and ductility of Al−Cu−Li alloy through cryogenic rolling followed by aging. *Trans. Nonferrous Met. Soc. China* **2017**, *27*, 1920–1927. [CrossRef]
21. Li, C.; Xiong, H.Q.; Bhatta, L.; Wang, L.; Zhang, Z.Y.; Wang, H.; Kong, C.; Yu, H.L. Microstructure evolution and mechanical properties of Al−3.6Cu−1Li alloy via cryorolling and aging. *Trans. Nonferrous Met. Soc. China* **2020**, *30*, 2904–2914. [CrossRef]
22. Saimoto, S. Deformation kinetics and constitutive relation analyses of bifurcation in work-hardening of face-centred cubic metals at cryogenic temperatures. *Acta Mater.* **2019**, *174*, 43–52. [CrossRef]

23. Jayaganthan, R.; Panigrahi, S.K. Effect of Cryorolling Strain on Precipitation Kinetics of Al 7075 Alloy. *Mater. Sci. Forum* **2008**, *584–586*, 911–916. [CrossRef]
24. Tanaka, Y.; Masumura, T.; Tsuchiyama, T.; Takaki, S. Effect of dislocation distribution on the yield stress in ferritic steel under identical dislocation density conditions. *Scr. Mater.* **2020**, *177*, 176–180. [CrossRef]
25. Ma, K.; Wen, H.; Hu, T.; Topping, T.; Isheim, D.; Seidman, D.; Lavernia, E.; Schoenung, J. Mechanical behavior and strengthening mechanisms in ultrafine grain precipitation-strengthened aluminum alloy. *Acta Mater.* **2014**, *62*, 141–155. [CrossRef]
26. Ma, Y.; Zhou, X.; Thompson, G.E.; Hashimoto, T.; Thomson, P.; Fowles, M. Distribution of intermetallics in an AA 2099-T8 aluminium alloy extrusion. *Mater. Chem. Phys.* **2011**, *126*, 46–53. [CrossRef]
27. Pathak, M.K.; Joshi, A.; Mer, K.K.S.; Jayaganthan, R. Mechanical properties and microstructural evolution of bulk UFG Al 2014 processed through cryorolling and warm rolling. *Acta Metall. Sin.* **2019**, *32*, 845–856. [CrossRef]
28. Kumar, P.; Singh, A. Investigation of fracture behaviour and low cycle fatigue properties of cryorolled Al-Mg alloy. *Theor. Appl. Fract. Mech.* **2018**, *98*, 78–94. [CrossRef]
29. Fu, B.; Fu, L.; Liu, S.; Wang, H.R.; Wang, W.; Shan, A. High strength-ductility nano-structured high manganese steel produced by cryogenic asymmetry-rolling. *J. Mater. Sci. Technol.* **2018**, *34*, 695–699. [CrossRef]
30. Li, R.; Zhang, S.; Kang, H.; Chen, Z.; Yang, F.; Wang, W.; Zou, C.; Li, T.; Wang, T. Microstructure and texture evolution in the cryorolled CuZr alloy. *J. Alloys Compd.* **2017**, *693*, 592–600. [CrossRef]

metals

MDPI

Article

Influence of Morphology of Intermetallic Particles on the Microstructure and Properties Evolution in Severely Deformed Al-Fe Alloys

Andrey Medvedev [1,*], Maxim Murashkin [1,2], Nariman Enikeev [1,2], Evgeniy Medvedev [1] and Xavier Sauvage [3]

1 Institute of Physics of Advanced Materials, Ufa State Aviation Technical University, 450008 Ufa, Russia; m.murashkin.70@gmail.com (M.M.); nariman.enikeev@gmail.com (N.E.); medvedev.ufa@mail.ru (E.M.)
2 Laboratory of Mechanics of Advanced Bulk Nanomaterials for Innovative Engineering Applications, Saint Petersburg State University, 198504 Saint Petersburg, Russia
3 Groupe de Physique des Matériaux, Normandie Université, UNIROUEN, INSA Rouen, CNRS, 76000 Rouen, France; xavier.sauvage@univ-rouen.fr
* Correspondence: medvedevandreyrf@gmail.com; Tel.: +7-917-456-9074

Abstract: This study focuses on the difference in microstructural features and physical properties of Al-2Fe and Al-4Fe alloys subjected to large plastic straining. The difference in the intermetallic particle morphology in the initial state is shown to be a key parameter influencing the particle and grain fragmentation process and, as a result, the properties of these two alloys. We demonstrate that the shape and average size of Al-Fe intermetallic particles provide stronger effect on the microstructure evolution during high pressure torsion (HPT) than their volume fraction. The formation of Fe supersaturated solid solution in Al in these two alloys during deformation is discussed in connection to the morphology of the intermetallic phase. The major microstructural attributes, responsible for the solid solution formation, are highlighted.

Keywords: Al-Fe alloys; severe plastic deformation; supersaturated solid solution; mechanical strength; electrical conductivity

Citation: Medvedev, A.; Murashkin, M.; Enikeev, N.; Medvedev, E.; Sauvage, X. Influence of Morphology of Intermetallic Particles on the Microstructure and Properties Evolution in Severely Deformed Al-Fe Alloys. *Metals* **2021**, *11*, 815. https://doi.org/10.3390/met11050815

Academic Editor: Babak Shalchi Amirkhiz

Received: 27 April 2021
Accepted: 14 May 2021
Published: 17 May 2021

1. Introduction

Efforts to increase aluminum alloys' strength without significant sacrifice in electrical conductivity is a current trend in industry. At the same time, the material used as an electrical conductor should demonstrate good thermal stability, which can be challenging in case of aluminum alloys [1]. With this in mind, finding ways to simultaneously improve mechanical and electrical properties is important. This combination is essential for creating new lightweight conductive materials for the electrical industry, and it can comprise ultrafine-grained Al-Fe alloys.

Aluminum alloys, particularly Al-Fe alloys, have several advantages as conductive materials. First, aluminum and iron are very common and cheap metals, which make them economically preferable. Second, the solubility of iron in aluminum for conventionally processed alloys, at temperatures ranging from room to near-melting, is close to zero [2]. This eliminates the major contribution to electrical resistance (i.e., solid solute atoms). The other contributions are grain boundaries, particles, and dislocation density [2].

Previous studies of Al-Fe alloys have found applications in conductive wires in automobiles [3–6] and as household cables. However, conventional approaches have reached the limit of increasing the mechanical strength and electrical conductivity of these materials. This is due to the absence of iron solubility in aluminum limiting the variations of precipitation morphology.

In conventionally produced aluminum alloys, there are usually two major intermetallic phases: $Al_{13}Fe_4$ and Al_6Fe [7–9]. These phases precipitate as needle- and plate-like particles of relatively large size (microns to hundreds of microns). Conventional approaches, such as drawing/rolling etc., are not able to reduce these particles to the nanoscale, which would provide increased mechanical strength. Nanoscale particles could easily be formed from solid solution by simple heat treatment, but, as was mentioned earlier, formation of solid solution of Fe in Al is difficult to achieve via conventional methods.

Another possibility is to obtain a supersaturated solid solution thanks to the introduction of severe plastic deformation (SPD) [10,11]. As well as leading to grain size refinement, particle shredding, and an increase of defect density, SPD is also known for the formation of supersaturated solid solution in systems, which is considered very hard or nearly impossible [5,10].

A number of studies were conducted involving SPD of Al-Fe alloys. The concentration of Fe in a solid solution was increased up to 2 wt.% or more [10–25]. Such solid solutions may give an opportunity to homogeneously precipitate Al-Fe intermetallic particles in a way that was previously impossible.

The formation of strain-induced supersaturated solid solutions depends on a number of parameters. It was previously demonstrated that it could be influenced not only by the second-phase volume fraction [12], but also by the particle morphology [13]. Larger interphase areas naturally give more opportunities for solute atoms to migrate into the matrix. In addition, the second-phase morphology directly influences the fragmentation process during deformation. This effect was demonstrated in [13], where differences in the particle morphology before deformation affected microstructures and the mechanical and physical properties of the alloy after SPD. However, the underlying mechanisms were not deeply investigated, and to optimize the second-phase fragmentation process and the solid solution formation, one needs a systematic study of the influence of the primary as-cast microstructure (volume fraction, intermetallic phase, particle size, morphology, and distribution). To achieve this goal, the present study focuses on two alloys (Al-2Fe and Al-4Fe) with very different initial microstructures.

2. Materials and Methods

The samples of Al-2 wt.% Fe and Al-4 wt.% Fe alloys were prepared on the basis of A99 grade primary aluminum (GOST 11069-2001). Melting was carried out in a GRAFICARBO GF 1100 electric resistance furnace (Graficarbo S.R.L., Zorlesco, Italy) in a graphite crucible at 830 °C at Russia's National University of Science and Technology "MISIS". Ingots with a diameter of 20 mm and a length of 200 mm were obtained by casting into a graphite mold (at the 20 K/s cooling rate). The total amount of trace materials, including Si, did not exceed 0.1 wt.% for either of the alloys.

The cylinders were then sliced into discs (1.5 mm thick) using a wire-cutting APTA-120 machine (Delta-Test, Fryazino, Russia). HPT was carried out by placing the disc-shaped sample between two rotating anvils and applying high hydrostatic pressure to the sample, inducing a simple shear type deformation. In this study, discs of as-cast alloy were subjected to 20 revolutions of HPT performed at room temperature (RT) under a pressure of 6 GPa and a speed of 1 rpm in closed anvils. Such a processing route was chosen based on primary experiments showing that these parameters provide uniform microstructures [26,27] and, possibly, supersaturated solid solutions.

Microstructural, mechanical, and electrical data were collected from a specific location in the middle of the sample's radius 5 mm, thus rendering it comparable with other research.

Transmission electron microscopes (TEMs, JEOL, Tokyo, Japan) using different analysis modes (i.e., BF bright-field analysis, SAED-selected area electron diffraction) and STEM (scanning transmission electron microscopy-JEOL ARM 200F) were employed for the microstructure analysis. Objects for TEM were prepared by twin-jet electro-polishing on a Struers Tenupol-5 (Copenhagen, Denmark) with 20% nitric acid in methanol below −20 °C at an operating voltage of 20 V. To ensure statistically reliable results, at least three foils were used for each state.

Scanning electron microscopy (SEM) was performed on a ThermoFisher Helios G4 PFIB DualBeam (Waltham, MA, USA).

X-ray diffraction (XRD) analysis was conducted with a Rigaku Ultima IV (Tokyo, Japan) diffractometer using CuKα radiation (30 kV and 20 mA). Values of lattice parameter a, coherent scattering domain (CSD), size D, and elastic microdistortion level <ε2>1/2 were calculated via the Rietveld refinement method using MAUD software (v.2.97, University of Trento, Trento, Italy) [28]. To calculate dislocation density (ρ), Equation (2) was used:

$$\rho = \frac{2\sqrt{3}\langle\varepsilon^2\rangle^{1/2}}{D \times b} \quad (1)$$

where $b = a\sqrt{2}/2$ is the Burgers vector for FCC metals, and D is the coherent scattering domain [29].

Tensile tests were carried out in triplicate using an Instron 5982 (Instron, Norwood, MA, USA) machine at RT and a strain rate of 10^{-3} s^{-1}. Yield stress ($\sigma_{0.2}$), ultimate tensile stress (σ_{UTS}), and ductility (δ), measured as elongation-to-failure, were obtained from the small samples with gauge dimensions of 1.0 mm × 1.0 mm × 4.0 mm, prepared by wire-cutting. The maximum error for sample dimensions was 0.02 mm.

Electrical conductivity (ω) was determined with a ±2% error, using the eddy current method [30]. Electrical conductivity relative to annealed copper (International Annealed Copper Standard, %IACS) was calculated according to Equation (3):

$$IACS = \frac{\omega_{Al}}{\omega_{Cu}} \times 100\% \quad (2)$$

where ω_{Al} is the measured electrical conductivity of aluminum alloy, and ω_{Cu} is the conductivity of annealed chemically pure copper (58 MS/m).

3. Results

3.1. Microstructure Evolution of the Alloys during HPT

3.1.1. SEM Analyses

Figure 1a,b shows the microstructure of the as-cast Al-2Fe and Al-4Fe samples. For the Al-2Fe alloy (Figure 1a), the microstructure consists of the net/web of eutectic phase particles, surrounding the dendritic aluminum phase. This could also be described as the ellipsoidal-shape areas of the aluminum phase, divided by aluminum-intermetallic conglomerates. The volume fraction of the intermetallic phase in the Al-2Fe alloy is 4.6%.

The microstructure of the Al-4Fe alloy is different from that of the Al-2Fe alloy. The amount of the intermetallic phase is higher than 7.7%. It comprises very coarse plate-like particles with a thickness/width up to 10 microns, and a length up to 50 microns (Figure 1b). In addition to these coarse particles, the microstructure also consists of smaller fractions of particles distributed between them. Smaller particles in the Al-4Fe alloy form a structure, similar to Al-2Fe alloy, only these particles are larger and are located sparsely in comparison.

The intermetallic particles in the alloys are the mix of the Al_6Fe and $Al_{13}Fe_4$ phases [21], formed during the crystallization [31–36].

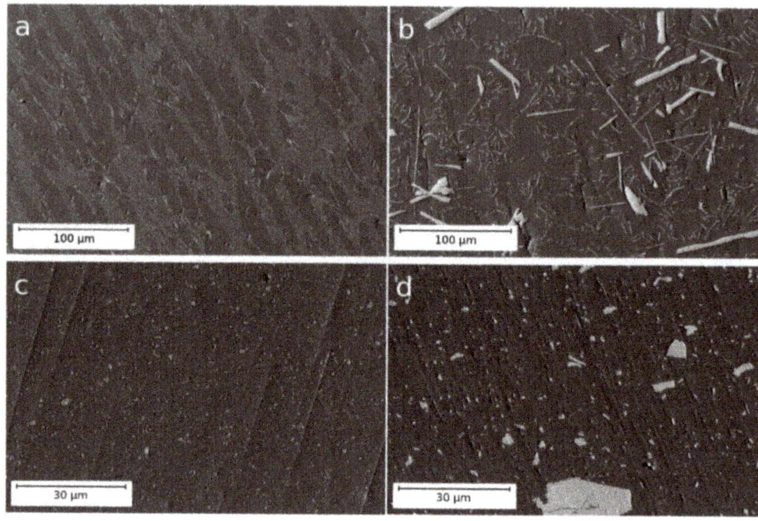

Figure 1. Backscattered SEM images of the samples of the Al-2Fe and Al-4FE alloys: (**a**) Al-2Fe as-cast, (**b**) Al-4Fe as-cast, (**c**) Al-2Fe after HPT, (**d**) Al-4Fe after HPT.

It can be seen that the Al-2Fe alloy has a casting microstructure very similar to that in Al-RE (rare earth) alloys [12,13] and other Al-Fe alloys [20–23].

Figure 1c,d show the microstructure formed in alloys during HPT. While the intermetallic particles in the Al-2Fe alloy are pretty small and hardly distinguishable (Figure 1c), the size of these particles in Al-4Fe after HPT varies from a fraction to tens of microns.

Since the intermetallic particles in the Al-2Fe alloy exhibit a narrow size distribution, we can affirm that the fragmentation of the particles occurred homogeneously. On the contrary, the refinement of the coarse particles in the Al-4Fe alloy was incomplete, as fragments of large particles are still exhibited in the microstructure (Figure 1d) and the particle size distribution is large.

3.1.2. XRD Analyses

XRD profiles corresponding to the as-cast and HPT-processed states of Al-2Fe and Al-4Fe alloys are presented in Figure 2, where peaks of different phases are indexed. For the phase analysis we have used the reference crystal structures corresponding to Al (FCC, $a = b = c = 4.049$ Å), Al_6Fe (orthorhombic, $a = 6.46$ Å, $b = 7.44$ Å, $c = 8.78$ Å), and $Al_{13}Fe_4$ (monoclinic, $a = 15.49$ Å, $b = 8.08$ Å, $c = 12.47$ Å, β=107.69°), where a, b, c are the lattice parameters. Thanks to the analysis of the XRD profiles we have revealed the phases with the crystal structure close to stoichiometry of the reference phases. Interestingly, the as-cast specimen of Al-2Fe alloy mainly contains a stable $Al_{13}Fe_4$ phase, while in the as-cast Al-4Fe alloy a metastable Al_6Fe phase is dominant. This can be explained by the chemical composition of the as-cast materials: it is known that, depending on the alloy composition and cooling rate, different types of Al-Fe intermetallic phases can be formed [36].

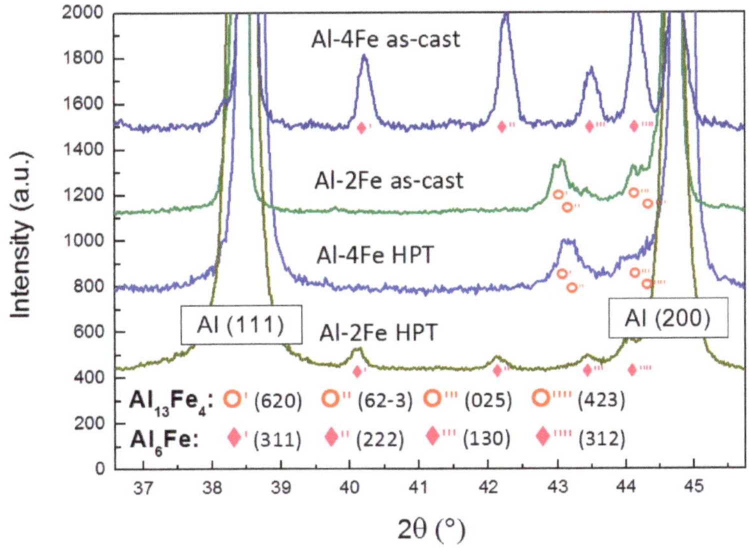

Figure 2. XRD profiles of the Al-2Fe and Al-4Fe alloys in the as-cast state and after HPT.

After HPT the opposite trend is observed: Al_6Fe and $Al_{13}Fe_4$ phases are formed in the Al-2Fe and Al-4Fe alloys, respectively, while the peaks of precursor particles disappear.

Revealing possible scenarios for phase transformations in Al-Fe alloys during HPT is beyond the scope of this study and will be reported later. The results of qualitative phase analysis and structural parameters evaluation are thus presented.

The precise quantitative phase analysis is complicated by texture effects especially in the coarse-grained states where statistically limited number of grains are oriented in the reflection position with respect to the scattering vector. Nevertheless, the XRD profile of Al-4Fe in the as-cast state demonstrates visibly the highest intensity of all major peaks related to the precipitate phase among all studied cases. It indicates that this state is characterized with the maximal volume fraction of precipitates.

HPT resulted in a noticeable shift in the positions of Al peaks to higher diffraction angles in both alloys. The calculated lattice parameter for Al matrix reduced by 0.078% for the Al-2Fe alloy and 0.035% for the Al-4Fe alloy (which is well consistent with the data reported in [36]). This indicates to a process of strain-induced dissolution of Fe in Al [36], which is more intensive in Al-2Fe. Besides, Al-2Fe has a smaller coherent domain size and higher dislocation density: about 90 nm and 2.2×10^{14} m^{-2}, respectively, versus 200 nm and 2.9×10^{13} m^{-2} for the Al-4Fe alloy. From the presented data it follows that HPT-induced structure refinement, defect accumulation, and Fe dissolution are more pronounced in Al-2Fe compared to Al-4Fe despite the difference in the alloying element concentration.

3.1.3. TEM and STEM Analyses

Figure 3a,b shows the bright-field TEM images, comparing the microstructures of the Al-2Fe and Al-4Fe alloys samples after HPT, respectively. To the right from BF images in Figure 3 the selected area diffraction (SAED) patterns obtained from a region of 1.5 μm size, are presented. The number of particles, contained within the aperture area of the SAED patterns, is quite small, thus the number of the reflections from particles is small as well. However, the number of reflections was just enough to index phases.

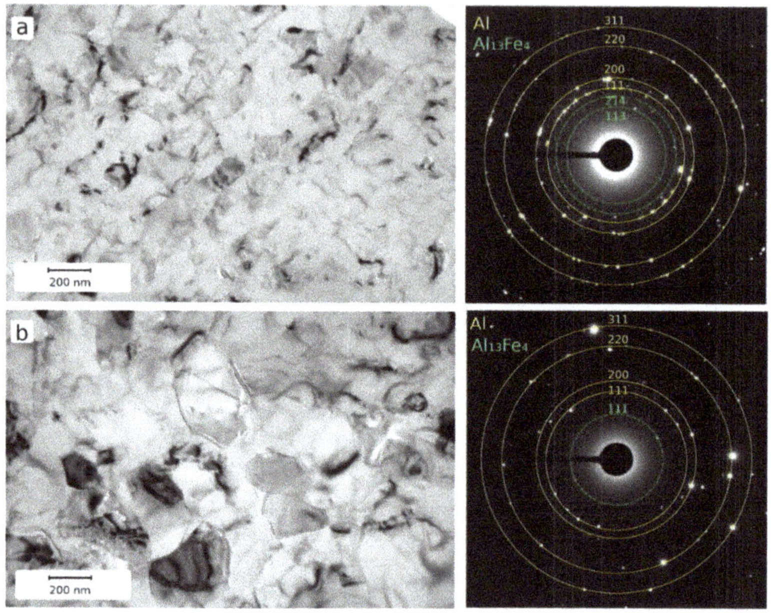

Figure 3. TEM bright-field (**left**) and SAED patterns (**right**) from the samples of the Al-2Fe (**a**) and Al-4Fe (**b**) after HPT. Indexed reflections are highlighted for the Al and Al13Fe4 phases with the corresponding color.

Analysis of the SAED patterns showed the presence of the Al and $Al_{13}Fe_4$ phase in both Al-2Fe and Al-4Fe alloys after HPT. Peaks that could be attributed to the Al_6Fe phase could not be observed in TEM. This is probably due to the fact, that the amount of Al_6Fe phase is quite small. Besides, overlap between the Al_6Fe (131), Al_6Fe (222) and Al_6Fe (132) peaks with Al (111) and Al (200) peaks makes them hardly distinguishable.

The number of reflections on the Al-4Fe SAED patterns is lower due to the larger mean grain size.

The microstructure of the Al-2Fe alloy after HPT is composed of equiaxed, ultrafine grains with a mean grain size ~125 ± 10 nm. Figure 3b shows that the microstructure of the Al-4Fe alloy after HPT exhibits the equiaxed grains with a mean size ~340 ± 20 nm. The microstructures of both alloys exhibit grains with high-angles misorientation.

Bright-field TEM does not provide sufficient information for the size, distribution, and morphology of the intermetallic particles. Intermetallic particles in these alloys contains a high amount of Fe, which should provide a clear contrast between the matrix and the particles in the z-contrast mode (HAADF) of STEM [24].

Figure 4 shows STEM images of Al-2Fe (Figure 4a,b) and Al-4Fe (Figure 4c,d) alloys after HPT. According to these data, the mean size of particles in the Al-4Fe alloy is significantly larger than in the Al-2Fe alloy (277 ± 16 nm versus 78 ± 4 nm). The difference in grain size of these two alloys (Figure 3a,b) is very likely to be connected with the difference in particle size since the deformation conditions for both alloys were similar.

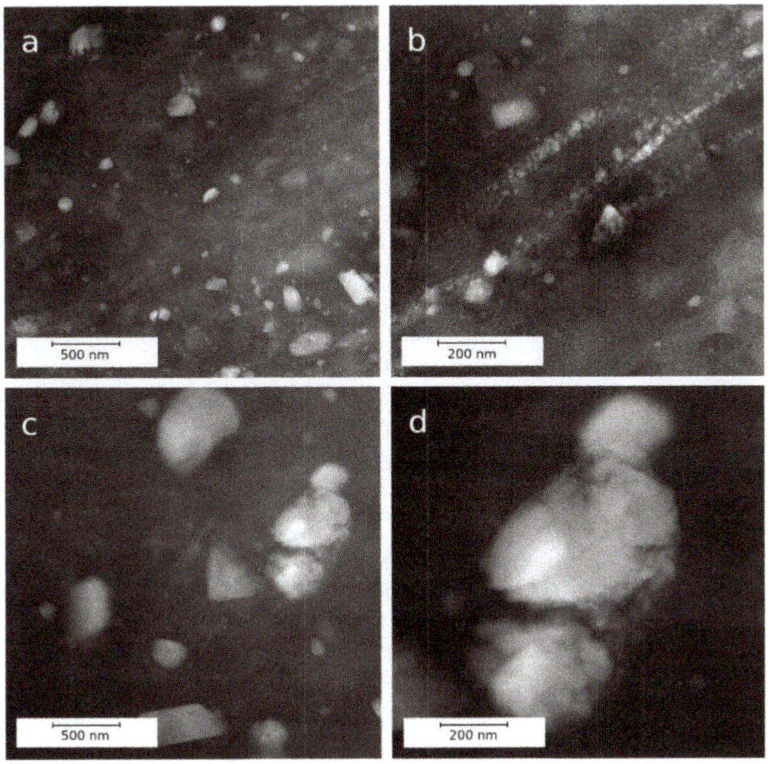

Figure 4. STEM HAADF images, obtained from the samples of the Al-2Fe alloy (**a**,**b**) and Al-4Fe alloy (**c**,**d**) HPT.

The origin of differences in the two alloys, subjected to the same deformation treatment, probably lies in the particle fragmentation process. In the Al-2Fe alloy, linear arrays of fine particles can be observed (Figure 4b). Such arrays are former intermetallic particles, fragmented and elongated during the deformation.

The particle fragmentation in the Al-4Fe alloy after HPT is different. During the observation, no similar arrays could be observed. Instead, "clouds" of finer particles surrounding coarse ones were exhibited (Figure 4d). This means that the amount of deformation induced by HPT was probably not large enough to cause cracks throughout the whole particle. Instead, small particles are fragmented from the surface region, while the core of the coarse particle remains intact. Thus, even after HPT, the coarse particles are still present, and fine particles are distributed non-uniformly. All this contributes to a reduced impact on the structure evolution during the HPT in comparison to the Al-2Fe alloy, as it will be discussed later

3.2. Physical and Mechanical Properties

Table 1 presents the results of the study of electrical and mechanical properties of Al-2Fe and Al-4Fe alloys in as-cast condition and after HPT.

Table 1. Mechanical and electrical properties of Al-Fe alloys in the as-cast state and after HPT.

Alloy	State	$\sigma_{0.2}$, MPa	σ_{UTS}, MPa	δ, %	ω, MS/m	IACS, %
Al-2Fe	As-cast	55 ± 6	100 ± 2	26 ± 2	32.4 ± 0.2	55.8 ± 0.3
	HPT	564 ± 12	649 ± 6	5.0 ± 0.3	23.4 ± 0.7	40.4 ± 1.7
Al-4Fe	As-cast	70 ± 6	88 ± 7	18 ± 2	31.2 ± 0.2	54.2 ± 1.2
	HPT	270 ± 19	340 ± 22	8.0 ± 0.5	27.8 ± 0.1	48.0 ± 0.6

Despite the difference in the iron content and the intermetallic particle morphology in the initial state, both alloys have a comparable ultimate tensile strength (~90 MPa). The presence of coarse intermetallic particles in the Al-4Fe alloy embrittles it Al-4Fe alloy elongation to failure is 8% lower than that of the Al-2Fe alloy in the initial state (Figure 5).

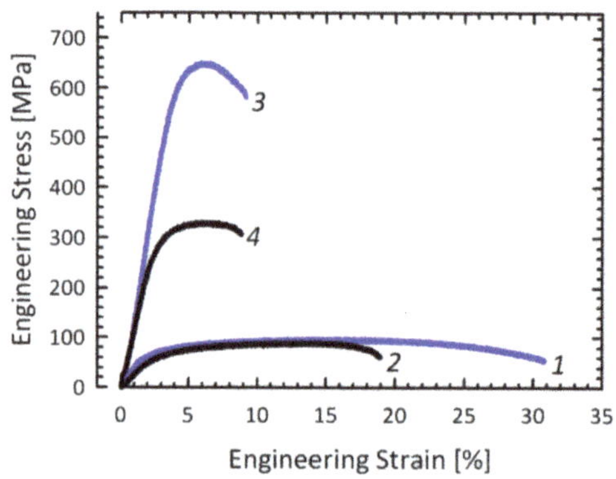

Figure 5. Engineering stress-strain curves obtained by tensile tests of Al-Fe alloys in the as-cast state (1: Al-2Fe, 2: Al-4Fe) and after HPT (3: Al-2Fe, 4: Al-4Fe). For the sake of clarity, only one curve is provided for each state.

The level of the electrical conductivity is higher for the Al-2Fe alloy in the as-cast state. This is due to the lower total amount of iron in the alloy, and due to the absence of coarse intermetallic particles, contrary to the Al-4Fe alloy.

As a result of microstructure refinement during HPT, the level of the ultimate tensile strength increases to ~650 MPa for the Al-2Fe alloy and to ~340 MPa for the Al-4Fe alloy. The increase in tensile strength is not as big for the Al-4Fe alloy as for the Al-2Fe alloy due to the larger intermetallic particle size and a nonuniform distribution of nanoscaled particles (Section 3.1.3). HPT also leads to a decrease in the ductility of the Al-2Fe and the Al-4Fe alloys. It should be noted, however, that after HPT both alloys still maintain a relatively high level of elongation to failure: ~5% for the Al-2Fe alloy and ~8% for the Al-4Fe alloy. Because the fragmentation of the intermetallic particles in the Al-4Fe alloy is not as complete as in the Al-2Fe alloy, the Al-4Fe alloy samples still contain areas free of the finely dispersed second phase particles. These areas improve the elongation to failure for the Al-4Fe alloy in comparison to the Al-2Fe alloy.

The electrical conductivity after HPT decreases for both Al-2Fe and Al-4Fe alloys. The decrease in the electrical conductivity is 15.4 %IACS for the Al-2Fe alloy and 6.2 %IACS for the Al-4Fe alloy (Table 1). The electrical conductivity is affected, among other things, by the presence of the finely dispersed second phase particles and the size of the grains. It was observed in Section 3.1.3 that Al-4Fe alloy after HPT has a higher mean grain size and also contains areas, free of the fine intermetallic particles, while Al-2Fe alloy after HPT is characterized by lower mean grain size and uniformly distributed fine second-phase particles. Such difference in microstructure may explain the difference in the level of electrical conductivity.

The electrical conductivity is known to be particularly sensitive to the presence of the solid solution, so the sharp decrease of it in the Al-2Fe alloy can be the indirect sign of the solid solution formation. This will be discussed in the detail in the Discussion section.

4. Discussion

4.1. Particle Morphology

The SEM data show that the particle fragmentation process is not the same in Al-2Fe and Al-4Fe alloys, presumably due to the morphology of the intermetallic particles in the as-cast state. The difference in the alloys' properties could be explained by the difference in the fragmentation process. To evaluate this difference, we take the overall interphase area parameter total surface of the "aluminum–particle" interface.

Corresponding calculations and measurements have been undertaken in order to find the overall interphase area between aluminum and intermetallic particles. The following assumption has been made: all particles have the shape of perpendicular prisms with a base of *AP* (area of particle) and a height of *h* (Figure 6).

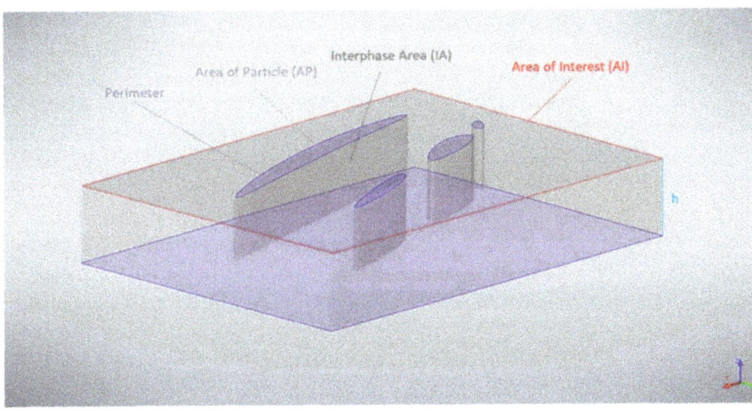

Figure 6. Particle morphology model describing the terms necessary for the calculations.

The area of interest (*AI*) is calculated via the markers on the SEM images. For statistical purposes, acquisition of data has been performed over multiple SEM images for each of the alloys.

The interphase area (*IA*) is calculated as a perimeter (*P*) multiplied by the *h*:

$$IA = P \times h \tag{3}$$

Relative interphase area (*RIA*) is calculated as the interphase area divided by the studied volume (area of interest multiplied by the *h*):

$$RIA = \frac{IA}{AI \times h} = \frac{P}{AI} \tag{4}$$

Table 2 represents the results of the statistical analysis of the data, collected from the Al-Fe alloys SEM images. The average particle size (area) in the as-cast state is smaller in the Al-2Fe alloy 0.26 μm^2 against 2.55 μm^2 in the Al-4Fe alloy.

Table 2. Analysis data of Al-Fe alloys' intermetallic particles, obtained from the SEM images.

Alloy	State	Average Particle Area, μm^2	Average Particle Perimeter, μm	Perimeter-to-Area Ratio, μm^{-1}	RIA, μm^{-1}
Al-2Fe	As-cast	0.26 ± 0.002	0.26 ± 0.002	1.00	0.396
	HPT	0.02 ± 0.001	0.49 ± 0.006	24.50	1.466
Al-4Fe	As-cast	2.55 ± 0.160	2.85 ± 0.180	1.10	0.234
	HPT	0.25 ± 0.040	1.24 ± 0.020	4.96	0.460

HPT leads to particle refinement, and thus to a decrease of their average area of 13 times for the Al-2Fe and of 10 times for the Al-4Fe alloy.

The average perimeter of particles is bigger in the Al-4Fe alloy in both the as-cast (0.49 μm against 0.26 μm) and HPT (1.24 μm against 0.49 μm) conditions, indicating the presence of the coarser particles in Al-4Fe both before and after deformation. It can be seen (Table 2) that the total perimeter of the particles increases by about two times for the Al-2Fe alloy after HPT, and decreases by about two times for the Al-4Fe alloy. For us, the perimeter-to-area ratio (*PtA*) is of greater importance, since it considers both of these parameters.

The perimeter-to-area ratio (*PtA*) can be more informative, since it combines these two parameters. The value of the *PtA* is the same for the alloys in the as-cast state, but changes drastically after HPT from 1 to 24.5 μm^{-1} for the Al-2Fe alloy and from 1.1 to 4.96 μm^{-1} for the Al-4Fe alloy. The higher the value of the *PtA*, the smaller the particles in the studied volume. Hence, the particle refinement in the Al-4Fe alloy is five times less intensive than in the Al-2Fe alloy.

Figure 7 shows the area distribution of particles in Al-Fe alloys depending on their size. In the as-cast state in the Al-2Fe alloy, the area distribution is close to normal, while the Al-4Fe alloy plot has a plateau, pointing to the presence of coarser particles. After HPT, both curves for the Al-2Fe and Al-4Fe alloys become skewed left, showing the decrease in the average area size of the particles. While the distribution of the particle size is characterized by a single peak, for the Al-2Fe alloy, the curve for the Al-4Fe alloy exhibits a second peak, signifying bimodal distribution.

The *RIA* parameter shows how many interphase boundaries are presented at a given volume. The higher the value of *RIA*, the higher the density of the interphase boundaries. In the as-cast state, *RIA* in the Al-2Fe alloy is higher than in the Al-4Fe alloy 0.396 μm^{-1} against 0.234 μm^{-1}. The difference is even more notable after HPT—the *RIA* in the HPT state in the Al-2Fe alloy is more than three times greater than in the Al-4Fe alloy.

The *RIA* parameter and the level of electrical conductivity in the as-cast Al-2Fe alloy are higher than in the Al-4Fe alloy. This means that in the as-cast condition, the surface amount of the interphase boundaries is not as significant as the total amount of the intermetallic phase. During the HPT, however, we observe a much more intense change of properties (Section 3.2), making the interphase area (and, thus, the morphology of the intermetallic phase in the as-cast state) more significant for the physical properties than its total amount.

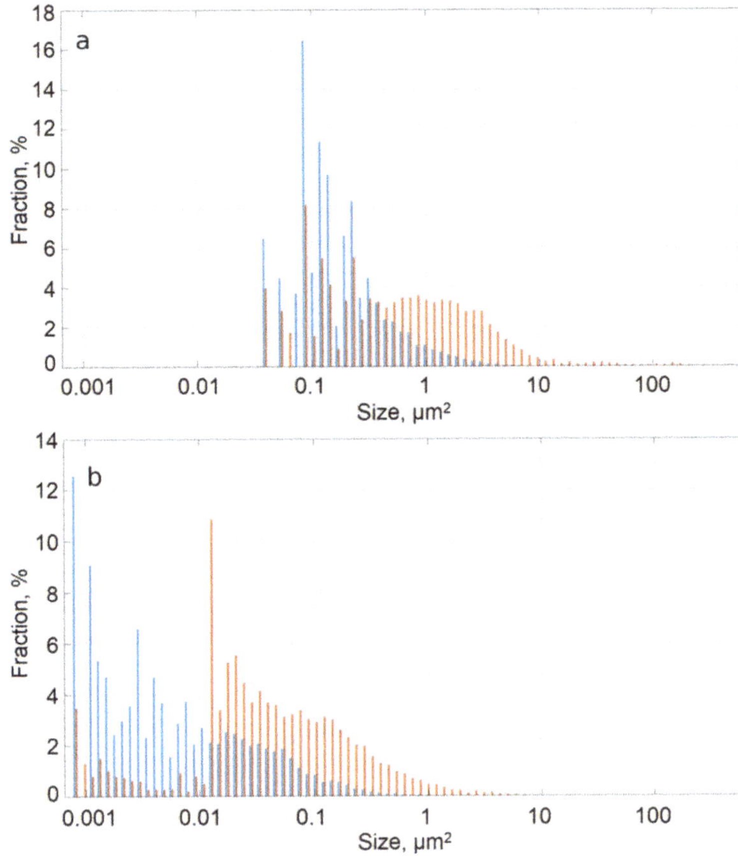

Figure 7. Particle area distribution: (**a**) in as-cast state, (**b**) after HPT. Histogram for Al-2Fe alloy is marked by blue color, for Al-4Fe alloy is marked by orange color. μm^2.

4.2. Formation of the Solid Solution

The differences in mechanical strength and electrical conductivity are linked to different structural parameters, such as dislocation density, grain size, size and density of particles and so on.

Among all, the negative effect of solid solution on the conductivity is much higher than that of the other crystal lattice defects [1]. A sharp decrease in the electrical conductivity after the HPT (Table 1) can indirectly indicate the strain-induced formation of supersaturated solid solution of Fe in Al.

The lattice parameters of the Al-2Fe and the Al-4Fe alloys in the as-cast condition are close to that of pure Al. It allows suggesting that nearly all Fe atoms are stored within the intermetallic phases which is consistent with the results presented in [25,36]. After HPT, the lattice parameter of both Al-2Fe and Al-4Fe alloys decreases. This change can be associated with the dissolution of Fe in Al [16,25,36]. Al-4Fe alloy, where the decrease of the lattice parameter is smaller than in Al-2Fe alloy, exhibits a tendency to form solid solution with a smaller concentration of Fe.

The discussion in Section 4.1 supports the HPT-induced dissolution of Fe in Al. Since Fe could only migrate to Al from intermetallic particles, and the *RIA* parameter after HPT is 3 times higher in the Al-2Fe alloy than in the Al-4Fe alloy, the microstructure of the as-cast Al-2Fe alloy provides wider opportunities for Fe to diffuse into the Al matrix by the higher density of interphase boundaries.

The difference in the lattice parameter, electrical conductivity, and density of the interphase boundaries before and after HPT in both alloys indicates more intensive formation of the strain-induced solid solution of Fe in Al in Al-2Fe alloy. Since intermetallic particles are still present after the HPT, we can safely assume that the concentration of Fe in the supersaturated solid solution is far less than 1 at. % (which is the total amount of Fe in Al-2Fe alloy). The solid solution is formed due to the partial dissolution of intermetallic particles. This observation is consistent with the previous studies, focused on the intermetallic phase transformation in Al-Fe alloys [14,15,25,36].

Higher dissolution rate of Fe in Al matrix in Al-2Fe alloy than in Al-4Fe alloy during HPT proves that the total amount of alloying element is not as important for the formation of the solid solution as the morphology of the intermetallic phase in the initial state.

5. Conclusions

This study shows that in Al-Fe alloys the morphology of the intermetallic phase in the cast-state has a notable effect on the structural evolution during 20 revolutions of HPT. This feature can be even more important, than the concentration of Fe in the alloy. The initial size and dispersity of the particles, formed during crystallization, their volume fraction, and, respectively, the length of the interphase boundaries plays an important role in forming mechanical and electrical properties by HPT. The following specific conclusions are drawn from the results presented in this study:

1. The initial microstructure of the Al-2Fe and the Al-4Fe alloys is composed of the mixture of Al and eutectic Al_6Fe and $Al_{13}Fe_4$ phases. In the Al-2Fe alloy the intermetallic particles are shaped as plates/needles with the length up to 150 nm, while Al-4Fe alloy contains, among the fine plate-like particles, coarser particles with the size up to tens of microns.
2. As a result of HPT, the intermetallic particles become fragmented and redistributed in the bulk of the Al-2Fe and Al-4Fe alloys. Intermetallic particles in the Al-2Fe alloy are in the comparable size range and are uniformly distributed within the Al matrix, while in the Al-4Fe alloy the size of the particles after HPT varies from tens of nanometers to few microns, and their distribution is bimodal.
3. Al-4Fe alloy after HPT is characterized by a larger grain size than Al-2Fe alloy. Intermetallic particles morphology in Al-2Fe alloy created the possibility for the solid solution formation, thus inhibiting the grain boundary migration. In addition, in Al-4Fe alloy intermetallic particles were distributed heterogeneously, and thus created regions in material's volume where the movement of the grain boundaries was not inhibited.
4. HPT processing leads to an increase of the UTS to ~650 MPa for the Al-2Fe alloy and to the decrease of its electrical conductivity to 40.4 %IACS. The changes in the Al-4Fe alloy have a similar character but are less pronounced due to the different morphology of intermetallic particles in the as-cast state.
5. We show that Al-2Fe alloy has a uniform size distribution of the intermetallic particles after HPT because of the different particle morphology in the initial state. In particularly, the uniformity of the particle size distribution is achieved due to the higher, relative to the Al-4Fe alloy, overall interphase area in the as-cast state.
6. Supersaturated solid solution of the Fe in Al is formed by HPT in Al-2Fe alloy. Al-4Fe alloy, due to lower density of interphase boundaries in the as-cast state, has fewer opportunities for the Fe to diffuse into the Al matrix, so the solid solution forms it with a significantly lower Fe concentration.
7. Predesigned morphology in low-solubility alloys, coupled with deformation techniques, could be effectively used as a tool for the production of high-strength materials with a unique set of properties.

Author Contributions: Conceptualization, M.M. and A.M.; methodology, X.S.; software, N.E.; writing—original draft preparation, A.M.; writing—review and editing, N.E., M.M. and X.S.; visualization, E.M. All authors have read and agreed to the published version of the manuscript.

Funding: This research was funded by Russian Science Foundation, Grant number 20-79-10133.

Institutional Review Board Statement: Not applicable.

Informed Consent Statement: Not applicable.

Data Availability Statement: Not applicable.

Acknowledgments: Authors are grateful to the personnel of the research and technology Joint Research Center, Nanotech, Ufa State Aviation Technical University, for their assistance with instrumental analysis.

Conflicts of Interest: The authors declare no conflict of interest.

References

1. Murashkin, M.Y.; Sabirov, I.; Sauvage, X.; Valiev, R.Z. Nanostructured Al and Cu alloys with superior strength and electrical conductivity. *J. Mater. Sci.* **2016**, *51*, 33–49. [CrossRef]
2. Polmear, I.J. *Light Alloys-From Traditional Alloys to Nanocrystalls*, 4th ed.; Monash University: Melbourne, Australia, 2006.
3. Jablonski, M.; Knych, T.; Smyrak, B. Effect of iron addition to aluminium on the structure and properties of wires used for electrical purposes. In Proceedings of the 5th International Conference on Light Metals Technology, Luneburg, Germany, 19–22 July 2011; Trans Tech Publications Ltd.: Luneburg, Germany, 2011; pp. 459–462.
4. Jablonski, M.; Knych, T.; Smyrak, B. New Aluminium Alloys for Electrical Wires of Fine Diameter for Automotive Industry. *Arch. Metall. Mater.* **2009**, *54*, 671–676.
5. Sekia, S.; Susai, K.; Takamura, S. Development of aluminium wire for automotive harnesses. In Proceedings of the 60th International Wire & Cable Symposium (IWCS), Charlotte Convention Center, Charlotte, NC, USA, 7–9 November 2011; pp. 445–449.
6. Horikoshi, T.; Koruda, H.; Shimizu, M.; Aoyama, S. Development of Aluminium Alloy Conductor with High Electrical Conductivity and Controlled Tensile Strength and Elongation. *Hitachi Cable Rev.* **2006**, *25*, 31–34.
7. Nasu, S.; Gonser, U.; Preston, R.S. Defects and phases of iron in aluminium. *J. Phys. Colloq.* **1980**, *41*, 385–386. [CrossRef]
8. Shabashov, V.A.; Brodova, I.G.; Mukoseev, A.G.; Sagaradze, V.V.; Litvinov, A.V. Mossbauer study of the dissolution of iron aluminides under strong cold deformation, Proceedings of the Russian Academy of Sciences. *Phys. Ser.* **2005**, *10*, 1459–1464.
9. Shabashov, V.A.; Brodova, I.G.; Mukoseev, A.G.; Sagaradze, V.V.; Litvinov, A.V. Structural transformations in the Al-Fe sys-tem under intense plastic deformation. *Phys. Met. Metall.* **2005**, *4*, 66–67.
10. Senkov, O.N.; Valiev, R.Z.; Pirzada, M.D.S.; Liu, J.; Froes, F.H. A new approach to hardening of Al-Fe alloys. Synthesis of Lightweight Metals: In proceedings of the 128th Annual Meeting & Exhibition of The Minerals, Metals & Materials Society, San Diego, CA, USA, 28 February–4 March 1999; pp. 129–134.
11. Stolyarov, V.V.; Brodova, I.G.; Yablonskikh, T.I.; Lapovok, R. The effect of backpressure on the structure and mechanical properties of the Al-5 wt % Fe alloy produced by equal-channel angular pressing. *Phys. Met. Metallogr.* **2005**, *100*, 182–191.
12. Medvedev, A.E.; Murashkin, M.Y.; Enikeev, N.A.; Valiev, R.Z.; Hodgson, P.D.; Lapovok, R. Enhancement of mechanical and electrical properties of Al-RE alloys by optimizing rare-earth concentration and thermo-mechanical treatment. *J. Alloys Compd.* **2018**, *745*, 696–704. [CrossRef]
13. Medvedev, A.E.; Murashkin, M.Y.; Enikeev, N.A.; Bikmukhametov, I.; Valiev, R.Z.; Hodgson, P.D.; Lapovok, R. Effect of the eutectic Al-(Ce,La) phase morphology on microstructure, mechanical properties, electrical conductivity and heat resistance of Al-4.5(Ce,La) alloy after SPD and subsequent annealing. *J. Alloys Compd.* **2019**, *796*, 321–330. [CrossRef]
14. Fadeeva, V.; Leonov, A. Amorphization and crystallization of Al-Fe alloys by mechanical alloying. *Mater. Sci. Eng. A* **1996**, *206*, 90–94. [CrossRef]
15. Fadeeva, V.; Leonov, A.V.; Khodina, L. Metastable Phases in Mechanically Alloyed Al-Fe System. *Mater. Sci. Forum* **1995**, *179–181*, 397–402. [CrossRef]
16. Mukhopadhyay, D.K.; Suryanarayana, C.; Froes, F.H.S. Structural evolution in mechanically alloyed Al-Fe powders. *Met. Mater. Trans. A* **1995**, *26*, 1939–1946. [CrossRef]
17. Senkov, O.N.; Froes, F.H.; Stolyarov, V.V.; Valiev, R.Z.; Liu, J. Non-equilibrium structures in aluminium-iron alloys subjected to severe plastic deformation. In Proceedings of the 1997 5th International Conference on Advanced Particulate Materials and Processes, West Palm Beach, FL, USA, 7–9 April 1997; pp. 95–102.
18. Senkov, O.; Froes, F.; Stolyarov, V.; Valiev, R.; Liu, J. Microstructure of Aluminum-Iron Alloys Subjected to Severe Plastic Deformation. *Scr. Mater.* **1998**, *38*, 1511–1516. [CrossRef]
19. Stolyarov, V.; Lapovok, R.; Brodova, I.; Thomson, P. Ultrafine-grained Al–5 wt.% Fe alloy processed by ECAP with backpressure. *Mater. Sci. Eng. A* **2003**, *357*, 159–167. [CrossRef]
20. Cubero-Sesin, J.M.; Horita, Z. Strengthening of Al through addition of Fe and by processing with high-pressure torsion. *J. Mater. Sci.* **2012**, *48*, 4713–4722. [CrossRef]
21. Cubero-Sesin, J.M.; Horita, Z. Age Hardening in Ultrafine-Grained Al-2PctFe Alloy Processed by High-Pressure Torsion. *Metall. Mater. Trans. A Phys. Metall. Mater. Sci.* **2015**, *46*, 2614–2624. [CrossRef]

22. Cubero-Sesin, J.M.; In, H.; Arita, M.; Iwaoka, H.; Horita, Z. High-pressure torsion for fabrication of high-strength and high-electrical conductivity Al micro-wires. In *Special Section: Ultrafinegrained Materials*, 19th ed.; Mathaudhu, S.N., Estrin, Y., Horita, Z., Lavernia, E., Liao, X.Z., Lu, L., Wei, Q., Wilde, G., Zhu, Y.T., Eds.; Kluwer Academic Publishers: Amsterdam, The Netherlands, 2014; pp. 6550–6557.
23. Cubero-Sesin, J.M.; Watanabe, M.; Arita, M.; Horita, Z.J. Aging and precipitation behavior in supersaturated Al-2%Fe alloy produced by high-pressure torsion. In Proceedings of the 14th International Conference on Aluminium Alloys, ICAA 2014, Trondheim, Norway, 15–19 June 2014; Trans Tech Publications Ltd.: Trondheim, Norway, 2014; pp. 766–771.
24. Duchaussoy, A.; Sauvage, X.; Edalati, K.; Horita, Z.; Renou, G.; Deschamps, A.; De Geuser, F. Structure and mechanical behavior of ultrafine-grained aluminum-iron alloy stabilized by nanoscaled intermetallic particles. *Acta Mater.* **2019**, *167*, 89–102. [CrossRef]
25. Medvedev, A.E.; Murashkin, M.Y.; Enikeev, N.A.; Valiev, R.Z.; Hodgson, P.D.; Lapovok, R. Optimization of Strength-Electrical Conductivity Properties in Al-2Fe Alloy by Severe Plastic Deformation and Heat Treatment. *Adv. Eng. Mater.* **2017**, *20*, 1700867. [CrossRef]
26. Valiev, R.Z.; Aleksandrov, I.V. *Bulk Nanostructural Metallic Materials*; Akademkniga: Moscow, Russia, 2007.
27. Vorhauer, A.; Pippan, R. On the homogeneity of deformation by high pressure torsion. *Scr. Mater.* **2004**, *51*, 921–925. [CrossRef]
28. Lutterotti, L.; Matthies, R.; Wenk, H.R.; Schultz, A.; Richardson, J. Combined texture and structure analysis of deformed lime-stone from time-of-flight neutron diffraction spectra. *J. Appl. Phys.* **1997**, *81*, 594–600. [CrossRef]
29. Williamson, G.K.; Smallman, R.E., III. Dislocation densities in some annealed and cold-worked metals from measurements on the X-ray debye-scherrer spectrum. *Philos. Mag.* **1956**, *1*, 34–46. [CrossRef]
30. ASTM E1004-09. *Standard Test Method for Determing Electrical Conductivity Using The Electromagnetic (Eddy-Current) Method*; ASTM International: West Conshohocken, PA, USA, 2009.
31. Saller, B.D.; Hu, T.; Ma, K.; Kurmanaeva, L.; Lavernia, E.J.; Schoenung, J.M. A comparative analysis of solubility, segregation, and phase formation in atomized and cryomilled Al–Fe alloy powders. *J. Mater. Sci.* **2015**, *50*, 4683–4697. [CrossRef]
32. Kim, D.H.; Cantor, B. Structure and decomposition behaviour of rapidly solidified Al-Fe alloys. *J. Mater. Sci.* **1994**, *29*, 2884–2892. [CrossRef]
33. Nayak, S.; Wollgarten, M.; Banhart, J.; Pabi, S.; Murty, B. Nanocomposites and an extremely hard nanocrystalline intermetallic of Al–Fe alloys prepared by mechanical alloying. *Mater. Sci. Eng. A* **2010**, *527*, 2370–2378. [CrossRef]
34. Sasaki, H.; Kita, K.; Nagahora, J.; Inoue, A. Nanostructures and mechanical properties of bulk Al-Fe alloys prepared by elec-tron-beam deposition. *Mater. Trans.* **2001**, *42*, 1561–1565. [CrossRef]
35. Sasaki, T.; Ohkubo, T.; Hono, K. Microstructure and mechanical properties of bulk nanocrystalline Al–Fe alloy processed by mechanical alloying and spark plasma sintering. *Acta Mater.* **2009**, *57*, 3529–3538. [CrossRef]
36. Shabashov, V.A.; Brodova, I.G.; Mukoseev, A.G.; Sagaradze, V.V.; Litvinov, A.V. Deformation-induced phase transformations in the Al–Fe system under intensive plastic deformation. *J. Phys. Condens. Matter* **2007**, *19*. [CrossRef]

metals

MDPI

Article

Effect of Severe Plastic Deformation on Structure Refinement and Mechanical Properties of the Al-Zn-Mg-Fe-Ni Alloy

Irina Brodova [1], Dmitriy Rasposienko [1], Irina Shirinkina [1], Anastasia Petrova [1,*], Torgom Akopyan [2] and Elena Bobruk [3,4]

[1] Mikheev Institute of Metal Physics, Ural Branch, Russian Academy of Sciences, 620108 Ekaterinburg, Russia; ibrodova@mail.ru (I.B.); dmitrijrasp@gmail.com (D.R.); shirinkina@imp.uran.ru (I.S.)
[2] Metal Forming Department, National University of Science and Technology MISiS, 119991 Moscow, Russia; aktorgom@gmail.com
[3] Institute of Physics of Advanced Materials, Ufa State Aviation Technical University, 450008 Ufa, Russia; e-bobruk@yandex.ru
[4] Laboratory for Mechanics of Bulk Nanostructured Materials, Saint Petersburg State University, 198504 Saint Petersburg, Russia
* Correspondence: petrovanastya@yahoo.com; Tel.: +7-950-647-3501

Abstract: This paper identifies the mechanisms of phase and structural transformations during severe plastic deformation by shearing under pressure (high-pressure torsion) of an Al-Zn-Mg-Fe-Ni-based aluminum alloy depending on different initial states of the material (an ingot after homogenizing annealing and a rod produced by radial-shear rolling). Scanning and transmission electron microscopy are used to determine the morphological and size characteristics of the structural constituents of the alloy after high-pressure torsion. It has been found that, irrespective of the history under high-pressure torsion, fragmentation and dynamic recrystallization results in a nanostructural alloy with a high microhardness of 2000 to 2600 MPa. Combined deformation processing (high-pressure torsion + radial-shear rolling) is shown to yield a nanocomposite reinforced with dispersed intermetallic phases of different origins, namely Al_9FeNi eutectic aluminides and $MgZn_2$, $Al_2Mg_3Zn_3$, and Al_3Zr secondary phases. The results of uniaxial tensile testing demonstrate good mechanical properties of the composite (ultimate tensile strength of 640 MPa, tensile yield strength of 628 MPa, and elongation of 5%).

Keywords: Al-Zn-Mg alloy; nanomaterials; nanocomposite; high pressure torsion; fragmentation; dynamic recrystallization; hardening; $MgZn_2$; $Al_2Mg_3Zn_3$; Al_3Zr; tensile test; transmission electron microscopy

Citation: Brodova, I.; Rasposienko, D.; Shirinkina, I.; Petrova, A.; Akopyan, T.; Bobruk, E. Effect of Severe Plastic Deformation on Structure Refinement and Mechanical Properties of the Al-Zn-Mg-Fe-Ni Alloy. *Metals* **2021**, *11*, 296. https://doi.org/10.3390/met11020296

Academic Editor: Christine Borchers

Received: 28 December 2020
Accepted: 31 January 2021
Published: 9 February 2021

1. Introduction

At present, Al-based composite materials are known to become increasingly important, for they have essential advantages over conventional commercial alloys due to a favorable combination of practical design characteristics and functional properties [1–4]. For example, aluminum-matrix composite materials reinforced with silicon carbide are used in automobile industry, aerospace engineering and other industries owing to high wear resistance, high specific strength, and heat conductivity [5–7]. It was shown that silicon carbide reinforced aluminum matrix composites are effectively additionally alloyed with the chromium [8]. The authors have found that the aluminum matrix composites with a varying weight percent of chromium (0–3 wt%) and a fixed percentage of silicon carbide (10%) were formed through the vortex casting. It was found that the chromium additive in the Al-SiC composites significantly improved the hardness, friction coefficient, and mechanical properties (strength, abrasion resistance, and wear resistance). Borides, nitrides, and oxides of refractory metals are often used as reinforcement particles for the Al matrix; the disperse phases of these particles reinforce the plastic Al matrix, thus creating barriers for the migration of dislocations and grain boundaries [1–3]. Besides artificial composites, which are produced by mechanical alloying or by powder metallurgy, there is a big class of natural aluminum-based eutectic composite materials. Typical examples of such alloys are represented by Al-Si alloys (silumins). New advanced alloys based on Al-Fe, Al-Ni, and Al-Fe-Ni eutectics have lately been developed and termed nickalins [9–11]. Well-known compositions of the 7xxx series—i.e., alloys belonging to the Al-Zn-Mg-Cu system—were proposed as the Al matrix of these alloys. Economical Al-Zn-Mg alloys with a total Fe and Ni content of 1 wt% are highly technological due to the eutectic constituent (Al + Al_9FeNi) and have high strength characteristics resulting from a dispersion-hardened Al matrix [10,11].

The state-of-the-art trend in improving the structure and properties of commercial alloys is the application of methods of severe plastic deformation, the best-studied ones being equal-channel angular pressing (ECAP) [12–14], equal channel angular extrusion (ECAE) [15–18], and torsion in Bridgman anvils (HPT) [19–21]. New SPD methods have been developed in recent years. For example, there are an accumulative roll bonding (ARB) [22,23] and a differential velocity sideways extrusion (DVSE) [24,25]. DVSE is a novel method for directly forming curved profiles from billets in one extrusion operation using two opposing punches. Modeling the process, authors [24] identified that a curvature of extrudate can be controlled and varied using a difference between the velocities of the two punches, defined by velocity ratio, as well as an extrusion ratio. The effectiveness of the new SPD method was tested during manufacturing a curved AA1050 bar widely used in the transport industry [25]. Due to the severe strains arising in the DVSE process, (greater than that for conventional pass of equal channel angular extrusion), significant grain refinement in the curved bar (grain size ~3 μm) from the original billet (grain size ~357 μm) was achieved in one extrusion operation. On the basis of structural studies, the authors found that grain refinement in the DVSE process is due mainly to continuous dynamic recrystallization. The refinement of the structure led to an increase in mechanical properties. Compared with the billet, the hardness, yield strength, and ultimate tensile strength of the formed curved bar increase by 134.8%, 354.0%, and 116.8% respectively, although the elongation to fracture was decreased by 60.0%. DVSE method includes hot extrusion (DVSE-HE), welding extrusion (DVSE-WE), and co-extrusion (DVSE-CE). DVSE-CE technology used for a forming curved Al/Mg sandwich bar was described in [26].

According to [14,17,18,27–29] submicrocrystalline and nanocrystalline alloys of 7XXX series exhibit a good combination of strength and plastic properties resulting from thermal deformation processing under different conditions.

The analysis of the available results on the design of high-strength Al-Zn-Mg-alloy-based composites raises a question of how the shape, size, and amount of Al_9FeNi eutectic aluminides change and how this affects the structure, phase composition, and properties of nickalins as a whole. The first studies along this line have been conducted in the last three years [30–32]. The effect of thermal pretreatment on the deformation behavior of nickalin under high-pressure torsion (HPT) was established [30], the sequence and kinetics of post-deformation processes under heating to 400 °C were determined [31], and the dynamic properties of an Al-Zn-Mg-Fe-Ni alloy before and after HPT in a wide range of strain rates, 10^3 to 10^5 s^{-1}, were measured [32]. The results have shown that the principal mechanisms of structure formation in a 7xxx series alloy under HPT remain the same, but the presence of Al_9FeNi eutectic aluminides in the composite structure affects the morphology and kinetics of the precipitation of secondary phases and increases the thermal stability of the composite and its strength.

It was reported in [33] that radial-shear rolling is an effective method for increasing the mechanical properties of nickalin through creating a microscale gradient structure and a uniform distribution of strengthening phases in it.

The purpose of this paper is to continue the study aimed at creating a composite material based on an Al-Zn-Mg-Fe-Ni alloy by combining thermal and deformation processing techniques. Particularly, the effect of complex deformation processing including RSR and HPT on the structural and phase transformations of the Al-Zn-Mg-Fe-Ni alloy and its mechanical properties is discussed. In view of the complex chemical composition of the alloy, it is obvious that the intermetallic phases of different origins make a great contribution to the formation of its structure and properties; therefore, the paper focuses on studying the evolution of their formation mechanisms under HPT with the application of high-resolution TEM methods.

2. Materials and Methods

A high-strength Al-Zn-Mg-Fe-Ni alloy (nickalin) was studied. The chemical composition of the experimental alloy is given in Table 1.

Table 1. Chemical composition (wt%) of the experimental alloy.

Al	Zn	Mg	Fe	Ni	Zr	Cu
Base	7.22	2.95	0.52	0.57	0.2	0.002

The ingot production conditions, and the raw material composition are described in detail in [11,33]. The melting was carried out in an electric resistance furnace in graphite -chased crucible based on high purity Al (99.99%). Zn and Mg were introduced into Al melt in the form of pure metals, whereas Fe, Ni and Zr in the form of binary master alloys (Al-10% Fe, Al-20% Ni, Al-1.5% Zr). The casting was carried out in a metal mold at a temperature of 780 °C (cooling rate during solidification was ~10 K/s). The ingots were subjected to two-stage homogenizing annealing at 450 °C and 540 °C with holding for three hours in each stage. The alloy was studied in the form of ingots, calibrated rods and disks produced by HPT at room temperature.

The ingots were turned to a diameter of 40 mm, heated in an oven to 480 °C, and subjected to two-pass radial shear rolling (RSR) in MiSiS 100 T and 14–40 mills [33]. The total rod elongation ratio was 2.78, the diameter of the initial billet was 31 mm, and the diameter of the resulting rod was 24 mm.

Severe plastic deformation by HPT was performed on 20-mm-diameter and 1-mm-thick specimens cut from an ingot (nickalin 1) and a rod after RSR (nickalin 2). The number of anvil revolutions was $n = 5, 10, 15$, for the true accumulated strains (with allowance for upsetting) $\varepsilon = 6.0, 6.7$, and 7.1, respectively. The mechanical properties were determined on flat microspecimens (5.7 mm long, 2 mm wide, and 1 mm thick) in a Shimadzu AGX-50 Plus (Kyoto, Japan) universal testing machine. To fix the microspecimens in the grips of the testing machine, we used a special clamping device described in detail in [34]. The strain rate was 5×10^{-4} s^{-1}.

The mechanical properties were estimated from the values of tensile yield strength YS, ultimate tensile strength UTS, and percentage elongation δ. Microhardness H_v was determined at a load of 0.2 N in a PMT-3 device (Moscow, Russia). The measurement error was at most 10%. Hardness was measured by the Brinell method, with a 5 mm ball, at a load of 250 kg. The specimen structure was studied at different scale levels by means of a Neophot-32 optical microscope (OM, Carl Zeiss, Jena, Germany), a Quanta 200 scanning electron microscope (SEM, FEI Company, Hillsboro, OR, USA), and a Tecnai G^2 30 Twin transmission electron microscope (TEM, FEI Company, Oregon, USA) with an accelerating voltage of 300 kV. X-ray diffraction (XRD) analysis was carried out using an Empyrean diffractometer (PANanalytical, Almelo, The Netherlands) at room temperature and Cu Kα radiation. Based on the X-ray diffraction data the lattice parameter of an Al-matrix was calculated. The specimens for ^{a}analysis were made with the use of mechanical polishing on a diamond suspension. Final smoothing was performed on a suspension of colloidal silicon dioxide. The specimens for thin structure examination were made by jet polishing in a Tenupol-5 (Struers, Denmark) in a 20% nitric acid solution and 80% methanol at a temperature of −25 °C and a voltage of 15 to 20 V. The size of the structural constituents was quantitatively analyzed from dark-field images with the use of the SIAMS 700 image analysis program (Russia).

3. Results

3.1. Structure and Phase Composition of the Alloy before HPT

Al-Zn-Mg-Fe-Ni alloy specimens of the same composition but different histories were the initial materials for severe plastic deformation (SPD). Batch I specimens (nickalin 1) were cut from the ingot and annealed in two stages at 450 and 540 °C with a 3 h holding at each stage. Their structure and properties were described in detail in [32]. Heat treatment forms a structure consisting of an Al matrix and Al_9FeNi eutectic aluminides located on the grain boundaries and dendritic cells. The grain size of the Al matrix is 380 μm and that of the eutectic aluminides is 2 to 3.5 μm. Besides, inside the grains of the Al solid solution there are precipitates of the secondary T-phase ($Al_2Zn_3Mg_3$) resulting from ingot cooling after homogenization.

A 24-mm-diameter rod produced by RSR at 480 °C after two passes was an object for HPT of batch II specimens (nickalin 2). Figure 1 shows the rod cross-section structure.

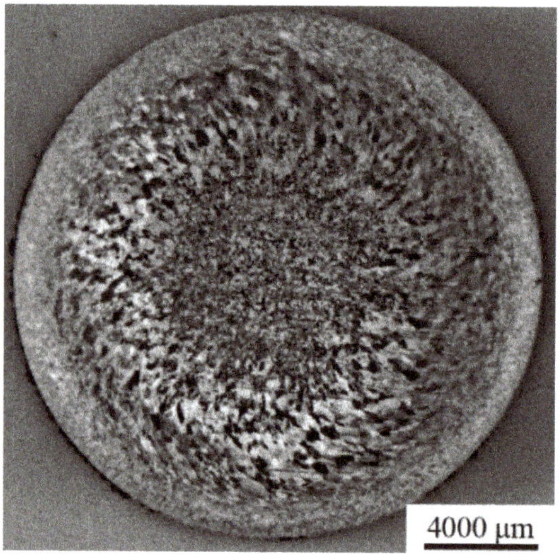

Figure 1. Microstructure in cross-section of the nickalin 2 rod after RSR (OM).

Three zones with a gradient structure typical for this RSR pattern are clearly visible. The width of the SPD zone on the rod periphery is ~2.5 to 3.0 mm; the middle zone at half the radius and the central zone have a width of 4.0 to 4.5 mm each. The gradient structure is characterized by different hardness values. The mid-radius zone has a hardness of 105 HB (105 HV), whereas the hardness in the central zone decreases to 97 HB (97 HV). In the peripheral zone, there is a dynamically recrystallized structure with the grain size increasing from 3 mm at the surface to 8 mm at the boundary with the second zone formed by the hot-deformed structure (Figure 2a). In the peripheral zone, the intermetallic particles become fragmented due to deformation and arrange themselves along the direction of rolling (Figure 2b). Elements of the cast structure of the matrix and eutectic conglomerates are preserved in the central part, the shape and size of aluminides being changed only slightly from the as-cast condition (Figure 2c,d). The EDS element analysis of these crystals has shown that they contain Fe and Ni, i.e., they belong to the Al_9FeNi phase (Figure 2e).

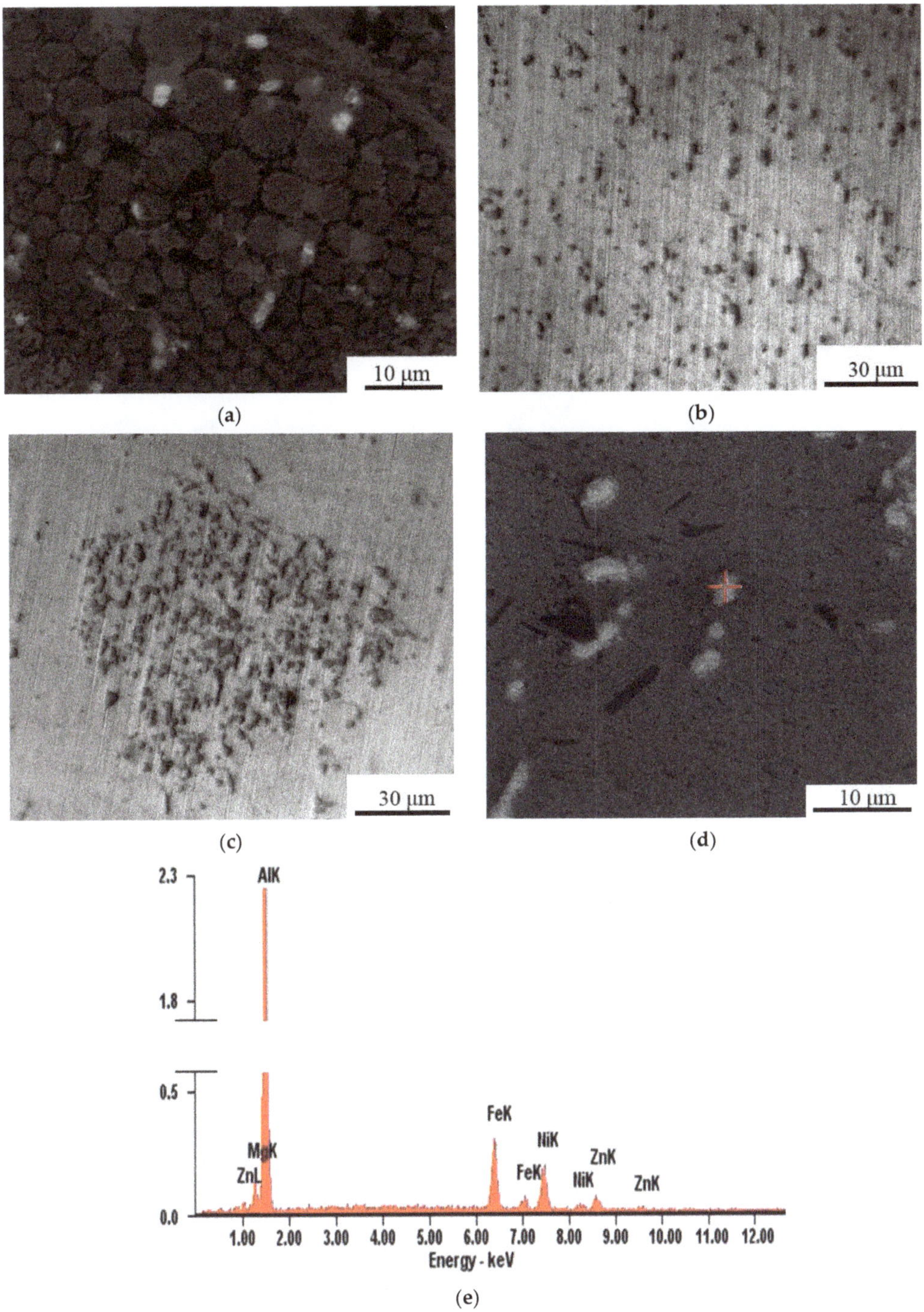

Figure 2. SEM microstructure showing the dynamically recrystallized structure (**a**) and the intermetallic particles (**b**,**d**) in peripheral zone and the intermetallic particles in the center zone (**c**) of cross-section of the nickalin 2 rod after RSR. The EDS element analysis of the Al_9FeNi phase (**e**).

The TEM results testify that, in the deformed region, the rod has a well-developed subgrain structure resulting from polygonization. The subgrains are mainly equiaxial and sized 2 to 6 μm. Along the subgrain boundaries there are disk- or stick-shaped (up to 1–5 μm) Al_9FeNi intermetallics of the crystallization origin. Globular and stick-shaped secondary phases have been found inside the subgrains. According to electron diffraction pattern, equiaxial nanoparticles 30 to 75 nm in diameter are the metastable Al_3Zr phase with an ordered cubic lattice $L1_2$ (Figure 3a,b). These dispersoids result from hot deformation. Sticks with a diameter of 5 nm and a length ranging between 50 and 100 nm (separate large particles reach a diameter of 20 nm and a length of 250 nm (Figure 3c,d)), belong to the $MgZn_2$ phase precipitating from the supersaturated Al solid solution during cooling after RSR.

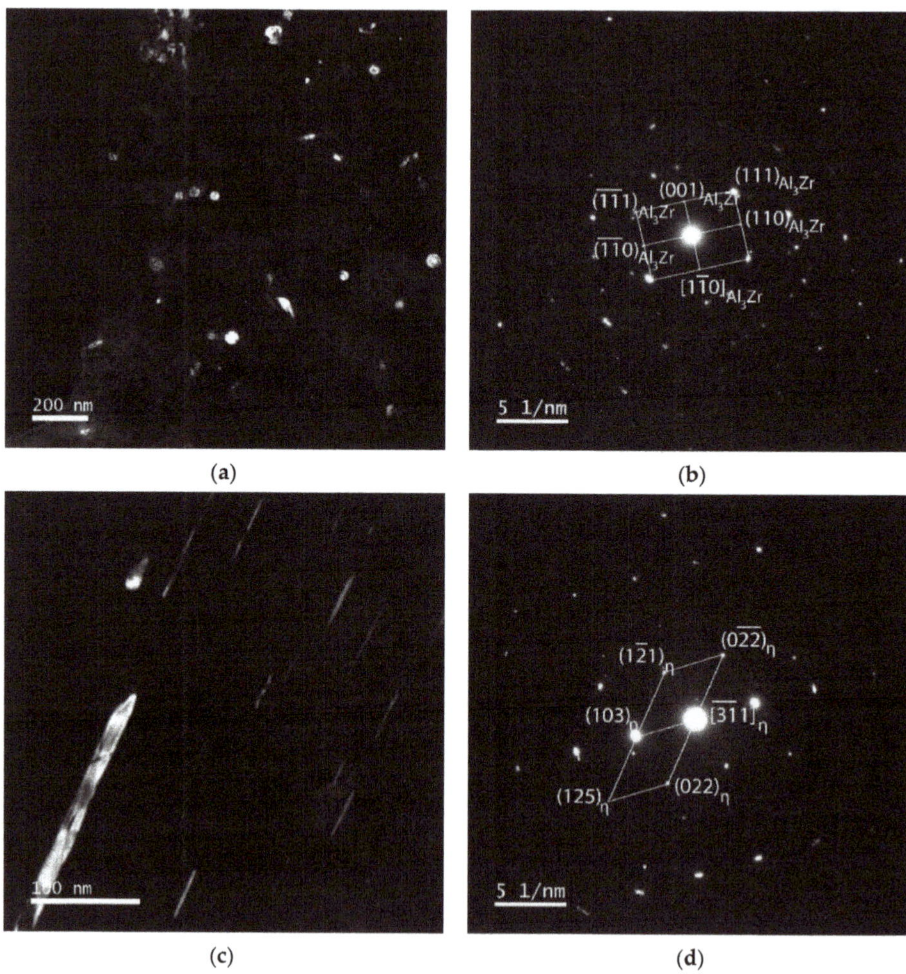

Figure 3. TEM dark-field images (**a**,**c**) and electron diffraction pattern (**b**,**d**) of intermetallic particles corresponding to the Al_3Zr (**a**,**b**) and $MgZn_2$ (**c**,**d**) phases in the nickalin 2 rod after RSR.

3.2. Structure and Phase Composition after HPT

The Al-Zn-Mg-Fe-Ni alloy in two structural states (nickalin 1 and nickalin 2) discussed in Section 3.1 was taken as macrocrystalline analogs for HPT specimens. The evolution of the structural and phase transformations in nickalin 1 and nickalin 2 was studied on specimens after HPT with different strains by varying the number of anvil revolutions $n = 5$, 10, and 15 ($\varepsilon = 6.0$, 6.7, and 7.1 respectively). Microhardness was measured in order to characterize the uniformity of their structure along the specimen radius. It follows from the measurements that microhardness variation from the edge to the center does not exceed 150 MPa. However, in order to compare correctly the structure and phase composition of the specimens produced with different numbers of revolutions, all the TEM studies were performed on foils cut from the mid-radius region.

As a result of deformation, coarse eutectic aluminides in nickalin 1 become fragmented, they break into particles sized ~1–2 μm. The Al matrix for $n = 5$ and 10 is represented by a fragmented structure with grain/subgrain diffusion boundaries and a nonuniform contrast inside the grains/subgrains, which is caused by a high level of internal stresses (Figure 4a). Fragmentation increases with the amount of strain and, according to the results of the quantitative analysis of the EDS element analysis, the average size of the grain/subgrain ranges from 120 to 160 nm at $n = 10$, with only 20% of them being larger than 200 nm. At $n = 15$, mainly high-angle boundaries (HABs) are formed in the structure, this being evidenced by the annular electron diffraction pattern with numerous discretely located point-like reflections (Figure 4b). The main array of grains/subgrains retains the strain-induced contrast, but in triple joints there are small (20 to 30 μm) dislocation-free grains formed through dynamic recrystallization. This mixed two-level structure decreases the average grain/subgrain size to the nanoscale (~100 nm).

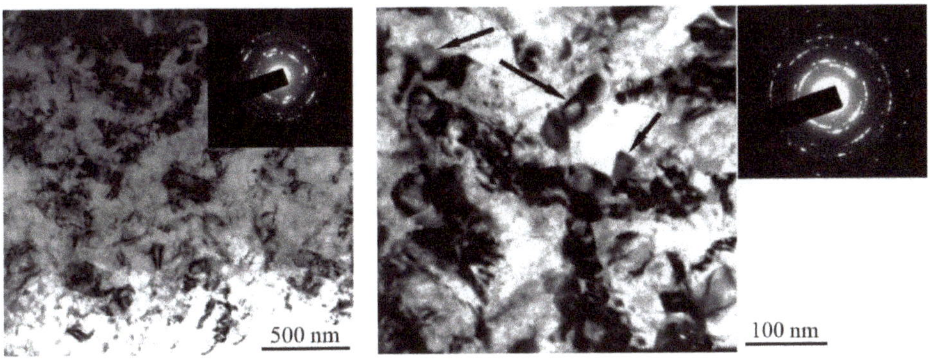

(a) (b)

Figure 4. TEM microstructure showing the deformation structure of the nickalin 1 after HPT, $n = 5$ (**a**) and $n = 15$ (**b**).

The phase composition of nickalin 1 changes with the amount of strain. When $n = 5$ and 10, the volume fraction of the particles of the secondary T-phase decreases due to their deformation dissolution, and when $n = 15$, there appears a strain-induced Al solid solution supersaturated with zinc and magnesium.

The five-revolution deformation nickalin 2 results in significant fragmentation of the polygonized structure of the alloy. When $\varepsilon = 6.0$, there appear variously sized subgrains, irregularly shaped or elongated. Most of them (~80%) have a size ranging from 50 to 250 nm, with separate large subgrains having a size of up to 500 nm (Figure 5a,b). The annular character of the electron diffraction patterns with diffuse reflexes and the analysis of the dark-field images of the microstructure demonstrate that the fragments are separated by both high- and low-angle boundaries (LABs), Figure 5c. Deformation changes the morphology and average size of intermetallic particles. Fragmentation of Al_9FeNi particles occurs, their characteristic size decreases to 0.6–3.5 μm, and the diameter of the Al_3Zr dispersoid precipitates also decreases to 30–40 nm, possibly, due to partial dissolution during HPT, as was noted in [30].

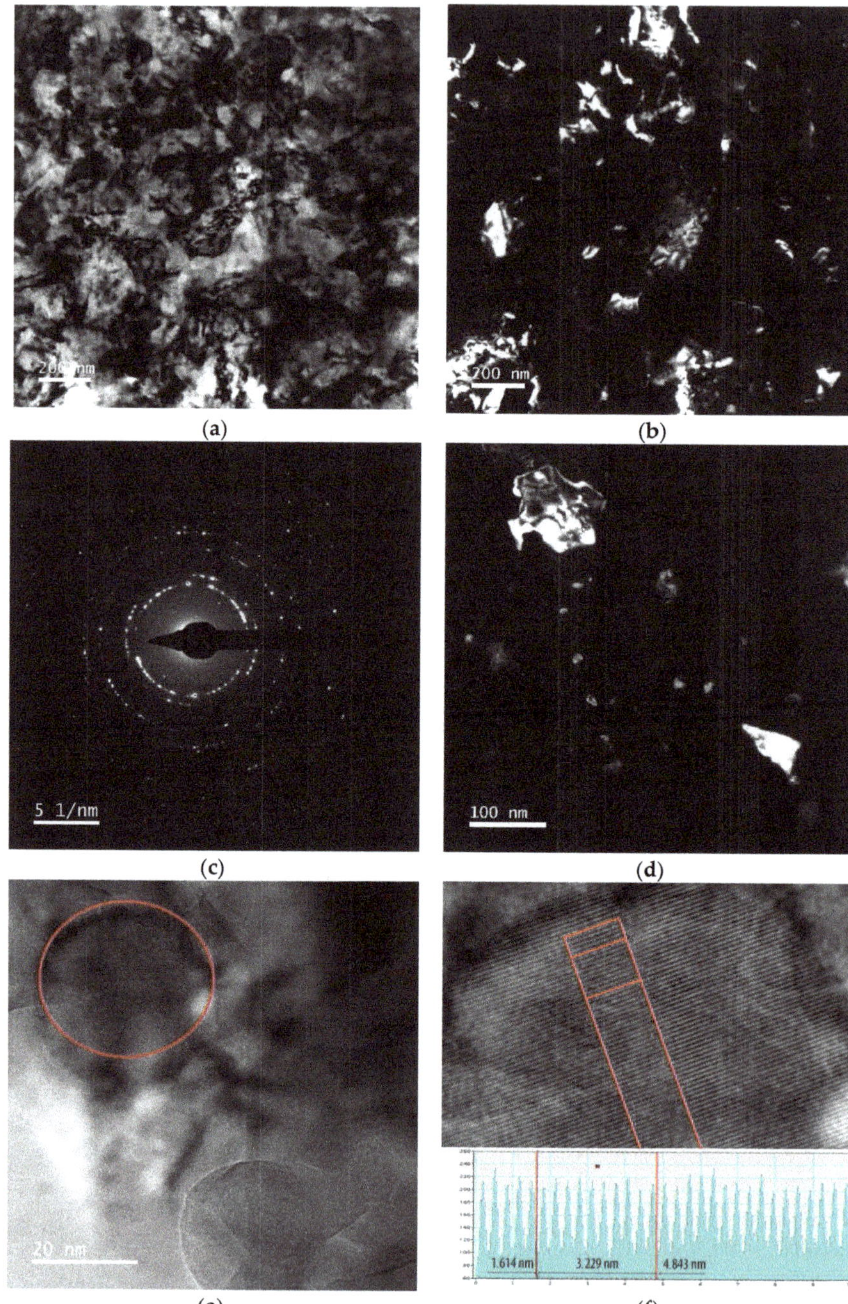

(a) (b) (c) (d) (e) (f)

Figure 5. TEM bright-field image (**a**), dark-field images (**b**,**d**) and electron diffraction pattern (**c**) of deformation structure (**a**–**c**) and intermetallic particles corresponding to the T (**d**,**e**) phase in the nickalin 2 after HPT, $n = 5$. (**e**) HRTEM image of T-phase (d = 0.3229 nm) particle. (**f**) An enlarged region of the particle shown in (**e**,**f**) and intensity histogram.

The analysis of the bright-field and dark-field images has shown that five-revolution deformation causes a complete dissolution of the zinc-magnesium phases, with the appearance of a strain-induced Al solid solution supersaturated with zinc and magnesium. On the other hand, the HPT process causes the decomposition the supersaturated solid solution and the formation of dispersed particles sized up to 20 nm along the boundaries of the matrix fragments (Figure 5d). The high dispersion and low volume fraction of these precipitates prevent them from being identified by electron microdiffraction patterns.

The calculation of the interplanar distances by the direct resolution of the particle lattice allows us to conclude that they are T-phase precipitates ($d_{(420)T}$ = 0.3222 nm, Figure 5e,f).

A characteristic feature of the structural state for n = 10 is the formation of a banded structure consisting of 40–120 nm grains/subgrains separated by low-angle and high-angle boundaries (Figure 6a–c).

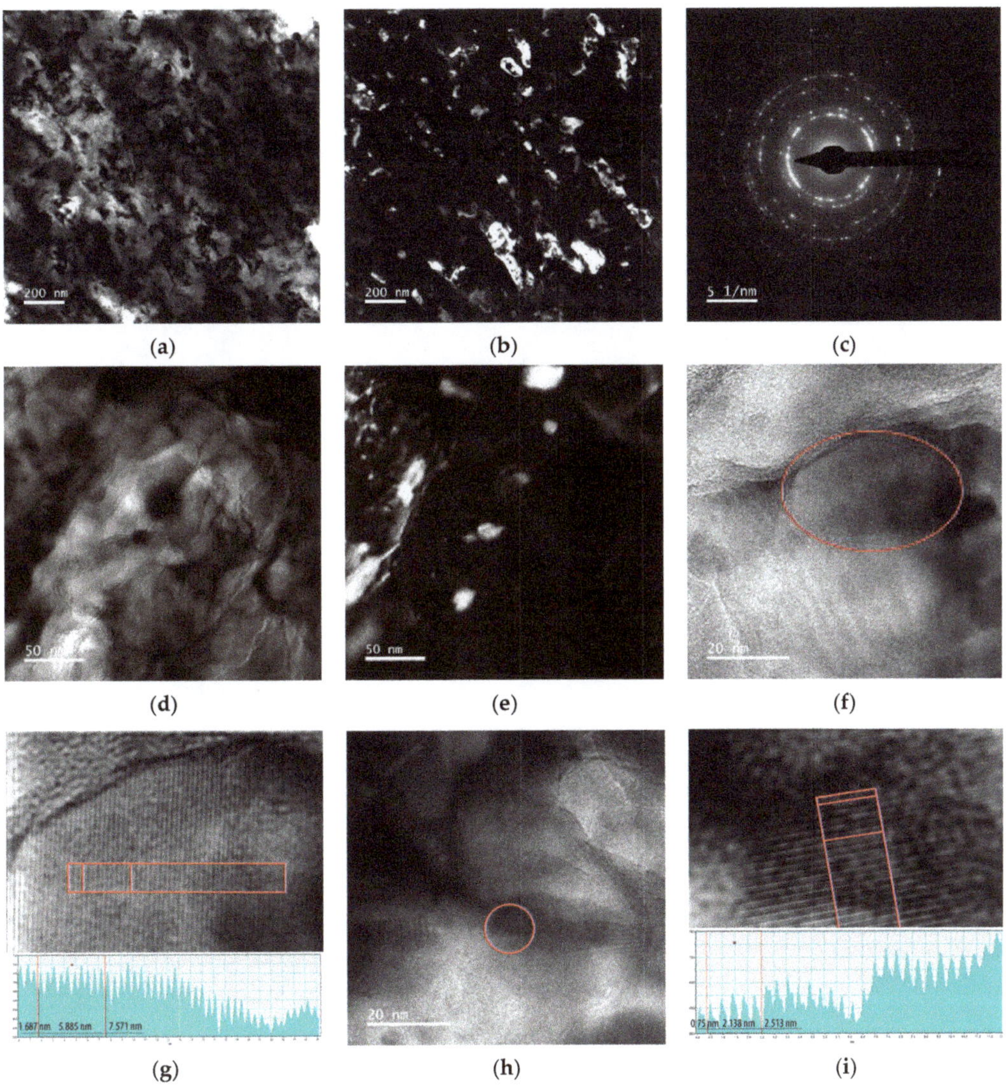

Figure 6. TEM bright-field image (**a**,**d**), dark-field images (**b**,**e**), and electron diffraction pattern (**c**) deformation structure (**a**–**c**) and intermetallic particles corresponding to the T and η (**d**–**g**) phases in the nickalin 2 after HPT, n = 10. HRTEM images of T (d = 0.5882 nm) (**f**,**g**) and η (d = 0.4276 nm) (**h**,**i**) phases particles. (**g**,**i**) Enlarged regions of the particle shown in (**f**,**h**) and intensity histograms.

According to the HRTEM data, the growing amount of strain initiates the decomposition of the supersaturated solid solution (Figure 6d,e); besides precipitates along the boundaries of T-phase fragments sized 15 to 40 nm ($d_{(210)T}$ = 0.5883 nm, Figure 6f,g), more dispersed 10 to 20 nm equiaxial particles of the η or η′-phase ($d_{(100)\eta'}$ = 0.4295 nm, $d_{(002)\eta}$ = 0.4283 nm, Figure 6h,i) are detectable inside the grains/subgrains. As the amount of strain increases, large Al_9FeNi inclusions continue decreasing in size to 1–1.5 μm, with individual fragments being nanosized; the Al_3Zr distribution density remains practically unchanged.

The maximum deformation with 15 revolutions causes results in a further refinement of the submicrocrystalline structure, this being clearly visible from the dark-field image (Figure 7a). The grains/subgrains size distribution becomes more uniform, with the average size reaching 30 to 90 nm, i.e., a nanostructured state is formed and grain misorientation angles increase, as follows from the annular electron diffraction pattern with numerous discrete point-like reflections (Figure 7b). Similarly to the structure of HPT nickalin 1, when n = 15, in the structure of HPT nickalin 2, in the triple joints and on the matrix-intermetallics interphase boundaries, there also appear recrystallized grains, i.e., two structure refinement processes occur simultaneously: fragmentation and in-situ recrystallization. On the background of the nanostructured matrix, there are intermetallic particles of different shapes and sizes, close to those found in HPT nickalin 2 after n = 10. In follows from the results of electron diffraction patterns and direct lattice resolution, they belong to four phases: Al_9FeNi, Al_3Zr, T, and η.

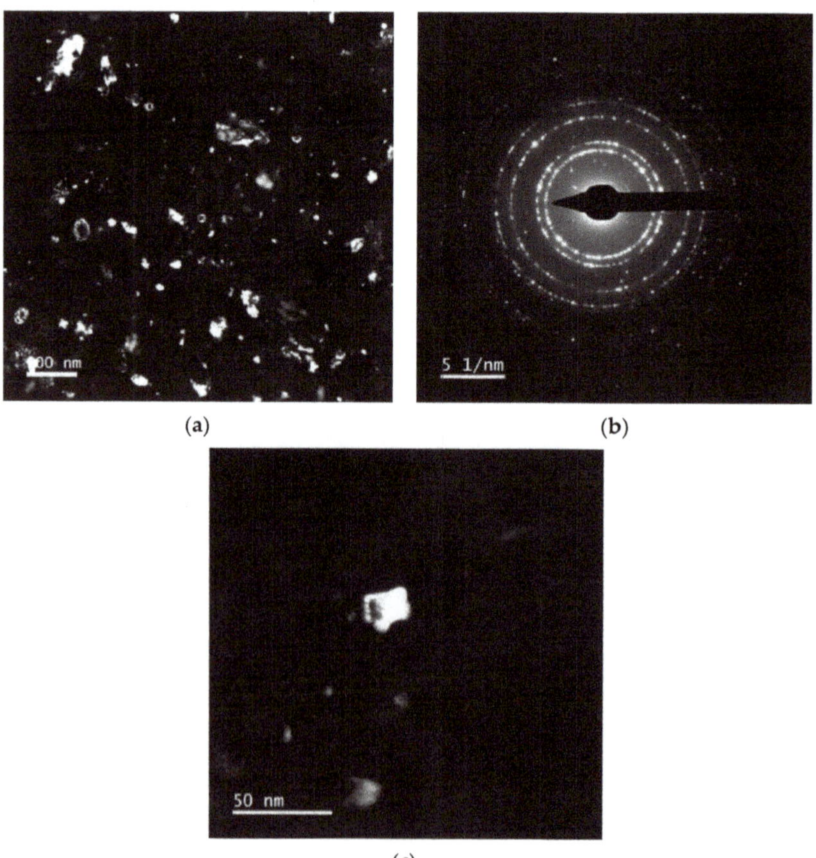

Figure 7. TEM dark-field images (**a**,**c**) and electron diffraction pattern (**b**) of deformed structure in the nickalin2 after HPT, n = 15.

Thus, complex deformation processing, namely RSR + HPT, yields an aluminum-matrix composite reinforced with dispersed intermetallics of different natures.

3.3. Hardness and Mechanical Properties of Nickalin

The size characteristics of the alloy microstructure before and after HPT are vividly demonstrated by hardness testing. It has been found that nanostructured HPT nickalin 2 has a microhardness of 1800 to 2000 MPa, which by a factor of 1.7–1.8 exceeds that measured at the mid-radius of the RSR rod. Nanocrystalline nickalin 1 produced by HPT with the same amount of strain (n = 15) also has double microhardness relative to the macrocrystalline analog.

The procedure discussed in Section 2 was used to measure the mechanical properties of nanocrystalline alloys that were obtained by HPT with n = 15. Typical stress–strain diagram is shown in Figure 8. The average values obtained by measuring the mechanical properties of two samples are shown in Table 2 in comparison with the properties of macrocrystalline analogs. Let us compare how mechanical characteristics of the cast alloy change depending on the type deformation processing. According to the Table 2, RSR increases the tensile yield strength (YS) of the cast alloy from 240 to 264 MPa. The combined treatment (RSR + HPT) causes an additional increase in YS to 628 MPa, i.e., 2.8 times relative to YS of an ingot and 1.8 times relative to ultimate tensile strength (UTS) of an ingot. It follows from the data that the strength properties of the nanostructured alloys after HPT and the combined treatment (RSR + HPT), exceed those for macrocrystalline analogs (an ingot, a rod). Tensile yield strength increases by a factor of 2.0–2.4 and ultimate tensile strength increases by a factor of 1.3–1.5. The different mechanical characteristics of nickalin 1 and nickalin 2 after HPT are determinates by the different structure, phase composition, and different structural strengthening mechanisms.

The higher values of the strength properties of HPT + RSR nickalin 2 (YS = 628 MPa, UTS = 640 MPa) then the strength properties of nickalin 1 (YS = 450 MPa, UTS = 470 MPa) are associated with a more dispersed structure and additional contribution of the strengthening from the dispersed particles of the secondary phase. The hardening mechanisms are described more detail in Section 4.

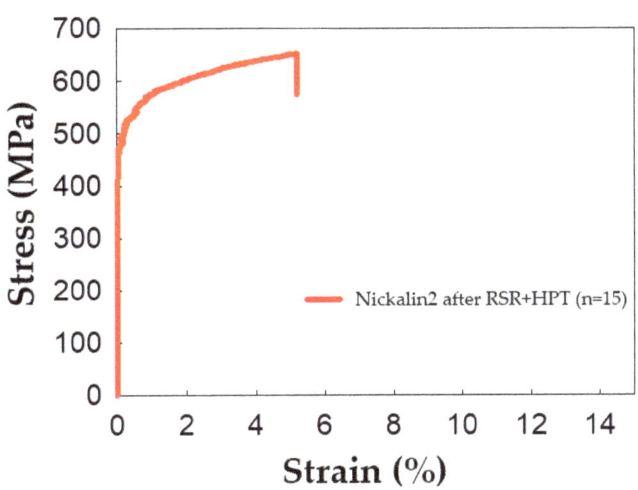

Figure 8. Stress–strain diagram of the Nickalin 2 after RSR + HP (n = 15).

Table 2. Uniaxial tensile tests for the nickalins after different treatments.

Condition	Tensile Yield Strength, MPa	Ultimate Tensile Strength, MPa	Elongation, %
Ingot after homogenization	220	350	6
Rod after RSR, two passes, (mid-radius)	264	436	10
Nickalin 1 (ingot) after HPT (n = 15)	450	470	7
Nickalin 2 (rod) after RSR+ HP (n = 15)	628	640	5

4. Discussion

This study shows the evolution of structural and phase transformations in the Al-Zn-Mg-Fe-Ni alloy with different histories during severe plastic deformation. An ingot after homogenization annealing (nickalin 1) and a rod after RSR (nickalin 2) served as initial specimens for HPT.

The microstructural studies, including SEM and high resolution TEM have shown that the structure formation mechanisms in common during HPT are as follows.

HPT results in an intensive refinement of the Al matrix structure to the nanocrystalline state. The formation of a dispersed structure follows two mechanisms: fragmentation at a true strain ε = 6.0 to 6.7 (n = 5, 10) and in-situ low-temperature dynamic recrystallization at ε = 7.1 (n = 15). The fragmented structure is characterized by the diffusion contrast of HABs, high dislocation density and the presence of LABs inside the grains/subgrains. The mixed structure consisting of deformation-induced fragments with LABs and HABs and recrystallized grains separated by HABs is bimodal, and this provides the best combination of strength and plastic properties. The average grain/subgrain size in HPT nickalin 1 is 100 nm, and it is 60 nm in HPT nickalin 2. Such nanostructures exhibit high hardness H_V = 2000 to 2600 MPa. The strengthening mechanisms for various materials after HPT were repeatedly proved in experiments reported in [19–21,35–37]; namely, according to the Hall–Petch equation, grain-boundary strengthening increases intensely, and the abundance of structural defects is responsible for the high value of the dislocation strengthening component.

The quantitative data obtained by TEM suggest that, during HPT, Al_9FeNi eutectic aluminides crash into fragments whose size becomes commensurate with the secondary phases; the fragments are uniformly distributed in the matrix volume. If the Al_9FeNi particles were micron-sized (2 to 5 μm) in the macrocrystalline analogs (the ingot and the rod), their size does not exceed 1 μm after HPT, and there appear nanoparticles when n = 15.

According to the TEM data, the phase composition of the alloy changes during HPT, the sequence and kinetics of the phase transformations being dependent on the history of the macrocrystalline analogs. As was noted in Section 3.2, during HPT of nickalin 1, at ε = 7.1, a supersaturated Al solid solution is formed due to the strain-induced dissolution of the secondary T-phase. Additional solid-solution hardening combined with grain-boundary and dislocation components provides high hardness to the material, H_V = 2500 MPa.

Additionally, the formation of a supersaturated Al solid solution during HPT can be proved by the results obtained during subsequent post-deformation heat treatment [31,32]. According to this data, low-temperature annealing (T = 120–160 °C) is accompanied by the decomposition of the supersaturated Al solid solution, which occurs simultaneously with the thermal softening of the alloy. Dispersed particles of the T and η strengthening phases precipitate as a result of artificial aging. Due to aging, the hardening achieved under HPT remains high. The presence of the nanostructure significantly reduces the duration of the maximum effect of Al solid solution decomposition from 3–6 h to 0.5–1.0 h. Consequently, the thermal deformation processing of nickalin 1 forms a two-phase material consisting of a nanostructured supersaturated Al solid solution and micron-sized Al_9FeNi aluminides of eutectic origin.

The phase transformations in nickalin 2 under HPT are more diverse. Prior to RSR deformation, the rod had a multiphase composition comprising, besides the Al matrix and Al_9FeNi aluminides, the Al_3Zr and $MgZn_2$ secondary phases. The conditions of their formation are described in Section 3.1.

Under HPT these phases alter their morphologies and sizes owing to two opposite processes induced by SPD, namely the strain-induced dissolution of intermetallics and the kinetic strain aging of the Al matrix. As a result of HPT (n = 5, ε = 6.0), the $MgZn_2$ phase dissolves with the accumulation of mobile dislocations, which, while moving, cut the particles and transfer magnesium and zinc atoms into the Al solid solution. An increase in true strain to ε = 6.7–7.1 (n = 10–15) intensifies the subsequent decomposition of the supersaturated solid solution under kinetic strain aging. This inversion of the matrix composition results in that the stick-shaped particles of the η phase disappear and that dispersed (40 nm) globular precipitates of the T-phase appear along the grain/subgrain boundaries of the Al matrix.

One of the possible reasons for the heterogeneous nucleation of secondary phases at the grain/subgrain boundaries is the appearance of abnormal grain-boundary segregations of the atoms of alloying elements under HPT, whose formation was proved theoretically [38] and experimentally [37,39]. In this connection, the precipitation of the secondary phases along the grain boundaries is easier, and they become dominant. Besides, the size of the particles and their discrete arrangement along the boundaries of the nanocrystalline structure does not decrease the alloy plasticity. Note that a distinctive feature of the morphology of the zinc-magnesium phases precipitating at low temperature under SPD is their globular shape.

As deformation with n = 10 increases (ε = 6.7), strain aging further develops, inside the grains/subgrains of the Al matrix there appear nanosized precipitates of the η' phase, the quantity of which increases with the amount of strain, ε = 7.1 (n = 15). The reasons for the alternation and competition of the T and η' phase precipitations under dynamic aging can be explained by referring to the results reported in [40], where it was found by atom probe tomography that, at an early stage of aging, clusters form in the Al-Zn-Mg-Cu alloy, with different Zn/Mg atomic ratios depending on Zn content in the supersaturated solid solution. With the atomic ratio Zn/Mg $<$ 1.3, the clusters transform into the T-phase, and when Zn/Mg $>$ 1.3 they turn into the η'-phase. As a result of the nonuniform deformation of the specimen volumes during HPT, the local regions of the strain-induced supersaturated Al solid solution have different combinations of alloying element concentrations. Accordingly, clusters appear with different Zn/Mg atomic ratios, which serve as centers for heterogeneous nucleation of strengthening phase precipitates.

The possibility of the occurrence of the above phase transformations during low-temperature SPD stems from a several orders of magnitude increase in the atomic diffusion coefficients of the alloying elements in the Al matrix due to excess vacancy concentration [41,42]. Besides, the formation of a strain-induced supersaturated solid solution may be due to the pinning and transfer of alloying element atoms on dislocations during their migration.

The XRD analysis confirms the TEM results and based on the change in the matrix lattice parameter before and after HPT from a = 0.4064 $\pm$ 0.0001 nm (n = 0) to a = 0.4067 $\pm$ 0.0002 nm (n = 15).

The analysis of the structural and strength characteristics of the Al-Zn-Mg-Fe-Ni alloy after HPT shows their interrelation. The difference between the phase compositions of nickalin 1 and nickalin 2 causes the different mechanical characteristics of these materials due to the action of different structural strengthening mechanisms. The total strength of nickalin 1 (YS = 450 MPa, UTS = 470 MPa) is the sum of grain-boundary strengthening by the Hall–Petch mechanism, dislocation strengthening resulting from high dislocation density, and solid-solution strengthening. The higher values of the strength properties of HPT nickalin 2 (YS = 628 MPa, UTS = 640 MPa), which is a multiphase nanocomposite, are caused by the additional contribution of dispersion strengthening from the dispersed particles of the secondary phase through the Orowan mechanism, as well as by the greatest contribution of the grain-boundary mechanism to the total strength due to the formation of smaller (half-size) grains. The comparison of the obtained strength characteristics with the corresponding properties of the Al-Zn-Mg-Fe-Ni alloy after four RSR cycles (YS = 410 MPa, UTS = 430 MPa) [33] shows the advantage of using combined deformation processing. Besides, the mechanical properties are comparable with those of other 7xxx series alloys after SPD and aging. For example, YS = 525 MPa, UTS = 547 MPa in the 7050 alloy after six ECAP passes (route Bc) [27] and YS = 520 MPa in the Al-Zn-Mg alloy after ECAP and aging [28].

5. Conclusions

1. The effect of combined deformation processing (RSR + HPT) on the structural and phase transformations and mechanical properties of the Al-Zn-Mg-Cu-Fe-Ni eutectic alloy has been studied. It has been found that severe plastic deformation is accompanied by the formation of a high-strength (Hv = 2000 to 2600 MPa) aluminum-matrix composite reinforced with dispersed intermetallic compounds of different origins.
2. With the application of high-resolution TEM methods, the morphological and size characteristics of the Al matrix, Al_9FeNi eutectic aluminides, and secondary phases has been determined. The mixed nanostructure of the matrix is formed by two mechanisms: fragmentation and continuous dynamic recrystallization; the average grain/subgrain size changes from 60 to 100 nm depending on the alloy history. Eutectic aluminides break down into micron-sized fragments and become uniformly distributed in the volume; the strengthening phases precipitate in the form of globules on the grain boundaries and inside the matrix grains.
3. The sequence of phase transformations induced by HPT, namely strain-induced dissolution of secondary phases and kinetic strain aging of the Al matrix, has been identified, as well as their evolution at different amounts of strain. The phase transformations induced by HPT are an additional channel of elastic energy relaxation. It has been demonstrated that the strain-induced dissolution of the initial Mg_2Zn phase and the formation of a supersaturated Al solid solution occur at $n = 5$. At $n = 10$–15 it is replaced by the decomposition of Al solid solution and the appearance of nanosized strengthening metastable T and η' phases.
4. The mechanical properties of the nanostructured composite (YS = 628 MPa, UTS = 640 MPa) have been determined; their comparison with the properties of 7xxx series alloys after SPD proves the expediency of using combined deformation processing for Al-Zn-Mg-Fe-Ni alloys.

Author Contributions: Conceptualization, methodology, and writing, I.B.; TEM investigation and writing, D.R.; SEM investigation, I.S.; TEM investigation, A.P.; Measurement of properties, T.A. and E.B. All authors have read and agreed to the published version of the manuscript.

Funding: This research was funded by the Ministry of Education and Science of the Russian Federation (a State Task on theme «Struktura», number AAAA-A18-118020190116-6).

Institutional Review Board Statement: Not applicable.

Informed Consent Statement: Not applicable.

Data Availability Statement: Data sharing not applicable.

Acknowledgments: The electron microscopic studies were performed at the Center of Collaborative Access Test Center of Nanotechnologies and Advanced Materials, Institute of Metal Physics, Ural Branch, Russian Academy of Sciences, Ekaterinburg, Russia).

Conflicts of Interest: The authors declare no conflict of interest.

References

1. Addzadeh, H.; Baharvandi, H.R.; Shirvani Moghaddam, K. Comparing the effect of processing temperature on microstructure and mechanical behavior of ($ZrSiO_4$ or TiB_2)/aluminum composites. *Mater. Sci. Eng. A* **2008**, *498*, 53–58. [CrossRef]

2. Abrosimova, G.E.; Aronin, A.S. Effect of the concentration of a rare-earth component on the parameters of the nanocrystalline structure in aluminum-based alloys. *Phys. Solid. State* **2009**, *51*, 1765–1771. [CrossRef]
3. Karbalaei, M.; Akbari, S.R.; Shirvani Moghaddam, K.; Baharvandi, H.R. Wear and friction behavior of nanosized TiB_2 and TiO_2 particle capture in matrix. *Compos. Mater.* **2015**, *49*, 3665–3681. [CrossRef]
4. Miracle, D.B. Metal matrix composites—From science to technological significance. *Compos. Sci. Technol.* **2005**, *65*, 2526–2540. [CrossRef]
5. Ma, Z.Y.; Tjong, S.C. High-temperature creep behaviour of SiC particulate reinforced Al–Fe–V–Si alloy composite. *Mater. Sci. Eng. A* **2000**, *278*, 5–15. [CrossRef]
6. Čadek, J.; Kuchařová, K.; Zhu, S.J. High temperature creep behaviour of an Al-8.5Fe-1.3V-1.7Si alloy reinforced with silicon carbide particulates. *Mater. Sci. Eng. A* **2000**, *283*, 172–180. [CrossRef]
7. Božić, D.; Vilotijević, M.; Rajković, V.; Gnjidić, Ž. Mechanical and Fracture Behaviour of a SiC-Particle-Reinforced Aluminum Alloy at High Temperature. *Mater. Sci. Forum* **2005**, *494*, 487–492. [CrossRef]
8. Kumara, J.; Singhb, D.; Kalsi, N.S.; Sharma, S.; Pruncu, C.I.; Pimenov, D.Y.; Rao, K.V.; Kapłonekh, W. Comparative study on the mechanical, tribological, morphological and structural properties of vortex casting processed, Al–SiC–Cr hybrid metal matrix composites for high strength wear-resistant applications: Fabrication and characterizations. *J. Mater. Res. Technol.* **2020**, *9*, 13607–13615. [CrossRef]
9. Belov, N.A. Quantitative phase analysis of the Al–Zn–Mg–Cu–Ni phase diagram in the region of compositions of high-strength nikalines. *Russ. J. Non-Ferrous Met.* **2010**, *51*, 243–249. [CrossRef]
10. Belov, N.A.; Alabin, A.N.; Sannikov, A.V.; Tabachkova, N.Y.; Deev, V.B. Effect of annealing on the structure and hardening of heat-resistant castable aluminum alloy AN2ZhMts. *Metal. Sci. Heat Treat.* **2014**, *56*, 353–358. [CrossRef]
11. Akopyan, T.K.; Belov, N.A. Approaches to the design of the new high-strength casting aluminum alloys of 7xxx series with high iron content. *Non-Ferrous Met.* **2016**, *1*, 20–26. [CrossRef]
12. Kim, Y.G.; Ko, Y.G.; Shin, D.H.; Lee, S. Effect of equal-channel angular pressing routes on high-strain-rate deformation behavior of ultra-fine-grained aluminum alloy. *Acta Mater.* **2010**, *58*, 2545–2554. [CrossRef]
13. Valiev, R.Z.; Langdon, T.G. Principles of equal-channel angular pressing as a processing tool for grain refinement. *Prog. Mater. Sci.* **2006**, *51*, 881–981. [CrossRef]
14. Shaeri, M.H.; Salehi, M.T.; Seyyedein, S.H.; Abutalebi, M.R.; Park, J.K. Microstructure and mechanical properties of Al-7075 alloy processed by equal channel angular pressing combined with aging treatment. *J. Mater. Des.* **2014**, *57*, 250–257. [CrossRef]
15. Segal, V.M. Engineering and commercialization of equal channel angular extrusion (ECAE). *Mater. Sci. Eng. A* **2004**, *386*, 269–276. [CrossRef]
16. Kulitskiy, V.; Malopheyev, S.; Mironov, S.; Kaibyshev, R. Grain refinement in an Al-Mg-Sc alloy: Equal-channel angular pressing versus friction-stir-processing. *Mater. Sci. Eng. A* **2016**, *674*, 480–490. [CrossRef]
17. Ning, Z.L.; Guo, S.; Cao, F.Y.; Wang, G.J.; Li, Z.C.; Sun, J.F. Microstructural evolution during extrusion and ECAP of a spray-deposited Al–Zn–Mg–Cu–Sc–Zr alloy. *J. Mater. Sci.* **2010**, *45*, 3023–3029. [CrossRef]
18. Yang, Y.; Zhao, Y.; Kai, X.; Zhang, Z.; Zhang, H.; Tao, R.; Chen, G.; Yin, H.; Wang, M. Effects of hot extrusion and heat treatment on microstructure and properties of industrial large-scale spray-deposited 7055 aluminum alloy. *Mater. Res. Express* **2018**, *5*. [CrossRef]
19. Zhu, Y.T.; Lowe, T.C.; Langdon, T.G. Performance and applications of nanostructured materials produced by severe plastic deformation. *Scr. Mater.* **2004**, *51*, 825–830. [CrossRef]
20. Valiev, R.Z.; Islamgaliev, R.K.; Alexandrov, I.V. Bulk nanostructured materials from severe plastic deformation. *Prog. Mater. Sci.* **2000**, *45*, 103–189. [CrossRef]
21. Estrin, Y.; Vinogradov, A. Extreme grain refinement by severe plastic deformation: A wealth of challenging science. *Acta Mater.* **2013**, *61*, 782–817. [CrossRef]
22. Mozaffari, A.; Danesh Manesh, H.; Janghorban, K. Evaluation of mechanical properties and structure of multilayered Al/Ni composites produced by accumulative roll bonding (ARB) process. *J. Alloys Compd.* **2010**, *489*, 103–109. [CrossRef]
23. Saito, Y.; Utsunomiya, H.; Tsuji, N.; Sakai, T. Novel Ultra-High Straining Process for Bulk Materials Development of the Accumulative Roll-Bonding (ARB) Process. *Acta Mater.* **1999**, *47*, 579–583. [CrossRef]
24. Zhou, W.; Lin, J.; Dean, T.A.; Wang, L. Feasibility studies of a novel extrusion process for curved profiles: Experimentation and modelling. *Int. J. Mach. Tools Manuf.* **2018**, *126*, 27–43. [CrossRef]
25. Zhou, W.; Lin, J.; Dean, T.A. Manufacturing a curved profile with fine grains and high strength by differential velocity sideways extrusion. *Int. J. Mach. Tools Manuf.* **2019**, *140*, 77–88. [CrossRef]
26. Yu, J.; Lin, J.; Dean, T.A. Development of novel differential velocity sideway extrusion tichniques to fabricate lightweight curved structural component. *Procedia Manuf.* **2020**, *50*, 125–128. [CrossRef]
27. Liddicoat, P.V.; Liao, X.-Z.; Zhao, Y.; Zhu, Y.; Murashkin, M.Y.; Lavernia, E.J.; Valiev, R.Z.; Ringer, S.P. Nanostructural hierarchy increases the strength of aluminium alloys. *Nat. Commun.* **2010**, *1*. [CrossRef] [PubMed]
28. Cardoso, K.R.; Travessa, D.N.; Botta, W.J.; Jorge, A.M., Jr. High Strength AA7050 Al alloy processed by ECAP: Microstructure and mechanical properties. *Mater. Sci. Eng. A* **2011**, *528*, 5804–5811. [CrossRef]
29. Elhefnawey, M.; Shuai, G.L.; Li, Z.; Nemat-All, M.; Zhang, D.T.; Li, L. On achieving superior strength for Al–Mg–Zn alloy adopting cold ECAP. *Vacuum* **2020**, *174*, 109191. [CrossRef]

30. Brodova, I.G.; Shirinkina, I.G.; Rasposienko, D.Y.; Akopyan, T.K. Structural Evolution in the Quenched Al–Zn–Mg–Fe–Ni Alloy during Severe Plastic Deformation and Annealing. *Phys. Met. Metallogr.* **2020**, *121*, 899–905. [CrossRef]
31. Shirinkina, I.I.G.; Brodova, I.G. Annealing-induced structural–phase transformations in an Al–Zn–Mg–Fe–Ni alloy after high pressure torsion. *Phys. Met. Metallogr.* **2020**, *121*, 344–351. [CrossRef]
32. Brodova, I.G.; Petrova, A.N.; Akopyan, T.K. The influence of severe plastic deformation on the structure and mechanical properties of eutectic Al–Zn–Mg–Fe–Ni alloy. *IOP Conf. Ser. Mater. Sci. Eng.* **2019**, *672*, 012022. [CrossRef]
33. Akopyan, T.K.; Belov, N.A.; Aleshchenko, A.S.; Galkin, S.P.; Gamin, Y.V.; Gorshenkov, M.V.; Cheverikin, V.V.; Shurkin, P.K. Formation of the gradient microstructure of a new Al alloy based on the Al-Zn-Mg-Fe-Ni system processed by radial-shear rolling. *Mater. Sci. Eng. A* **2019**, *746*, 134–144. [CrossRef]
34. Lomakin, I.V.; Arutyunyan, A.R.; Valiev, R.R.; Gadzhiev, F.A.; Murashkin, M.Y. Design and evaluation of an experimental technique for mechanical and fatigue testing of sub-sized samples. *Exp. Tech.* **2017**, *42*, 261–270. [CrossRef]
35. Zhang, Y.D.; Jin, S.B.; Trimby, P.; Liao, X.Z.; Murashkin, M.Y.; Valiev, R.; Sha, G. Strengthening mechanisms in an ultrafine-grained Al–Zn–Mg–Cu alloy processed by high pressure torsion at different temperatures. *Mater. Sci. Eng. A* **2019**, *752*, 223–232. [CrossRef]
36. Sauvage, X.; Murashkin, M.Y.; Straumal, B.B.; Bobruk, E.V.; Valiev, R.Z. Ultrafine grained structures resulting from SPD-induced phase transformation in Al–Zn alloys. *Adv. Eng. Mater.* **2015**, *17*, 1821–1827. [CrossRef]
37. Nurislamova, G.; Sauvage, X.; Murashkin, M.; Islamgaliev, R.; Valiev, R. Nanostructure and related mechanical properties of an Al–Mg–Si alloy processed by severe plastic deformation. *Philos. Mag. Lett.* **2008**, *88*, 459–466. [CrossRef]
38. Razumov, I.K. Nonequilibrian grain boundary segregation in alloys subjected to severe plastic deformation. *Condens. Matter Interphases* **2010**, *12*, 394–399.
39. Valiev, R.Z.; Enikeev, N.A.; Murashkin, M.Y.; Kazykhanov, V.U.; Sauvage, X. One origin of the extremely high strength of ultrafine-grained Al alloys produced by severe plastic deformation. *Scr. Mater.* **2010**, *63*, 949–952. [CrossRef]
40. Zou, Y.; Wu, X.; Tang, S.; Zhu, Q.; Song, H.; Cao, L. Co-precipitation of T′ and η′ phase in Al-Zn-Mg-Cu alloys. *Mater. Charact.* **2020**, *169*, 110610. [CrossRef]
41. Sauvage, X.; Enikeev, N.A.; Valiev, R.Z.; Nazedkina, Y.; Murashkin, M.Y. Atomic-scale analysis of o the segregation and precipitations mechanism in a severely deformed Al-Mg alloys. *Acta Mater.* **2014**, *72*, 125–136. [CrossRef]
42. Kazantzis, A.V.; Chen, Z.G.; De Hosson, J.T.M. Deformation mechanism of aluminum-magnesium alloys at elevated temperatures. *J. Mater. Sci.* **2013**, *48*, 7399–7408. [CrossRef]

metals

MDPI

Article

Investigation of Stresses Induced Due to the Mismatch of the Coefficients of Thermal Expansion of the Matrix and the Strengthening Particle in Aluminum-Based Composites

Oleg Matvienko [1,2], Olga Daneyko [1,2], Tatiana Kovalevskaya [1,2], Anton Khrustalyov [1,*], Ilya Zhukov [1] and Alexander Vorozhtsov [1]

1 Faculty of Physics and Engineering, National Research Tomsk State University, 634050 Tomsk, Russia; matvolegv@mail.ru (O.M.); olya_dan@mail.ru (O.D.); takov47@mail.ru (T.K.); gofra930@gmail.com (I.Z.); abv@mail.tomsknet.ru (A.V.)

2 Department of Physics, Chemistry and Theoretical Mechanics, Tomsk State University of Architecture and Building, 634003 Tomsk, Russia

* Correspondence: tofik0014@gmail.com; Tel.: +7-952-1555-568

Abstract: An experimental and theoretical investigation of the strength properties of aluminum alloys strengthened by dispersed nanoparticles, as well as the determination of the significance of various mechanisms responsible for the strengthening of the material, was carried out. Results of experimental investigation demonstrate that the hardening of aluminum alloy A356 by Al_2O_3 and ScF_3 nanoparticles leads to an increase in the yield strength, ultimate tensile strength, and plasticity. Despite the similar size of Al_2O_3 and ScF_3 nanoparticles, the physicomechanical properties of nanoparticles significantly affect the possibility of increasing the mechanical properties of the A356 aluminum alloy. A physicomathematical model of the occurrence of thermal stresses was developed caused by the mismatch of the coefficients of thermal expansion (CTEs) of the matrix and strengthening particles on the basis of the fundamental principles of mechanics of a deformable solid and taking into account the elastic properties of not only the matrix, but also the particle. The forming of thermal stresses induced due to the mismatch of the coefficients of thermal expansion of the matrix and the strengthening particle in aluminum-based composites was investigated. In the case of thermal deformation of dispersion-hardened alloys, when the CTE of the matrix and particles noticeably differ, an additional stress field is created in the vicinity of the strengthening particle. Thermal stresses increase the effective particle size. This phenomenon can significantly affect the result of the assessment of the yield strength. The strengthening caused by thermal mismatch makes the largest contribution to the yield strength improvement. The yield strength increments due to Nardon×Prewo and Orowan mechanisms are much lower.

Keywords: deformation; dispersion-hardening; nanoparticles; stress; strain; plastic deformation; thermal stresses; aluminum

Citation: Matvienko, O.; Daneyko, O.; Kovalevskaya, T.; Khrustalyov, A.; Zhukov, I.; Vorozhtsov, A. Investigation of Stresses Induced Due to the Mismatch of the Coefficients of Thermal Expansion of the Matrix and the Strengthening Particle in Aluminum-Based Composites. *Metals* **2021**, *11*, 279. https://doi.org/10.3390/met11020279

Received: 9 December 2020
Accepted: 29 December 2020
Published: 5 February 2021

1. Introduction

Aluminum-based alloys are widely used in the automobile industry not only because they are easily castable in complex forms but also because of their excellent wear resistance, light weight, and good strength, which make them widely used in the manufacture of numerous automotive parts, such as engine blocks and wheels. Additionally, aluminum-based alloys alloy are used as a matrix for obtaining composites [1,2], which have an enhanced wear resistance, favorable mechanical properties, and higher strength, Young's modulus, and fatigue resistance [3,4] at room and elevated temperatures. Dispersion-hardened alloys are characterized by isotropic mechanical properties, high plasticity, and high strength [5,6]. The physical theory of plasticity and strain hardening was developed by Orowan [7], Ashby [8], Hirsch, and Humphreys [9,10].

According to continuum linear elastic dislocation theory, Orowan [7] proposed that the interaction between dislocation and hardening particles leads to the formation of shear loops around the particles. This dislocation interaction scenario leads to what is known as Orowan strain hardening. Humphreys and Hirsch [11] proposed an alternative mechanism in which, if all dislocations are edge-oriented, each dislocation/particle interaction leads to two pairs of prismatic loops, one on each side of the particle. This multi-dislocation interaction, which involves the cross-slip of screw components of dislocations, is known as Hirsch looping. The material hardening depends on the types of particles, their volume fraction and homogeneous distribution in the matrix, and the interaction between matrix and strengthening particles.

The stress–strain curves at room temperature and at 400 °C for various types of dispersion-hardened aluminum alloys were determined in [12]. The results of investigations demonstrated that the effect of particles on the flow stress is in agreement with Orowan's theory [7]. A model to describe the creep of FCC metals has been used [13] for alloys and composites strengthened by the dispersion of nanosized particles. According to investigation results, the volume fraction and size of the nanosized particulate assumes a key role in determining the creep response of these materials. The results of investigations of the mechanical properties of aluminum alloys indicate that the hardness, yield strength, and ultimate tensile strength of the Al–2 vol.% C_{60} nanocomposites are higher than those of the monolithic Al samples.

The investigation in [14] dealt with aspects of the deformation behavior of dispersion-hardened aluminum materials during uniaxial creep in a broad temperature range. Results of the investigation demonstrated that the threshold stress decreases considerably with increasing temperature.

The authors of [15] focused on the examination of materials of submicron matrix grain size strengthened with nanosized particles. The results obtained within the scope of this work showed the possibility for significant improvement of the microcomposite properties.

The results of the studies of the effect of nanodiamond addition on the physicomechanical properties of the drilling tool metal matrix were discussed in [16]. The optimal concentration of nanodiamond powder additives in the composition of a hard alloy mixture was defined. The addition of 1–2 wt.% nanodispersed diamonds was shown to contribute to binding oxygen in the thermo-oxidation of diamond in the course of the tool manufacture, which makes it possible to reduce the number of pores in the matrix material structure and to increase its density by 1.7% and hardness by 24.3%.

In [17], the effect of the aspect ratio of carbon nanotubes (CNTs) on strengthening aluminum metal matrix composites (Al MMCs) was investigated. The tensile results showed that the CNTs exhibited a strong strengthening effect in the composites regardless of their aspect ratios. However, the post-loading examination and quantitative analysis indicated that there was a strengthening mechanism transition for CNTs, which was closely associated with the aspect ratio or length of CNTs. The origin of such a transition was explored from the viewpoint of the dislocation–CNT interaction under loading.

The effect of Al_4C_3 formation on the tensile properties of multiwalled carbon nanotubes which reinforce an Al metal matrix was investigated in [18]. It was concluded that the formation of Al_4C_3 could effectively enhance the load transfer in composites.

Mathematical modeling of composites hardened by nanoparticles with a metal matrix was considered in [19]. The authors covered in detail the effect of the bulk content, distribution, and morphology of nanoparticles per average path length of a dislocation in the alloy (L). The authors revealed that the uniform distribution of grains in the matrix gave the smallest value of L. An increase in the bulk content of nanoparticles from 1 to 7 vol.% decreased the average path length of the dislocation by 50%, and, if their fraction exceeded 7 vol.%, the value of L was almost invariable.

In [20–23], a physical model of plastic deformation was suggested. The modeling of plastic deformation and strain hardening is based on the concept that the plastic slip in FCC alloys with incoherent nanoparticles leads to the formation of a defect structure with shear-forming dislocations, prismatic dislocation loops of vacancy and interstitial types, dislocation dipoles of vacancy and interstitial types, and interstitial atoms, monovacancies, and bivacancies.

An approach which combines methods of plasticity physics and mechanics of deformable solid was used in [24–26] to explore the limits of elastic and plastic resistance of the tube from dispersion-hardened aluminum alloy subjected to internal and external pressure.

It is known that, in composites, there is a large difference in the coefficients of thermal expansion between the matrix and the strengthening particles [27–29]. Thermally induced stresses could be generated as the result of a global heating and cooling of constructions, as well as local heating and cooling effects when a temperature gradient develops.

Therefore, the problem of thermally induced stresses due to a mismatch of the coefficients of thermal expansion between the matrix and strengthening particles has implications for the determination of mechanical properties of composite materials. The stress–strain state in plastic area around an oversized spherical particle which was introduced into a spherical hole in the matrix was analyzed in [30]. The analysis of stress–strain state must consider the contribution of thermally induced stresses in addition to the mechanically induced stresses [31]. These thermal stresses may be higher than the yield strength of the matrix and induce either debonding between particles and matrix or crack propagation leading to damage accumulation in the matrix [32,33].

The thermal stresses may be sufficient to generate new dislocations at the interfaces between the matrix and the reinforcement. In [29,34], the authors proposed that the increased strength observed in Al–SiC composites could be accounted for by a high dislocation density in the aluminum matrix, as observed in transmission electron microscopy (TEM). An increase in the density of newly created dislocations near reinforcement fibers was calculated in [24,35]. According to [36], when the composite is heated or cooled, misfit strains which are sufficient to generate dislocations occur because of differential thermal contraction at the Al–SiC interface. The investigations of the elastic and plastic stresses and strains generated about an inclusion, which has a coefficient of thermal expansion different from that of the matrix as a result of heating or cooling, was carried out in [34]. Furthermore, an in situ transmission electron microscopy investigation was undertaken into dislocation generation at the composite due to the differential thermal contraction. A simple model based on results of experimental investigation was developed to account for the relative dislocation density due to the differential thermal contraction.

Thus, it is very important to analyze the thermal stresses on the particle–matrix interface in order to predict a possible modification of the composite properties [37].

The purpose of this work was to experimentally and theoretically investigate the strength properties of aluminum alloys strengthened by nanoparticles, as well as to determine the significance of various mechanisms responsible for strengthening the material. Particular attention is paid to the development of a physicomathematical model of the occurrence of thermal stresses caused by a mismatch between the coefficients of thermal expansion (CTEs) of the matrix and strengthening particles, according to the fundamental principles of solid mechanics and taking into account, in contrast to previously existing models, the elastic properties of not only the matrix, but also the particle.

2. Methods

2.1. Experimental Procedure

As initial materials, we used nanopowders of ScF_3 (~80 nm), obtained via the sol–gel method [38], and alumina (~80 nm), obtained via the method of electric explosion of a conductor (EEW) [39], aluminum micropowder (<20 μm) [39], and A356 aluminum alloy. Figure 1 shows TEM images of the initial powders.

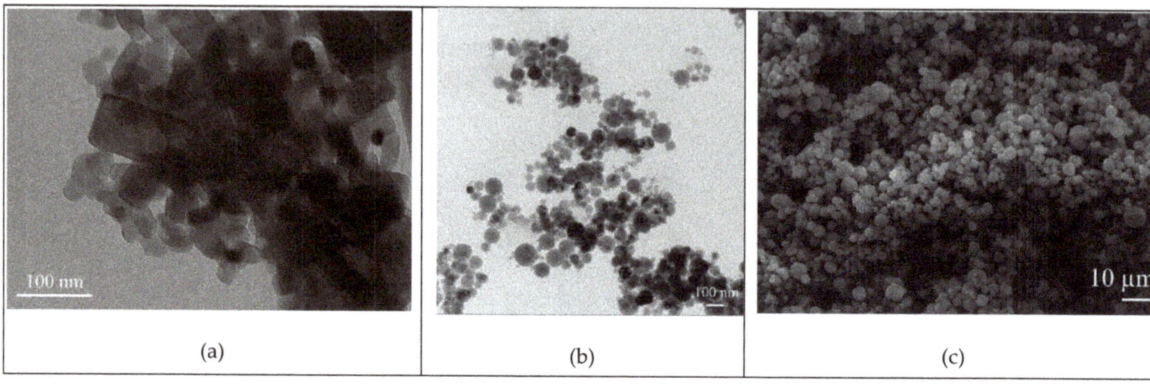

Figure 1. Transmission electron microscopy (TEM) images of initial powders: ScF_3 (**a**); Al_2O_3 (**b**); Al (**c**).

As initial components for the synthesis of ScF_3 nanoparticles, scandium oxide, hydrofluoric acid, and hydrochloric acid were used, which makes the process safer, without the use of poisonous fluorine, which is used in the direct synthesis of ScF_3 from metallic scandium.

For the introduction of nanoparticles, a powder mixture of Al–5 wt.% nanoparticles was prepared. For deagglomeration and distribution of nanoparticles in the powder mixture, stearic acid was used as a surfactant. First, 200 mL of petroleum ether and 1.5 wt.% stearic acid were added to the powder mixture. Then, for 20 min, the powder mixture was mechanically mixed. After mechanical mixing, the aluminum–nanoparticle powder mixture was air-dried (90 °C) and sieved. The resulting powder mixture was introduced into the melt using ultrasonic treatment, which allows degassing and refining the melt, preventing deagglomeration, and evenly distributing nanoparticles in the volume [40].

Aluminum alloy A356 was melted in a graphite crucible with a total melt volume of 500 g. Ultrasonic treatment was carried out using a water-cooled magnetostrictive transducer (power—5 kW, frequency—17.5 kHz, RELTECH, Saint-Petersburg, Russia). The waveguide of the ultrasonic equipment was made in the form of a niobium alloy (VN2AE) cone, the working amplitude of which was 30 μm. At a melt temperature of 730 °C, ultrasonic degassing was carried out for 1 min. After degassing, an aluminum–nanoparticle master alloy (5 wt.%) was introduced into the melt with ultrasonic treatment. After dissolution of the master alloy in the melt, ultrasonic treatment lasted for 2 min (temperature, 730 °C). The melt obtained at a temperature of 710 °C was poured into a steel chill mold. The amount of ScF_3 and Al_2O_3 nanoparticles in the A356 aluminum alloy ranged from 0 to 1 wt.%. The A356 alloy without particles was obtained with similar melt processing parameters. Mechanical tests were performed on a universal testing machine (Instron European Headquarters, High Wycombe, UK), Instron 3369, at a speed of 0.2 mm/min. The samples were tensioned at room temperature (25 °C), and the number of samples for each composition was at least four pieces. The structures of the obtained materials were investigated through optical microscopy, Olympus GX71 (Olympus Scientific Solutions Americas, Waltham, MA, USA). Samples were subjected to preliminary mechanical polishing, electrolytical etching, and anodization. The electrochemical oxidation of the metallographic specimen surface in a 5% solution of hydrofluoric acid (HBF4) at a voltage of 20 V and a current of 2 A was carried out to identify grain boundaries.

2.2. Strengthening Mechanism

It is generally accepted that nanocomposite hardening occurs due to the load transfer, Orowan mechanism, and CTE mismatch mechanism [41]. The total strengthening can be calculated as a superposition of the individual strengthening mechanisms.

$$\sigma_Y = \sigma_{Ym} + \sigma_{NP} + \sigma_{HP} + \sigma_{Or} + \sigma_{CTE}, \quad (1)$$

where σ_{Ym} is the yield strength of the matrix, σ_{NP} is the stress caused by the transfer of the load from the matrix to the particles, σ_{Or} is the Orowan stress, σ_{CTE} is the thermal stress arising from the difference in the coefficients of thermal expansion, elastic modulus, and shear modulus of the matrix and the particle.

The load transfer mechanism is the most accepted strengthening mechanism. The transfer of the load from the soft matrix to the hard particles when an external load is applied helps to harden the material, as proposed by Nardon and Prewo [42].

$$\sigma_{NP} = 0.5 f_p \sigma_{Ym}, \tag{2}$$

where f_p is the volume fraction of particles.

The grain size has an influence on the strength of the alloy, since the grain boundaries can impede the movement of dislocations. This is due to the different orientations of adjacent grains and to the high lattice disorder characteristic of these regions, which prevents the glide of dislocations in a continuous slip plane. The Hall–Petch equation relates stress to average grain size. There are empirical models available for predicting the yield strength change due to the change in the grain size in metal matrix composites by extending the Hall–Petch [43,44] relationship as follows:

$$\sigma_{HP} = k_Y \left(\frac{1}{\sqrt{d_{gr}}} - \frac{1}{\sqrt{d_{gr0}}} \right), \tag{3}$$

where σ_{HP} is the change in tensile strength due to the contribution of the Hall–Petch law, k_Y is the hardening coefficient (constant for each material), and d_{gr} and d_{gr0} are the average grain sizes of the obtained alloys. This equation assumes that the Hall–Petch parameters $k_Y \approx 68$ MPa$\cdot(\mu m)^{1/2}$ for Al alloys [45].

Particles perform a fundamental role in the final grain size of the matrix of composites as they can interact with grain boundaries and act as nuclei of crystallization, slowing or stopping grain growth. An increase in the volume fraction f_p and a decrease in the particle diameter δ_p lead to a finer-grained structure, which is theoretically modeled by the Zener–Smith equation [46].

$$d_{gr} = k_z \frac{\delta}{f_p}, \tag{4}$$

where k_z coefficient values are in the range $0.1 < k_z < 1$. Equation (5) is theoretically derived to describe the deceleration of migrating grain boundaries by particles.

The Orowan mechanism is based on the interaction of nanoparticles with dislocations. Solid "noncutting" particles impede the movement of dislocations, which leads to bending of the dislocation line around the particles (Orowan loops) under the action of an external load [47]. The Orowan mechanism is very important in metal matrix composites, hardening by fine particles when the interparticle distance is not large. An increase in the composite yield strength by Orowan stress may be expressed as follows [47]:

$$\sigma_{Or} = G_m \frac{b}{\Lambda_p - 2R_p}, \tag{5}$$

where G_m is the matrix shear modulus, b is the Burgers vector of the matrix ($b = 0.202$ nm), and Λ_p is the distance between particles.

When a composite is subjected to heating or cooling, the difference in the coefficients of thermal expansion, elastic modulus, and shear modulus between the matrix and hardening particles produce internal stress state changes. The improvement in yield strength from thermal mismatch can be calculated using the equation suggested in [48–50].

$$\sigma_{CTE} = 6G_m \sqrt{(\alpha_m - \alpha_p) \Delta T \frac{b}{\delta_p} \frac{f_p}{1 - f_p}}. \tag{6}$$

Equation (6) was obtained on the basis of the simplest geometric estimates under the assumption that the volume mismatch between the matrix and the reinforcing particles, arising from the difference in thermal expansion coefficients, leads to the appearance of geometrically necessary dislocations around the reinforcing particles. The disadvantages of the above approach include the fact that the stresses in Equation (6) do not depend on the elastic properties of the strengthening particles.

A more rigorous assessment of the thermal stresses arising from the difference in the coefficients of thermal expansion, elastic modulus, and shear modulus of the matrix and particle can be carried out using the methods of solid mechanics.

2.3. *Mathematical Model of the Stresses Caused by the Thermal Expansion Mismatch between the Matrix and Strengthening Particles*

Let us consider the stress–strain state that arises as a result of a change in the temperature of a spherical particle with a radius R_p surrounded by a matrix. The coefficients of linear thermal expansion of materials are assumed to be different.

The equation of equilibrium of an elastic medium in spherical coordinates under the assumption of spherical symmetry can be written in the following form:

$$\frac{\partial \sigma_{rr}}{\partial r} + \frac{2\sigma_{rr} - \sigma_{\varphi\varphi} - \sigma_{\theta\theta}}{r} = 0 \tag{7}$$

The relationship between stresses σ_{ij} and strains ε_{ij}, expressing the generalized Hooke's law, under nonisothermal conditions, according to the von Neumann hypothesis, has the following form:

$$\sigma_{ij} = \lambda\varepsilon_{kk}\delta_{ij} + 2G\varepsilon_{ij} - 3K\alpha\Delta T\delta_{ij}, \tag{8}$$

where λ is the the Lame coefficient, G is the shear modulus, K is the bulk strain modulus, α is the coefficient of linear thermal expansion, $\Delta T = T_{test} - T_{room}$ is the temperature difference between the testing temperature T_{test} and room temperature T_{room}, and δ_{ij} is the Kronecker symbol.

The components of the strain tensor under spherical symmetry are

$$\varepsilon_{rr} = \frac{\partial u}{\partial r},\ \varepsilon_{\varphi\varphi} = \frac{u}{r},\ \varepsilon_{\theta\theta} = \frac{u}{r},\ \varepsilon_{r\varphi} = 0,\ \varepsilon_{r\theta} = 0,\ \varepsilon_{\varphi\theta} = 0. \tag{9}$$

Substituting into the equilibrium equation stresses expressed in terms of displacements, we obtain

$$\frac{\partial}{\partial r}\left[\frac{1}{r^2}\frac{\partial}{\partial r}(ur^2)\right] = 0. \tag{10}$$

First, we integrate Equation (11) for a spherical particle for the following boundary conditions:

$$r = 0:\quad u = 0;\quad r = R_p:\quad u = U\ . \tag{11}$$

As a result of simple calculations, we obtain the displacement field in the particle.

$$u = U\frac{r}{R_p}. \tag{12}$$

The stress field in a particle using the generalized Hooke's law has the following form:

$$\sigma_{rr} = (3\lambda_p + 2G_p)\frac{U}{R_p} - 3K_p\alpha_p\Delta T, \tag{13}$$

$$\sigma_{\varphi\varphi} = (3\lambda_p + 2G_p)\frac{U}{R_p} - 3K_p\alpha_p\Delta T, \tag{14}$$

$$\sigma_{\theta\theta} = (3\lambda_p + 2G_p)\frac{U}{R_p} - 3K_p\alpha_p\Delta T. \tag{15}$$

The displacement of the matrix material can be defined as

$$u = \alpha_{\mathrm{m}} \Delta T r + \frac{U R_{\mathrm{p}}^2 - \alpha_{\mathrm{m}} \Delta T R_{\mathrm{p}}^3}{r^2}. \quad (16)$$

The parameter U is determined from the condition of continuity of the radial stresses at the boundary between the particle and the matrix: $\sigma_{rr}|_{-} = \sigma_{rr}|_{+}$. As a result, we obtain

$$U = \frac{3K_{\mathrm{p}}\alpha_{\mathrm{p}} + 4G_{\mathrm{m}}\alpha_{\mathrm{m}}}{3K_{\mathrm{p}} + 4G_{\mathrm{m}}} \Delta T R_{\mathrm{p}}. \quad (17)$$

Using the generalized Hooke's law, one can determine the stress field in the matrix.

$$\sigma_{rr} = -12 \frac{K_{\mathrm{p}} G_{\mathrm{m}}}{3K_{\mathrm{p}} + 4G_{\mathrm{m}}} (\alpha_{\mathrm{p}} - \alpha_{\mathrm{m}}) \Delta T \frac{R_{\mathrm{p}}^3}{r^3}, \quad (18)$$

$$\sigma_{\varphi\varphi} = 6 \frac{K_{\mathrm{p}} G_{\mathrm{m}}}{3K_{\mathrm{p}} + 4G_{\mathrm{m}}} (\alpha_{\mathrm{p}} - \alpha_{\mathrm{m}}) \Delta T \frac{R_{\mathrm{p}}^3}{r^3}, \quad (19)$$

$$\sigma_{\theta\theta} = 6 \frac{K_{\mathrm{p}} G_{\mathrm{m}}}{3K_{\mathrm{p}} + 4G_{\mathrm{m}}} (\alpha_{\mathrm{p}} - \alpha_{\mathrm{m}}) \Delta T \frac{R_{\mathrm{p}}^3}{r^3}. \quad (20)$$

We now turn from considering the stresses created by a single particle to the stresses caused by an ensemble of particles. The stress intensity characterizing the stress state of the material is determined by the following equation:

$$\sigma_{\mathrm{int}} = \sqrt{\frac{1}{2}\left(\sigma_{rr}^2 + \sigma_{\varphi\varphi}^2 + \sigma_{\theta\theta}^2 - 3(\sigma_{rr} + \sigma_{\varphi\varphi} + \sigma_{\theta\theta})^2\right)}. \quad (21)$$

Taking into account Equations (13)–(15) and (18)–(20), the intensity of stresses caused by the difference in the coefficients of thermal expansion of the strengthening particles and the matrix is equal to

$$\sigma_{\mathrm{int}} = 6\sqrt{3} \frac{K_{\mathrm{p}} G_{\mathrm{m}}}{3K_{\mathrm{p}} + 4G_{\mathrm{m}}} |\alpha_{\mathrm{p}} - \alpha_{\mathrm{m}}| \Delta T \min\left[1, \frac{R_p^3}{r^3}\right]. \quad (22)$$

The average value of the thermal stress arising from the difference in the coefficients of thermal expansion, elastic modulus, and shear modulus of the matrix and the particle is determined using the traditional procedure of averaging over a spherical volume with a radius $\Lambda_{\mathrm{p}}/2$.

$$\sigma_{\mathrm{CTE}} = \frac{24}{\Lambda_{\mathrm{p}}^3} \int_0^{\Lambda_{\mathrm{p}}/2} \sigma_{\mathrm{int}} r^2 dr. \quad (23)$$

Thus, the average value of thermal stresses can be estimated as

$$\sigma_{\mathrm{CTE}} = 6\sqrt{3} \frac{K_{\mathrm{p}} G_{\mathrm{m}}}{3K_{\mathrm{p}} + 4G_{\mathrm{m}}} f_{\mathrm{p}} |\alpha_{\mathrm{p}} - \alpha_{\mathrm{m}}| \Delta T (1 - \ln f_{\mathrm{p}}). \quad (24)$$

3. Results and Discussion

3.1. Experimental Results

Figure 2 shows SEM images of the microstructure of A356 aluminum alloys. The structure of the initial A356 alloy does not differ significantly from the alloys with ScF_3 nanoparticles. Silicon inclusions are present in the structure of all alloys (Figure 2a,b). At the same time, ScF_3 nanoparticles are concentrated in the A356 aluminum alloy around the silicon inclusions (Figure 2c,d) and along the grain boundaries. The nonregular distribution of nanoparticles in the structure of the A356 alloy occurs due to the force of attraction between nanosized inclusions, as well as under the influence of the solidification front, which displaces them to the boundaries when the melt is cooled.

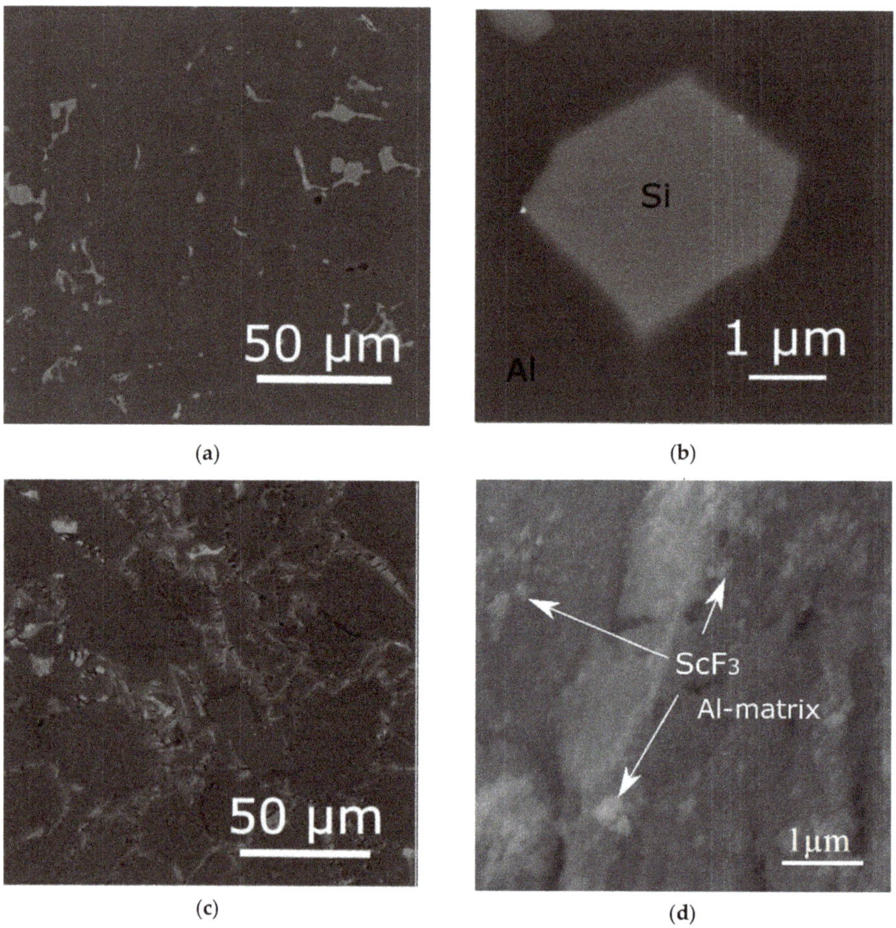

Figure 2. SEM images of the structure A356 alloys (**a**,**b**) and A356–1% ScF_3 (**c**,**d**).

The microstructures of the initial aluminum alloy A356, A356 + 0.2% ScF_3, and A356 + 1% ScF_3 are shown in Figure 3. It can be seen that the introduction of 0.2% ScF_3 nanoparticles into the aluminum alloy led to a decrease in the average grain size from 310 to 190 μm. An increase in the amount of nanoparticles in the A356 alloy to 1% led to a decrease in the grain size to 100 μm.

Figure 4 shows tensile diagrams, and Table 1 shows data on the mechanical properties of aluminum alloys strengthened with ScF_3 and Al_2O_3 nanoparticles. The results were obtained at room temperature ($T_{room} = 293$ K).

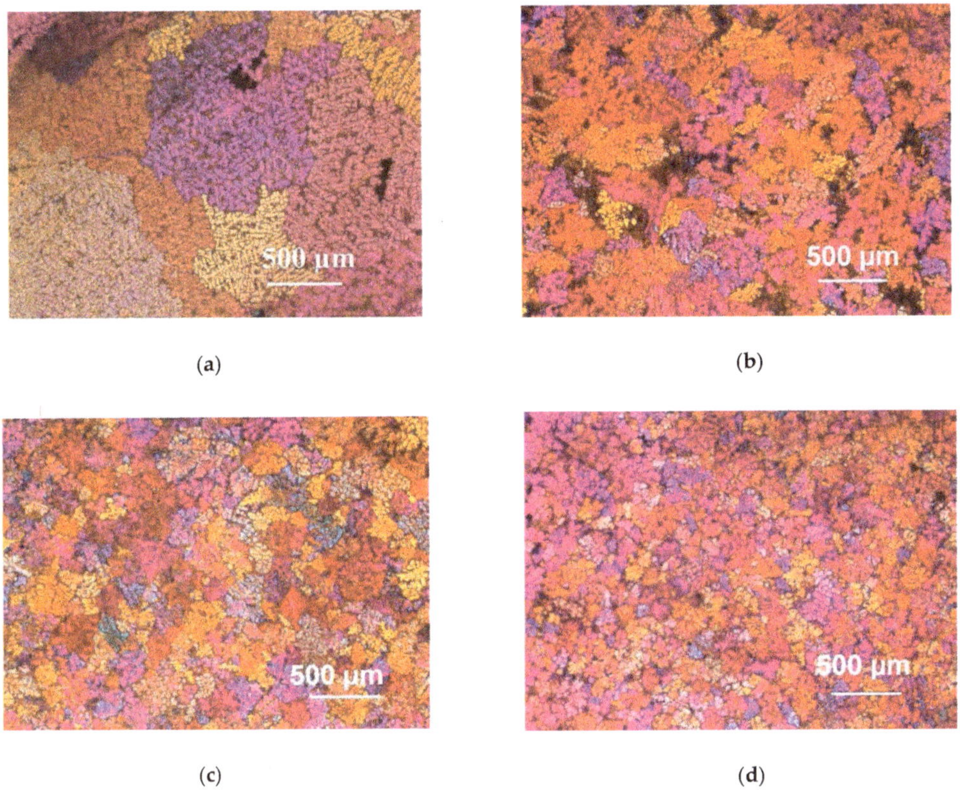

Figure 3. Microstructure of A356 alloy without treatment (**a**), A356 after ultrasonic treatment (**b**), A356–0.2% ScF_3 (**c**), and A356–1% ScF_3 (**d**).

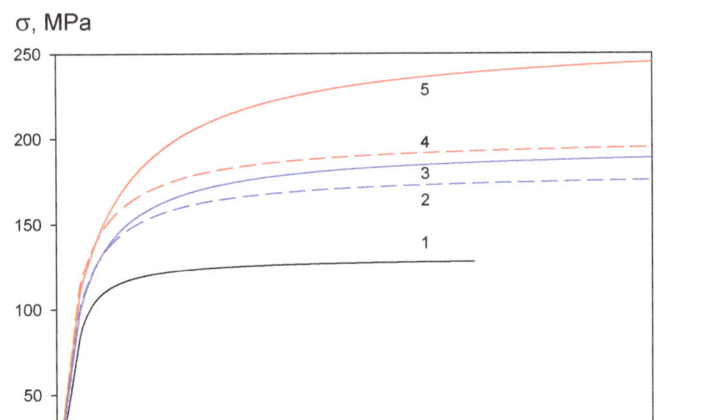

Figure 4. Loading diagrams of aluminum alloys: 1—A356; 2—A356–0.2% Al_2O_3; 3—A356–0.2% ScF_3; 4—A356–1% Al_2O_3; 5—A356–1% ScF_3.

Table 1. Mechanical properties of aluminum alloys.

Alloy	$\sigma_{0.2}$ (MPa)	σ_B (MPa)	Plasticity (%)
A356	85 ± 8	130 ± 7	3.5 ± 0.1
A356–0.2% ScF_3	98 ± 6	190 ± 11	4.3 ± 0.3
A356–1% ScF_3	109 ± 8	250 ± 9	4.2 ± 0.1
A356–0.2% Al_2O_3	100 ± 4	180 ± 7	4.4 ± 0.2
A356–1% Al_2O_3	113 ± 6	195 ± 4	4.1 ± 0.3

Figure 4 and Table 1 demonstrate that the hardening of aluminum alloy A356 by Al_2O_3 and ScF_3 nanoparticles led to an increase in the yield strength, ultimate tensile strength, and plasticity. The introduction of 0.2 wt.% Al_2O_3 nanoparticles made it possible to increase the yield strength, ultimate tensile strength, and plasticity from 85 to 100 MPa, from 130 to 180 MPa, and from 3.5% to 4.1%, respectively, and an increase in the content of Al_2O_3 nanoparticles allowed an increase in the yield strength and ultimate tensile strength of the alloy to 113 MPa and 195 MPa, respectively.

The use of 0.2 wt.% ScF_3 nanoparticles increased the yield strength, ultimate tensile strength, and ductility of the A356 aluminum alloy to 98, 190 MPa, and 4.3%, respectively, and an increase in the content of ScF_3 nanoparticles made it possible to increase the yield strength and ultimate tensile strength of the alloy to 109 MPa and 250 MPa, respectively. Despite the similar size of Al_2O_3 and ScF_3 nanoparticles (~80 nm), the physicomechanical properties of nanoparticles significantly affected the possibility of increasing the mechanical properties of the A356 aluminum alloy.

The approximation of the obtained experimental stress–strain curve allowed us to obtain the function of $\sigma(\varepsilon)$ with an error not exceeding 0.1%.

$$\sigma = \begin{cases} G\varepsilon, & \text{if} \quad \varepsilon \leq \tau_0/G \\ \tau_0 + \tau_1 \dfrac{\varepsilon - \tau_0/G}{\varepsilon_* + \varepsilon}, & \text{if} \quad \tau_0/G < \varepsilon \end{cases}, \tag{25}$$

where τ_0 is the yield strength, $\tau_1 = \tau_\infty - \tau_0$ is the hardening stress, which characterizes the maximum increase of the flow stress during the plastic deformation, and ε_* is an empirical parameter that determines the rate at which the flow curve reaches the asymptote.

The values of the material constants: τ_{0*}, τ_{1*}, and ε_* for various volume fractions of scandium fluoride particles are presented in Table 2.

Table 2. The material constants τ_{0*}, τ_{1*}, and ε_*.

Alloy	τ_{0*} (MPa)	τ_{1*} (MPa)	ε_*
A356–0.2% ScF_3	98	92	0.011
A356–1% ScF_3	109	141	0.013
A356–0.2% Al_2O_3	102	78	0.012
A356–1% Al_2O_3	114	81	0.013

3.2. Results of Theoretical Investigation

The theoretical investigations were conducted for a matrix of aluminum Al 5083 alloy hardened by reinforcement particles. The main calculations were performed for the following parameters [51]: α_m = 2.3 × 10^{-5} K^{-1}, Young modulus of 73 GPa, matrix shear modulus of G_m = 28.08 GPa, and yield strength of $\sigma_{Ym} = 85$ MPa.

A variety of oxides, carbides, borides, and fluorides were utilized as the reinforcement phase in aluminum alloys. Some selected physical and mechanical properties of commonly used reinforcements are summarized in Table 3.

Table 3. Characteristics of reinforcement particles.

Reinforcement	Density (g/cm^3)	Young Modulus (GPa)	Shear Modulus (GPa)	Coefficient Thermal Expansion (10^{-6} K)
SiC	3.15–3.20	450–480	90–131	3.9–4.3
B_4C	2.35–2.55	440–472	180–195	3.2–3.4
ScF_3	2.51–2.54	98–101	37–39	−(1.4–1.2)
TiB_2	4.43–4.52	500–545	182–191	4.6–4.7
Al_2O_3	3.94–3.96	450–460	88–162	7.7–8.5
TiO_2	3.97–4.05	276–288	108–114	8.4–11.8

Let us consider the contribution of various mechanisms to material hardening.

The intensity of the stresses caused by the transfer of the load from the matrix to the particles σ_{NP} is determined by the volume fraction of the hardening particles and, at $f_p = 0.01$, is 0.5% of the yield strength of the matrix material. For $\sigma_{Ym} = 85$ MPa, the calculated value of $\sigma_{NP} = 0.425$ MPa, which indicates an insignificant contribution of this mechanism to the hardening of the material.

The influence of the average grain size on the stress intensity is be rated by the Hall–Petch law, which describes the hardening of a material through the retardation of dislocations by grain boundaries in a polycrystal. The values were calculated (Equation (4)) taking into account the average grain size d_{gr} obtained from optical images of the structure (Figure 3). The calculated values σ_{HP} for the alloys were as follows: A356—2.5 MPa, A356–0.2%—3.2 MPa, and A356–1%—3.7 MPa. Despite a significant decrease in the average grain size, this mechanism had little effect on the increase in the mechanical properties of the A356 alloy. Taking this into account, the effect of the difference in CTE should be considered.

Let us proceed to the analysis of the effect of Orowan stress on the material hardening process. For convenience of analysis, we rewrite Equation (5) in the following form:

$$\sigma_{Or} = G_m \frac{b}{\Lambda_p - 2R_p} = G_m \frac{b}{\delta_p} \frac{\sqrt[3]{f_p}}{1 - \sqrt[3]{f_p}}. \quad (26)$$

Figure 5 shows the dependence of Orowan stresses on the volume fraction of hardening particles, calculated according to Equation (26).

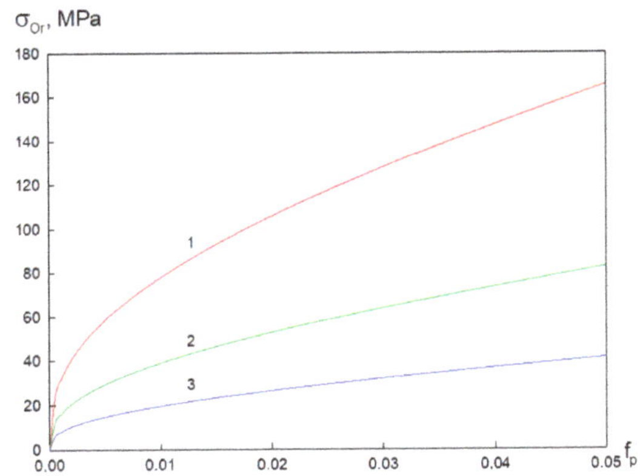

Figure 5. Dependence of Orowan stresses σ_{Or} in alloy A356 strengthened by ScF_3 particles on the volume fraction of hardening particles, calculated for $\Delta T = 100$ K. Particle size: 1—$\delta_p = 20$ nm; 2—$\delta_p = 40$ nm; 3—$\delta_p = 80$ nm.

It can be seen from Figure 5 that an increase in the volume fraction of hardening particles at the same size led to an increase in Orowan stresses. A similar effect was observed with an increase in the size of particles at the same volume fraction. This effect was associated with a decrease in the minimum distance between particles $\Lambda_p - \delta_p$. Dislocations under the influence of the applied external influence bypass the particles, leaving rings around them ("Orowan rings"). If this process occurs when the amount of bend is small, then the required increase in energy will be less than in the case when the dislocation line must completely loop around the particles before it is released. As the distance between the particles decreases, the length of the dislocation line increases significantly. As a result, the efficiency of particles as obstacles to the movement of dislocations increases, and the Orowan stresses increase.

Let us proceed to the analysis of the effect of thermal stresses arising from the difference in the coefficients of thermal expansion, elastic modulus, and shear modulus of the matrix and the particle on the hardening of the material.

In Figure 6, the dependence of the radial stresses σ_{rr} in the particle and matrix on the radial coordinate is shown when the composite was heated from room temperature (293 K). A region of tensile stresses is formed due to the difference in elastic properties. The radial stresses in the particle material have constant values. In the matrix material, these stresses decrease quite sharply with increasing distance from the particle and matrix interface and become negligible at a distance on the order of $5R_p$.

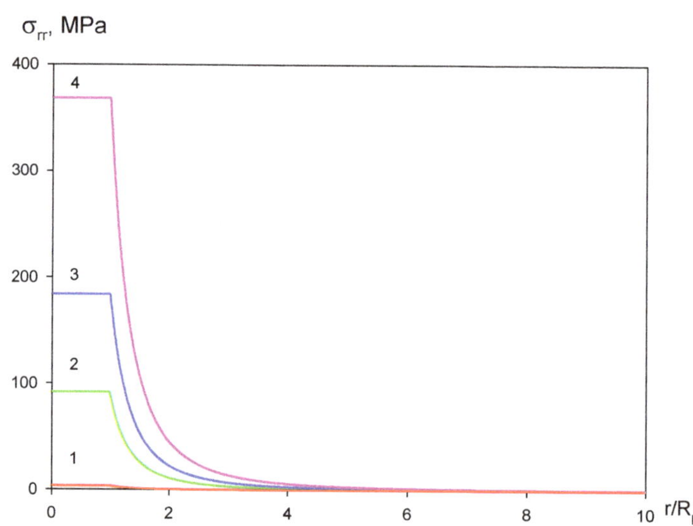

Figure 6. The distribution of the radial stresses σ_{rr} in the particle ScF_3 and matrix for various values of the temperature difference: 1—ΔT = 10 K; 2—50 K; 3—100 K; 4—200 K.

Figure 7 shows the radial distribution of tangential stresses when the composite was heated from room temperature (293 K). In a particle, these stresses have constant positive values, which are associated with stretching due to the impact of the matrix. A sharp jump in tangential stress occurs at the boundary between the particle and the matrix. In this case, tangential stresses become compressive. The tangential stresses decrease sharply with an increase in the distance from the interface and become negligible at a distance on the order of $5R_p$.

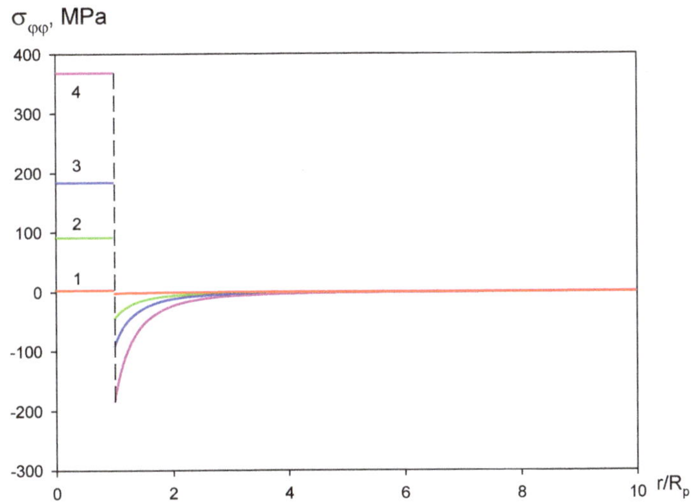

Figure 7. The distribution of the tangential stresses σ_{rr} in the particle and matrix for various values of the temperature difference: 1—ΔT = 10 K; 2—50 K; 3—100 K; 4—200 K.

Note that, due to spherical symmetry, tangential and meridional stresses are equal: $\sigma_{\varphi\varphi} = \sigma_{\theta\theta}$.

An analysis of the effect of the temperature difference ΔT on the stress state allows us to conclude that, with growth, there is an increase in stresses in the particle and the adjacent part of the matrix. In this case, to a first approximation, the magnitude of the arising stresses is proportional to ΔT.

Let us determine the maximum shear stresses $\tau_{\max} = \frac{1}{2}|\sigma_{rr} - \sigma_{\varphi\varphi}|$ arising in a dispersion-strengthened material as a result of the temperature change. Figure 8 shows the dependence of maximal shear stresses on the radial coordinate. In the hardened particle, $\tau_{\max} = 0$. A sharp increase in $\tau_{\max} = 0$ occurs at the interface between the particle and the matrix. Then, as the distance from the particle boundary increases, the value of the maximum shear stresses $\tau_{\max} = 0$ monotonously decreases and becomes vanishingly small.

According to the condition of Saint Venant and Tresca, the plastic deformation begins when the maximal shear stress reaches half of the yield strength. The mathematical formulation of this condition has the following form [52]:

$$\tau_{\max} = \frac{1}{2}\tau_0. \tag{27}$$

It is very critical for an engineer to locate and evaluate the maximum shear stress in a material in order to design the construction in such a way to resist failure.

On the basis of the dependences in Equations (11) and (12), it is possible to determine the magnitude of the temperature difference leading to the occurrence of plastic deformation.

$$\Delta T_{\rm pl} = \frac{3K_{\rm p} + 4\mu_{\rm m}}{9K_{\rm p}\mu_{\rm m}(\alpha_{\rm p} - \alpha_{\rm m})}\tau_0. \tag{28}$$

According to Equation (22), the plastic deformation due to thermal stresses of the aluminum matrix with strengthening scandium fluoride particles occurs when the temperature difference is approximately equal to 72 K. Therefore, above thermal stress, consideration should be given when designing technological constructions for a considerable temperature range.

Figure 9 shows the dimensionless stress intensity $\tau_* = \tau_{\rm int}/\tau_{\rm scale}$ in the vicinity of nine particles. The scale used here is the stress intensity at the particle–matrix interface.

$$\tau_{\text{scale}} = 6\sqrt{3}\frac{K_p \mu_m}{3K_p + 4\mu_m}(\alpha_p - \alpha_m)\Delta T. \quad (29)$$

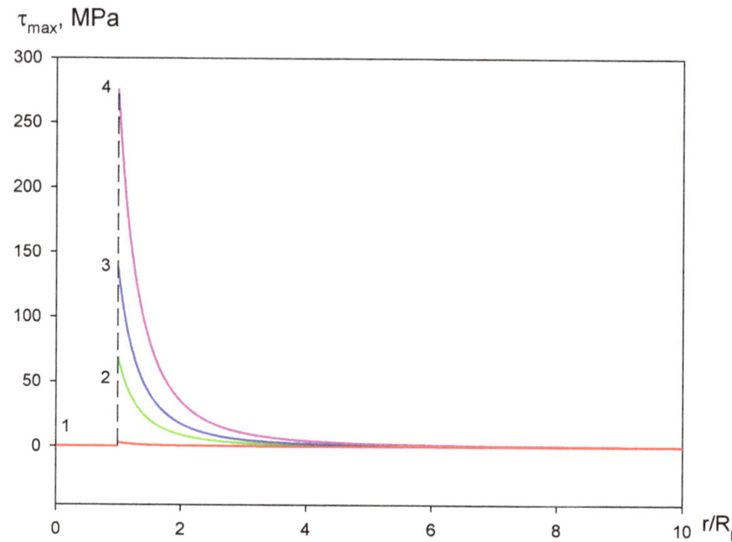

Figure 8. The distribution of the maximal shear stresses σ_{rr} in the particle and matrix for various values of the temperature difference: 1—ΔT = 10 K; 2—50 K; 3—100 K; 4—200 K.

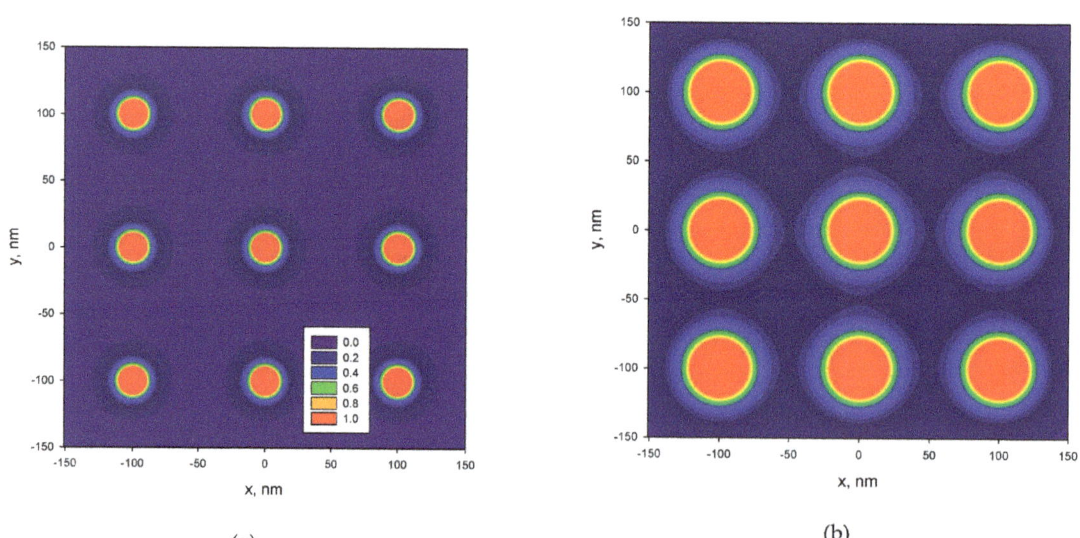

Figure 9. Dimensionless intensity of thermal stresses. Distance between hardening particles $\Lambda_p = 100$ nm: (**a**) $R_p = 5$ nm; (**b**) R_p = 20 nm.

The highest stress values were observed in particles. With distance from the particles, the stress intensity sharply decreased. With an increase in the size R_p of particles at the same distance between them Λ_p, the region in which thermal stresses were observed caused by the difference in the coefficients of thermal expansion of the matrix and the particle increased. The influence of neighboring particles at $R_p < 0.2\Lambda_p$ was insignificant.

Thus, with a small volume fraction $f = (2R_p/\Lambda_p)^3 < 0.06$, the analysis of the stress–strain state of the dispersion-hardened material caused by the difference in thermal expansion coefficients could be carried out without taking up to the effects of the collective interaction of particles and the matrix.

Figure 10 shows the dependence of σ_{CTE} on the volume fraction of hardening particles, calculated for different values of ΔT.

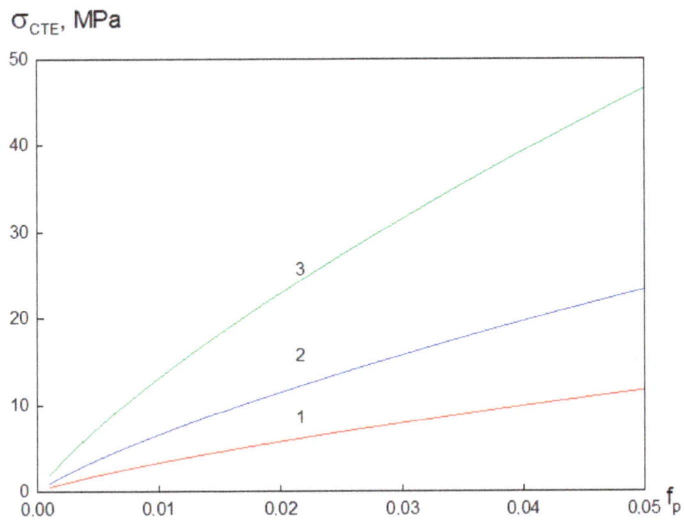

Figure 10. Dependence of thermal stresses σ_{CTE} in the A356 alloy hardened by ScF_3 particles on the volume fraction of hardening particles, calculated for various values of ΔT. Particle size $\delta_p = 80$ nm: 1—$\Delta T = 50$ K; 2—$\Delta T = 100$ K; 3—$\Delta T = 200$ K.

Figure 10 shows that, with an increase in the volume fraction of hardening particles, an increase in thermal stresses occurred σ_{CTE}. This fact was associated with an increase in the number of hardening particles in the alloy and, accordingly, with an increase in their contribution to the hardening of the material. When the material was heated or cooled, as a result of the mismatch between the thermal expansion coefficients and the elastic constants of the matrix and the particles, thermal stresses increased. The calculation results show that an increase in the volume fraction of the strengthening ScF_3 particles in the A356 alloy from 0.1% to 5% at $\Delta T = 50$ K led to an increase in σ_{CTE} from 0.46 MPa to 11.8 MPa, and, at $\Delta T = 200$ K, it led to an increase in σ_{CTE} from 1.84 MPa to 46.538 MPa.

Figure 11 shows the dependence σ_{CTE} on the volume fraction of particles in alloys, hardened by particles from different materials with a temperature change $\Delta T = 100$ K.

The qualitative behavior of all curves in Figure 11 coincides with an increase in the volume fraction of the hardening phase, while an increase in stresses occurs σ_{CTE}. However, the intensity of the stresses depends on the material of the hardening particles. Thus, with a volume fraction $f_p = 5\%$ and a temperature difference $\Delta T = 100$ in alloys hardened with ScF_3 particles, the thermal stresses are $\sigma_{CTE} = 23.61$ MPa; when hardened with Al_2O_3 particles, the magnitude of thermal stresses is $\sigma_{CTE} = 37.64$ MPa; when hardened with TiO_2, it is $\sigma_{CTE} = 40.53$ MPa. A comparison of calculations performed according to Equation (25) with calculations performed according to Equation (7) shows that, for small volume fractions of the hardening phase, Equation (7) underestimates σ_{CTE}, and, for large values of f_p in predicting σ_{CTE} using Equation (7), one can consider an average estimate for particles of different composition.

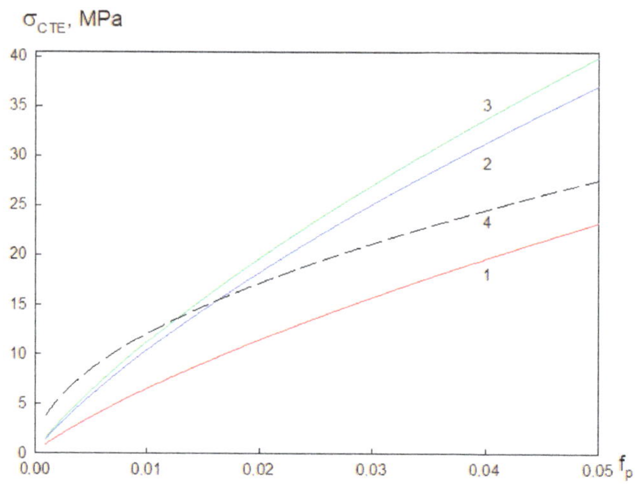

Figure 11. Dependence of thermal stresses σ_{CTE} in the A356 alloy on the volume fraction of hardening particles; particle size $\delta_p = 80$ nm, $\Delta T = 100$ K. Curves 1–3 correspond to the calculation using Equation (25), while curve 4 corresponds to the calculation using Equation (7): 1—ScF_3; 2 —TiO_2; 3—Al_2O_3.

The contributions of the described mechanisms to the yield strength of the composite calculated from Equations (3), (4), (6), and (25) are presented in Figure 12. The results demonstrate that theoretical values are very close to the experimental data. The strengthening caused by thermal mismatch makes the largest contribution to the yield strength improvement. The yield strength increments due to Nardon–Prewo and Orowan mechanisms are much lower.

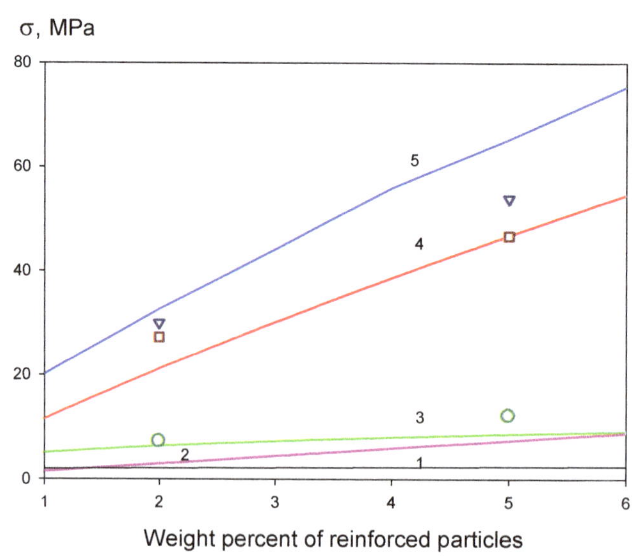

Figure 12. Dependence of different strengthening mechanisms on the mass fraction of the hardening particles: 1—Hall-Petch stresses; 2—Nardon and Prewo stresses; 3—Orowan stresses; 4—thermal stresses; 5—strengthening stresses $\sigma_{Str} = \sigma_{NP} + \sigma_{HP} + \sigma_{Or} + \sigma_{CTE}$; ○—Orowan stresses [53]; □—thermal stresses [53]; ▽—strengthening stresses [53].

The predicted values of the total improvement in yield strength due to various strengthening mechanisms were 31 MPa and 61 MPa for the alloys with 2% and 5% mass fraction of Al_2O_3. Experimental results [53] showed that the strengthening stresses are equal to 30 MPa and 45 MPa, respectively. The difference between the experimental and theoretical results for the alloy with the 5% mass fraction of Al_2O_3 may be explained by the agglomeration of the hardening particles in the composite.

Figure 13 shows the dependence of the yield strength on the mass fraction of the second phase in alloys reinforced with particles of different materials with a change in temperature $\Delta T = 100$. An increase in the mass fraction of particles led to an increase in the limiting shear stress for all considered cases. The smallest values τ_0 were observed for alloys hardened with titanium oxide. At low values of the mass fraction (less than 2%), the highest values τ_0 were achieved in alloys hardened with scandium fluoride; at large values of the mass fraction of the second phase, the highest values were achieved in alloys hardened with alumina.

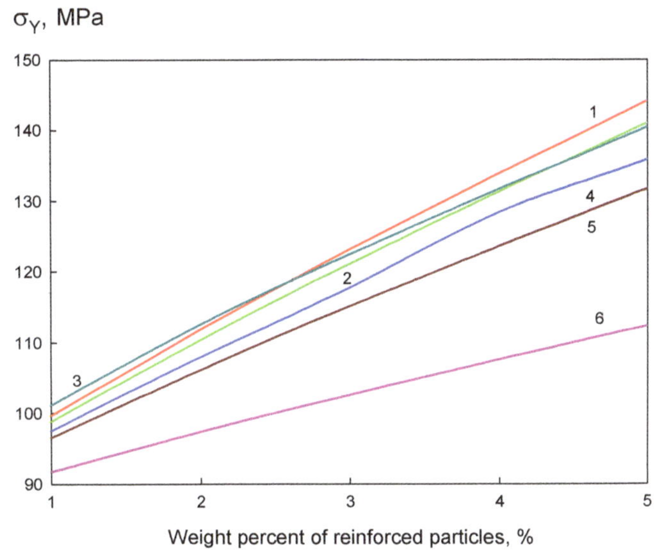

Figure 13. Dependence of the yield stress on the mass fraction of the second phase in alloys strengthened by particles of different materials with temperature difference $\Delta T = 100$ K: 1—SiC; 2 —B_4C; 3—ScF_3; 4—TiB_2; 5—Al_2O_3; 6—TiO_2.

3.3. Verification of the Results

Verification of the modeling results was carried out by comparison with experimental data. Table 4 presents the mechanical properties of the A356 alloy and A356-based composites.

Table 4 shows both experimental results and theoretical predictions of the yield strength σ_Y for different temperatures. The yield strength of composites was greatly enhanced with an increase in reinforcement ratio for all tested conditions. A considerable improvement in yield strength of the composite was recorded with integration into the matrix of the 1% Al_2O_3 and 1% ScF_3 disperse phase. The enhancement in strain hardening capacity of a composite at elevated temperature led to decreased variations in the yield strength of composites compared with aluminum alloy A356.

Table 4. Yield strength of A356 alloy and A356-based composites.

Alloy	Room Temperature, $\Delta T = 0$		$\Delta T = 50$ K		$\Delta T = 100$ K	
	σ_Y, MPA, Experiment	σ_Y, MPA, Results of Modeling	σ_Y, MPA, Experiment	σ_Y, MPA, Results of Modeling	σ_Y, MPA, Experiment	σ_Y, MPA, Results of Modeling
A356	85 ± 8	-	74 ± 4	-	61 ± 6	-
A356–1%Al_2O_3	113 ± 6	107	106 ± 5	101	100 ± 6	97
A356–1%ScF_3	109 ± 8	114	107 ± 4	111	98 ± 6	102

Upon comparing the experimental and theoretical values, one can see that, in general, the results of the predictions were fairly close to the experimental data. The good correlation between the experimental measurements and simulation results validates the correct methods and approaches for the simulation of processes of plastic deformation.

4. Conclusions

An experimental and theoretical investigation of the strength properties of aluminum alloys strengthened by nanoparticles, as well as a determination of the significance of various mechanisms responsible for the strengthening of the material, was carried out.

Results of experimental investigation demonstrated that the hardening of aluminum alloy A356 by Al_2O_3 and ScF_3 nanoparticles led to an increase in the yield strength, ultimate tensile strength, and plasticity. The introduction of 0.2 wt.% Al_2O_3 nanoparticles made it possible to increase the yield strength, ultimate tensile strength, and plasticity from 85 to 100 MPa, from 130 to 180 MPa, and from 3.5% to 4.1%, respectively, and an increase in the content of Al_2O_3 nanoparticles allowed an increase in the yield strength and ultimate tensile strength of the alloy to 113 MPa and 195 MPa, respectively.

The use of 0.2 wt.% ScF_3 nanoparticles increased the yield strength, ultimate tensile strength, and ductility of the A356 aluminum alloy to 98, 190 MPa, and 4.3%, respectively, and an increase in the content of ScF_3 nanoparticles made it possible to increase the yield strength and ultimate tensile strength of the alloy to 109 MPa and 250 MPa, respectively. Despite the similar size of Al_2O_3 and ScF_3 nanoparticles (~80 nm), the physicomechanical properties of nanoparticles significantly affected the possibility of increasing the mechanical properties of the A356 aluminum alloy.

A physicomathematical model of the occurrence of thermal stresses was developed caused by the mismatch of the coefficients of thermal expansion of the matrix and strengthening particles, on the basis of the fundamental principles of mechanics of a deformable solid, and in contrast to existing models, taking into account the elastic properties of not only the matrix, but also the particle.

In the case of thermal deformation of dispersion-hardened alloys, when the CTE of the matrix and particles noticeably differ, an additional stress field is created in the vicinity of the strengthening particle. Thermal stresses increase the effective particle size. This phenomenon can significantly affect the result of the assessment of the yield strength.

In the particle, the stresses are constant. In the matrix material, these stresses decrease quite sharply. As the distance from the particle increases, the value of the shear stresses monotonously decreases and becomes vanishingly small, when the distance from the particle boundary exceeds five particle diameters.

The growth of temperature difference leads to an increase in contact pressure between the matrix and the particle. On the inner surface of the matrix, plastic flow begins when the maximum shear stresses exceed the yield strength of the material at a given temperature. The plastic deformation due to thermal stresses of the aluminum matrix with strengthening scandium fluoride particles occurs when the temperature difference is equal to approximately 72 K.

The strengthening caused by thermal mismatch makes the largest contribution to the yield strength improvement. The yield strength increments due to Nardon–Prewo and Orowan mechanisms are much lower.

Author Contributions: Conceptualization, O.M., O.D. and A.K.; methodology, O.M. and I.Z.; software, O.M.; validation, O.D. and T.K.; formal analysis, O.D. and T.K.; investigation, O.D. and A.K.; writing, O.M. and O.D.; supervision and funding acquisition, A.V. All authors read and agreed to the published version of the manuscript.

Funding: This work was financially supported by Grant N 17-13-01252 from the Russian Science Foundation and Ministry of Science and Higher Education of the Russian Federation, project No 0721-2020-0028.

Data Availability Statement: The data presented in this study are available in the article.

Acknowledgments: The authors acknowledge the Russian Science Foundation under Grant N 17-13-01252. I.Z. and A.K. acknowledge the financials support by the Ministry of Science and Higher Education of the Russian Federation, project No 0721-2020-0028.

Conflicts of Interest: The authors declare no conflict of interest.

References

1. Chawla, N.; Chawla, K.K. *Metal Matrix Composites*; Springer: New York, NY, USA, 2006; 211p, eBook ISBN 978-0-387-28567-2, Hardcover ISBN 978-0-387-23306-2, Softcover ISBN 978-1-4614-9843-8. [CrossRef]
2. Natarajan, N.; Vijayarangan, S.; Rajendran, I. Wear behaviour of A356/25SiCp aluminium matrix composites sliding against automobile friction material. *Wear* **2006**, *261*, 812–822. [CrossRef]
3. Weinert, K.; Lange, M.; Schoer, M. Machining of Light-metal Matrix Composites. In Proceedings of the 6th International Conference Magnesium Alloys and Their Applications, Frankfurt, Germany, 20 September 2000; ISBN 3-527-30282-4. [CrossRef]
4. Karabasov, Y.S. *Advanced Materials*; MISIS: Moscow, Russia, 2002; 736p.
5. Ashby, M.F.; Johnson, K. *Materials and Design, the Art and Science of Materials Selection in Product Design*; Butterworth Heinemann: Oxford, UK, 2002; 602p, ISBN 0750655542/978-0750655545.
6. Arnhold, V.; Hummert, K. *New Materials by Mechanical Alloying Techniques*; Arzt, E., Schult, L., Eds.; DGM Informationsgeselischaft Verlag: Oberursel, Germany, 1989; 263p, ISBN 3-88355-133-3.
7. Orowan, E. Discussion on internal stresses. In *Symposium on Internal Stresses in Metals and Alloys*; Institute of Metals: London, UK, 1948; pp. 451–453.
8. Ashby, M.F. Work hardening of dispersion-hardened crystals. *Philos. Mag.* **1966**, *14*, 1157–1178. [CrossRef]
9. Hirsch, P.B.; Humphreys, F.J. Comment on "Dispersion hardening in metals" by E.W. Hart. *Scr. Metall.* **1973**, *7*, 259–267. [CrossRef]
10. Hazzledine, P.M.; Hirsch, P.B. A coplanar Orowan loops model for dispersion hardening. *Philos. Mag.* **1974**, *30*, 1331–1351. [CrossRef]
11. Humphreys, F.J.; Hirsch, P.B. Work-hardening and recovery of dispersion hardened alloys. *Philos. Mag.* **1976**, *34*, 373–390. [CrossRef]
12. Hansen, N. Dispersion strengthening of aluminium-aluminium-oxide products. *Acta Met.* **1970**, *18*, 137–145. [CrossRef]
13. Spigarelli, S.; Paoletti, C. A new model for the description of creep behaviour of aluminium-based composites reinforced with nanosized particles. *App. Sci. Manuf.* **2018**, *112*, 346–355. [CrossRef]
14. Kröpfl, I.; Vöhringer, O.; Macherauch, E. Creep Behavior of Dispersion-Hardened Aluminum Materials. *Mechan. Time Depend. Mater.* **1999**, *3*, 1–13. [CrossRef]
15. Stobrawa, P.; Rdzawski, Z.M.; Głuchowski, W. Structure and properties of dispersion hardened submicron grained copper. *J. Achiev. Mater. Manufact. Eng.* **2007**, *20*, 195–198.
16. Novikov, N.V.; Bogatyreva, G.P.; Bogdanov, R.K.; Il'nitskaya, G.D.; Isonkin, M.; Marinich, M.A.; Tkach, V.N.; Tsysar', M.A.; Zaitseva, N. The effect of the additions of nanodispersed diamonds on physico-mechanical properties of the drilling tool metal matrix. *J. Superhard Mater.* **2011**, *33*, 268–273. [CrossRef]
17. Chen, B.; Shen, J.; Ye, X.; Jia, L.; Li, S.; Umeda, J.; Takahashi, M.; Kondoh, K. Length effect of carbon nanotubes on the strengthening mechanisms in metal matrix composites. *Acta Mater.* **2017**, *140*, 317–325. [CrossRef]
18. Zhou, W.; Yamaguchi, T.; Kikuchi, K.; Nomura, N.; Kawasaki, A. Effectively enhanced load transfer by interfacial reactions in multi-walled carbon nanotube reinforced Al matrix composites. *Acta Mater.* **2017**, *125*, 369–376. [CrossRef]
19. Cao, G.; Chen, X.; Kysar, J.W.; Lee, D.; Gan, Y.X. The mean free path of dislocations in nanoparticle and nanorod reinforced metal composites and implication for strengthening mechanisms. *Mech. Res. Commun.* **2007**, *34*, 275–282. [CrossRef]
20. Daneyko, O.I.; Kovalevskaya, T.A.; Matvienko, O.V. The influence of incoherent nanoparticles on thermal stability of aluminum alloys. *Russ. Phys. J.* **2018**, *61*, 1229–1235. [CrossRef]
21. Daneyko, O.I.; Kovalevskaya, T.A.; Kulaeva, N.A. Modeling of plastic deformation of dispersion-hardened materials with $L1_2$ superstructure particles. *Russ. Phys. J.* **2017**, *60*, 508–514. [CrossRef]
22. Daneyko, O.I.; Kovalevskaya, T.A. Temperature effect on stress-strain properties of dispersion-hardened crystalline materials with incoherent nanoparticles. *Russ. Phys. J.* **2019**, *61*, 1687–1694. [CrossRef]
23. Kovalevskaya, T.A.; Daneyko, O.I. The influence of scale parameters of strengthening phase on plastic shear zone in heterophase alloys with incoherent nanoparticles. *Russ. Phys. J.* **2020**, *62*, 2247–2254. [CrossRef]

24. Matvienko, O.; Daneyko, O.; Kovalevskaya, T. Mathematical modeling of plastic deformation of a tube from dispersion-hardened aluminum alloy. *MATEC Web Conf.* **2018**, *243*, 8. [CrossRef]
25. Matvienko, O.V.; Daneyko, O.I.; Kovalevskaya, T.A. Elastoplastic deformation of dispersion-hardened aluminum tube under external pressure. *Russ. Phys. J.* **2018**, *61*, 1520–1528. [CrossRef]
26. Matvienko, O.V.; Daneyko, O.I.; Kovalevskaya, T.A. Elastoplastic deformation of dispersion-hardened aluminum tube under external and internal pressure. *Russ. Phys. J.* **2019**, *62*, 720–728. [CrossRef]
27. Huang, Y.D.; Hort, N.; Dieringa, H.; Kainer, K.U. Analysis of instantaneous thermal expansion coefficient curve during thermal cycling in short fiber reinforced AlSi12CuMgNi composites. *Compos. Sci. Technol.* **2005**, *65*, 137–147. [CrossRef]
28. Vaidya, R.U.; Chawla, K.K. Thermal expansion of metal−matrix composites. *Comp. Sci. Technol* **1994**, *50*, 13–22. [CrossRef]
29. Arsenault, R.J.; Shi, N. Dislocations generation due to differences between the coefficients of thermal expansion. *Mater. Sci. Eng.* **1986**, *81*, 151–187. [CrossRef]
30. Lee, J.K.; Earmme, Y.Y.; Aaronson, H.I.; Russell, K.C. Plastic relaxation of the transformation strain energy of a misfitting spherical precipitate: Ideal plastic behavior. *Metall. Mater. Trans. A* **1980**, *11*, 1837–1847. [CrossRef]
31. Yang, C.; Sun, W.; Seneviratne, W.; Shashidhar, A. Thermally induced loads of fastened hybrid composite/aluminum structures. *J. Aircr.* **2008**, *45*, 569–580. [CrossRef]
32. Parrini, L.; Schaller, R. Thermal stresses in metal matrix composites studied by internal friction. *Acta Mater.* **1996**, *44*, 4881–4888. [CrossRef]
33. Drozd, Z.; Trojanová, Z.; Lukáč, P. Temperature deformation of the ax41 composites studied by the dilatometer measurements. In Proceedings of the ECCM15—15th European Conference on Composite Materials, Venice, Italy, 24–28 June 2012; pp. 1–8.
34. Vogelsang, M.; Arsenault, R.J.; Fisher, R.M. An in situ HVEM study of dislocation generation at Al/SiC interfaces in metal matrix composites. *Metall. Mater. Trans. A* **1986**, *17*, 379–389. [CrossRef]
35. Dunand, D.C.; Mortensen, A. On plastic relaxation of thermal stresses in reinforced metals. *Acta Metall. Mater.* **1991**, *39*, 127–139. [CrossRef]
36. Taylor, A.; Jones, R.M. *Silicon Carbide*; O'Connor, J.R., Smiltens, J., Eds.; Pergamon: Oxford, UK, 1960; pp. 147–154.
37. Chowdhury, A.S.M.F.; Mari, D.; Schaller, R. Thermal stress relaxation in magnesium matrix composites studied by mechanical spectroscopy. *Phys. Stat. Sol.* **2008**, *205*, 249–254. [CrossRef]
38. Brinker, C.J.; Scherer, G.W. *Sol-Gel Science*; Academic Press: New York, NY, USA, 1990; 881p.
39. Vorozhtsov, S.A.; Eskin, D.G.; Tamayo, J.; Vorozhtsov, A.B.; Promakhov, V.V.; Averin, A.A.; Khrustalyov, A.P. The application of external fields to the manufacturing of novel dense composite master alloys and aluminum-based nanocomposites. *Metall. Mater. Trans. A* **2015**, *46*, 2870–2875. [CrossRef]
40. Puga, H.; Costa, S.; Barbosa, J.; Ribeiro, S.; Prokic, M. Influence of ultrasonic melttreatment on microstructure and mechanical properties of AlSi9Cu3 alloy. *J. Mater. Process. Technol.* **2011**, *211*, 1729–1735. [CrossRef]
41. Bisht, A.; Srivastava, M.; Manoj Kumar, R.; Lahir, I.; Lahiri, D. Strengthening mechanism in grapheme nanoplatelets reinforced aluminum composite fabricated through spark plasma sintering. *Mater. Sci. Eng. A* **2017**, *695*, 20–28. [CrossRef]
42. Nardone, V.C.; Prewo, K.M. On the strength of discontinuous silicon carbide reinforced aluminum composites. *Scr. Metall.* **1986**, *20*, 43–48. [CrossRef]
43. Yu, H.; Xin, Y.; Wang, M.; Liu, Q. Hall-Petch relationship in Mg alloys: A review. *J. Mater. Sci. Tech.* **2018**, *34*, 248–256. [CrossRef]
44. Goh, C.; Wei, J.; Lee, L.C.; Gupta, M. Properties and deformation behaviour of Mg–Y_2O_3 nanocomposites. *Acta Mater.* **2007**, *55*, 5115–5121. [CrossRef]
45. Neite, G.; Kubota, K.; Higashi, K.; Hehmann, F. Magnesium. *Mater. Sci. Tech.* **2005**, *8*, 115.
46. Sanaty-Zadeh, A. Comparison between current models for the strength of particlate-reinforced et al matrix nanocomposites with emphasis on consideration of Hall–Petch effect. *Mater. Sci. Eng. A* **2012**, *531*, 112–118. [CrossRef]
47. Ma, F.C.; Zhou, J.J.; Liu, P.; Li, W.; Liu, X.K.; Pan, D.; Lu, W.J.; Zhang, D.; Wu, L.Z.; Wei, X.Q. Strengthening effects of TiC particles and microstructure refinement in insitu TiC-reinforced Ti matrix composites. *Mater. Charact.* **2017**, *127*, 27–34. [CrossRef]
48. Sree Manu, K.M.; Arun Kumar, S.; Rajan, T.P.D.; Riyas Mohammed, M.; Pai, B.C. Effect of alumina nanoparticle on strengthening of Al-Si alloy through dendrite refinement, interfacial bonding and dislocation bowing. *J. Alloys Compd.* **2017**, *712*, 394–405. [CrossRef]
49. Li, J.; Liu, B.; Fang, Q.H.; Huang, Z.W.; Liu, Y.W. Atomic-scale strengthening mechanism of dislocation-obstacle interaction in silicon carbide particle- reinforced copper matrix nanocomposites. *Ceram. Int.* **2017**, *43*, 3839–3846. [CrossRef]
50. Zhao, X.; Lu, C.; Tieu, A.K.; Pei, L.Q.; Zhang, L.; Cheng, K.Y.; Huang, M.H. Strengthening mechanisms and dislocation processes in ⟨111⟩ textured nanotwinned copper. *Mater. Sci. Eng. A* **2016**, *676*, 474–486. [CrossRef]
51. Ma, K.; Lavernia, E.J.; Schoenung, J.M. Particulate reinforced aluminum alloy matrix composites—A review on the effect of microconstituent. *Rev. Adv. Mater. Sci.* **2017**, *48*, 91–104.
52. Timoshenko, S.P.; Goodier, J.N. *Theory of Elasticity*; McGraw Hill: New York, NY, USA, 2010; 567p.
53. Ma, P.; Jia, Y.; Prashanth, K.; Yu, Z.; Yang, S.; Zhao, J.; Li, C. Effect of Al2O3 Nanoparticles as Reinforcement on the Tensile Behavior of Al-12Si Composites. *Metals* **2017**, *7*, 359. [CrossRef]

metals

MDPI

Article

Structure and Properties of Ca and Zr Containing Heat Resistant Wire Aluminum Alloy Manufactured by Electromagnetic Casting

Nikolay Belov [1,*], Torgom Akopyan [1], Natalia Korotkova [1], Maxim Murashkin [2,3], Victor Timofeev [4] and Anastasiya Fortuna [5]

1 Department of Metal Forming, National University of Science and Technology MISiS, Leninsky Prospekt 4, 119049 Moscow, Russia; nemiroffandtor@yandex.ru (T.A.); kruglova.natalie@gmail.com (N.K.)
2 Institute of Physics of Advanced Materials, Ufa State Aviation Technical University, K. Marx 12, 450008 Ufa, Russia; murashkin.maksim-y@net.ugatu.su
3 Laboratory for Mechanics of Bulk Nanostructured Materials, St. Petersburg State University, Universitetskaya Nab. 7/9, 199034 St. Petersburg, Russia
4 Department of Electrical Engineering, Siberian Federal University, 79 Svobodnyy Prospekt, 660041 Krasnoyarsk, Russia; viktortim0807@mail.ru
5 Department of Physical Materials Science, National University of Science and Technology MISiS, Leninsky Prospekt 4, 119049 Moscow, Russia; anastasya_fortuna@mail.ru
* Correspondence: nikolay-belov@yandex.ru; Tel.: +7-9154145945

Abstract: Experimental aluminum alloy containing 0.8% Ca, 0.5% Zr, 0.5% Fe and 0.25% Si (wt.%), in the form of a long-length rod 12 mm in diameter was manufactured using an electromagnetic casting (EMC) technique. The extremely high cooling rate during alloy solidification ($\approx 10^4$ K/s) caused the formation of a favorable microstructure in the ingot characterized by a small size of the dendritic cells, fine eutectic particles of Ca-containing phases and full dissolution of Zr in Al the solid solution. Due to the microstructure obtained the ingots possess high manufacturability during cold forming (both drawing and rolling). Analysis of the electrical conductivity (EC) and microhardness of the cold rolled strip and cold drawn wire revealed that their temperature dependences are very close. The best combination of hardness and EC in the cold rolled strip was reached after annealing at 450 °C. TEM study of structure evolution revealed that the annealing mode used leads to the formation of $L1_2$ type Al_3Zr phase precipitates with an average diameter of 10 nm and a high number density. Experimental wire alloy has the best combination of ultimate tensile strength (UTS), electrical conductivity (EC) (200 MPa and 54.8% IACS, respectively) and thermal stability (up to 450 °C) as compared with alloys based on the Al–Zr and Al– rare-earth metal (REM) systems. In addition, it is shown that the presence of calcium in the model alloy increases the electrical conductivity after cold forming operations (both drawing and rolling).

Keywords: aluminum–zirconium wire alloys; electromagnetic casting (EMC); drawing; electrical conductivity (EC); phase composition; nanoparticles

Citation: Belov, N.; Akopyan, T.; Korotkova, N.; Murashkin, M.; Timofeev, V.; Fortuna, A. Structure and Properties of Ca and Zr Containing Heat Resistant Wire Aluminum Alloy Manufactured by Electromagnetic Casting. *Metals* **2021**, *11*, 236. https://doi.org/10.3390/met11020236

Academic Editor: Qudong Wang

Received: 20 December 2020
Accepted: 26 January 2021
Published: 1 February 2021

1. Introduction

For many years, electrical engineering applications have been recognized as one of the main uses of aluminum in terms of industry economics, which is constantly developing under the current widespread tendency for replacing copper conductors [1]. Most conductive Al alloys are related to 1xxx (Al-Fe-Si), 6xxx (Al-Mg-Si) and 8xxx (Al-Fe) families. For manufacturing electrical parts with appropriate parameters, continuous casting and rolling are currently used for shaping and deformation strengthening. However, the latter is highly reduced after heating to above ~250 °C due to recrystallization.

For the sake of both improving heat resistance and operating temperature, Al-Zr alloys have been developed for particular applications, like overhead cables for long-distance power transmission [2–7]. Transition element Zr provides a far higher recrystallization temperature (up to 400 °C) due to the formation of the nanoscale metastable Al_3Zr phase with a $L1_2$ crystal lattice [8–13] while heating to above 350 °C and decay of solid solution (Al) [14–19]. However, if not controlling the foregoing parameters of melting and casting, the unfavorable Al_3Zr phase with a $D0_{23}$ structure may appear. According to the binary Al-Zr diagram [6], the liquidus temperature rises dramatically with Zr, hence, the temperature of the melt must be considered depending on the concentration. Additionally, for achieving metastable solubility, a relatively high cooling rate must be provided. The latter may be achieved by techniques like electromagnetic casting (EMC) that is superior to direct chill casting (DC). EMC allows producing long rod billets with far higher surface quality and an exceptionally fine microstructure due to support of the liquid pool by electromagnetic forces, continuous stirring, and increased cooling rates up to 10^4 K/s [20]. Based on the cooling rates, close to rapid solidification techniques [1,21], it may be argued that EMC is a very appropriate method for obtaining high-solute alloys, include those alloyed with transition elements. It was recently demonstrated [22] that Al alloy containing 0.6% Zr, 0.4% Fe and 0.4% Si (wt.%) may be successfully manufactured by EMC to long-length 12 mm circular cross-section rods with a microstructure containing Zr-rich solid solution. This structure provided excellent processability in cold drawing down to 3 mm, and the wire product showed a remarkable combination of strength (ultimate tensile strength (UTS) ~230 MPa) and electrical conductivity (55.6% IACS), both remaining stable after heating up to 400 °C.

In addition to Fe and Si used, subsequent development can be considered in alloying by rare-earth metal (REM) [21,23] and calcium [24,25]. They are both recognized to be very efficient for improving the heat resistance of Al alloys. However, while rare earth metals (REM) are more established for Al alloys, unconventional calcium may bring far higher cost efficiency combined with favoring the physical and mechanical properties. Particularly, it was shown that the Al wire alloy containing Al-4% Ca-1% Fe-0.6% Si-0.2% Zr-0.1% Sc manufactured from DC cast 150 mm billet has an improved combination of properties compared with Al–REM alloys [24]. This is due to its structure combining an Al matrix strengthened by $Al_3(Zr,Sc)$-$L1_2$ phase particles and uniformly distributed submicron particles of Ca-bearing phases. Simulation of DC casting and flat rolling for the Al-0.5% Ca-0.5% Fe-0.25% Si-0.2% Zr-0.1% Sc alloy also showed the possibility for achieving such a combination of basic characteristics [25]. In respect to the foregoing compositions, scandium alloying may bring excessive costs of products. Using EMC technology, it is possible to introduce up to 0.6% Zr, that can be as efficient as the combination of Zr and Sc. Hence, we find it expedient to consider the Al–Ca–Zr–Fe–Si system as the base one. According to our earlier results [5,22,24–26], the target structure can be achieved in the alloy containing ~1% Ca, 0.5% Zr, 0.5% Fe, 0.25% Si (wt.%). Based on above, the goal of the present work is to study the structure, mechanical properties and electrical conductivity of the aluminum wire alloy of the foregoing composition manufactured by direct cold drawing from an as-cast 12 mm circular cross-section EMC rod and subjected to various heat treatment routes.

2. Experimental Methods

Experimental aluminum alloy in the form of a long-length rod 12 mm in diameter and ≈20 m in length was manufactured using electromagnetic casting (EMC) technique (equipment of the Research and Production Centre of Magnetic Hydrodynamics (Krasnoyarsk, Russia) [27]). The alloy was prepared from commercial aluminum (99.5 wt.%). Calcium, zirconium, iron and silicon were introduced into the aluminum melt at 880–900 °C in the form of master alloys. The casting was carried out at a temperature of ≈830 °C which was higher than the alloy liquidus (810 °C). The fragment of long rod billet (Figure 1a) was used for experimental study. According to spectral analysis made using Oxford Instruments, the actual chemical composition of the experimental alloy (hereinafter referred to as ACZ) was close to the target one (Table 1). For a more detailed analysis of the quaternary Al-Ca-Fe-Si system, additional studies of the microstructure of the ACZ alloy obtained during slow solidification (melt cooling rate is ~0.01 K/min) were carried out. From this, we can expect the formation of the phase composition close to the equilibrium one.

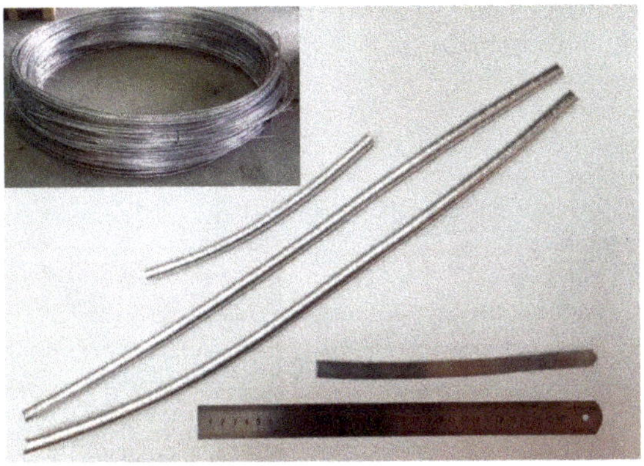

(**a**)

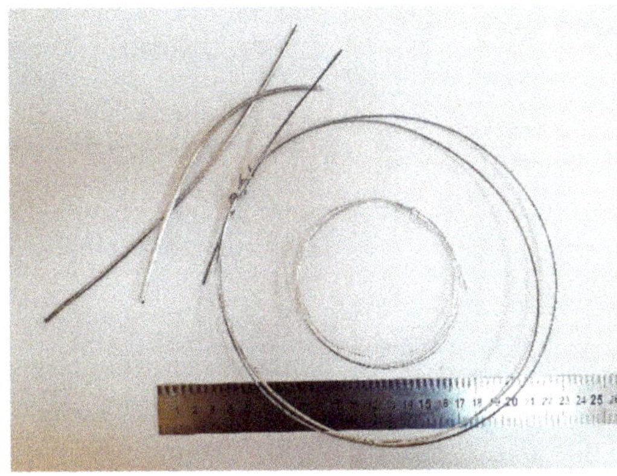

(**b**)

Figure 1. As-cast wire rod (12 mm) and cold rolled strip (**a**), and (**b**) cold drawn wire of ACZ alloy.

Table 1. Chemical compositions of experimental alloys.

Designation	Concentration, wt.% (at.%)				
	Ca	Fe	Si	Zr	Al
ACZ	0.81 ± 0.02 (0.55)	0.48 ± 0.03 (0.23)	0.24 ± 0.02 (0.23)	0.46 ± 0.05 (0.14)	balance
0 Ca [1]	–	0.68 ± 0.04 (0.40)	0.23 ± 0.03 (0.22)	–	balance
0.1 Ca [1]	0.13 ± 0.02 (0.09)	0.67 ± 0.03 (0.33)	0.22 ± 0.03 (0.21)	–	balance
0.2 Ca [1]	0.19 ± 0.02 (0.10)	0.65 ± 0.04 (0.33)	0.21 ± 0.03 (0.20)	–	balance
0.75 Ca [1]	0.73 ± 0.03 (0.42)	0.41 ± 0.04 (0.20)	0.15 ± 0.03 (0.14)	–	balance
1 Ca [1]	1.03 ± 0.04 (0.69)	0.41 ± 0.05 (0.20)	0.16 ± 0.04 (0.16)	–	balance

[1] Zr-free alloys (0–1 wt.% Ca) were prepared in a resistance furnace (GRAFICARBO) in a form of flat ingots (10 mm × 40 mm × 180 mm in size) and considered in section "Discussion".

The as-cast EMC rod was processed using drawing (the reduction ratio was 94%) to obtain the wire (3 mm diameter, see in Figure 1b). The obtained wire was subjected to annealing at 350–450 °C together with the 2 mm strip (Figure 1a) made additionally from the as-cast EMC rod by cold rolling (using a Chinetti LM160 laboratory-scale rolling mill machine). This strip was prepped for the analysis of the influence of deformation on the decomposition of (Al) during annealing. The stepwise annealing modes (Table 2) used for all experimental samples (EMC rod, wire and strip) were previously substantiated [22]. The Vickers hardness and electrical conductivity (EC) of the samples were measured at each annealing step. The measurement was carried out at room temperature.

Besides, the ternary and 4 quaternary alloys containing 0.5% Fe, 0.25% Si and up to 1% Ca were prepared to estimate the effect of calcium on the electrical conductivity. These alloys were poured at 750 °C into a flat graphite mold (of 10 mm × 40 mm × 180 mm in size). The cold rolled sheet products of 2 mm thickness were prepared from as-cast ingots.

The microstructure was studied using scanning electron microscopy (SEM, TESCAN VEGA 3, Tescan Orsay Holding, Brno, Czech Republic), electron microprobe analysis (EMPA, OXFORD Aztec, Oxford Instruments, Oxford shire, UK), and transmission electron microscopy (TEM, JEM 2100, JEOL, Tokyo, Japan). Mechanical polishing was used together with electrolytic polishing, which was carried out at a voltage of 15 V at a temperature of −25 °C in an electrolyte containing 20% nitric acid and 80% methanol. The thin foils for

TEM were prepared by ion thinning with a PIPS (Precision Ion Polishing System, Gatan, Pleasanton, CA, USA) machine and studied at 160 kV.

Tensile tests for as-processed wire specimens were conducted by using a universal testing machine, model Zwick Z250 (Zwick Roell AG, Ulm, Germany). The Vickers hardness (HV) was measured using a MetkonDuroline MH-6 (METKON Instruments Inc., Bursa, Turkey) universal tester. A load of 1 kg and a holding time of 10 s were used to determine the Vickers hardness. The hardness was measured at least five times at each point.

The specific electrical conductivity (EC) of the EMC rod and the cold rolled strip was determined using the eddy current method with a VE-26NP eddy structures instrument (CJSC Research institute of introscopy SPEKTR, Moscow, Russia). The electrical resistivity of the cold drawn wire was measured for straightened samples of at least 1 m in length in the rectified part (in accordance with IEC 60468:1974 standard [28]).

Table 2. Annealing regimes for electromagnetic casting (EMC) rod, strip and wire.

Designation	Regime Treatment
R-EMC casting rod (diameter 12 mm)/S-cold rolled strip (thickness 2 mm)	
R/S	As-cast/as-cold rolled
R300/S300	Annealing at 300 °C, 3 h
R350/S350	R300/S300 + annealing at 350 °C, 3 h
R400/S400	R350/S350 + annealing at 400 °C, 3 h
R450/S450	R400/S400 + annealing at 450 °C, 3 h
R500/S500	R450/S450 + annealing at 500 °C, 3 h
R550/S550	R500/S500 + annealing at 550 °C, 3 h
R600/S600	R550/S550 + annealing at 600 °C, 3 h
W–Wire (diameter 3 mm) manufactured by cold drawing of as-cast EMC rod	
W	As-drawn
W350	W300 + annealing at 350 °C, 3 h
W400	W350 + annealing at 400 °C, 3 h
W450	W450 + annealing at 450 °C, 3 h

3. Results

3.1. Characterization of as-Cast Structure

The extremely high cooling rate during ACZ alloy solidification caused the formation of a favorable microstructure (Figure 2a) characterized by small size of dendritic cells, fine eutectic particles of Ca-containing phases and full dissolution of Zr in Al solid solution. The measured average size of the dendritic cells is ~4 μm (Figure 2b) which, according to other studies [22,29,30], corresponds to a cooling rate of ~10^4 K/s. Calcium-bearing eutectic particles corresponding to the quaternary eutectic (Al) + Al_4Ca + $Al_{10}CaFe_2$ + Al_2CaSi_2 [26] are detected in the form of thin veins located along the boundaries of the aluminum dendritic cells. It should be noted that the as-cast structure does not contain needle-shaped inclusions, for example, iron-containing ones.

Remelting of EMC rod followed by pouring into a graphite mold (cooling velocity about 20 K/s) leads to coarsening of structure and the formation of some needle-like particles (Figure 2c). On the other hand, the structure of the alloy after slow solidification (in a furnace) differs considerably from that of the initial EMC rod. Along with the expected general coarsening of the structure, the phase composition of the alloy changes. In particular, needle-shaped Al_3Fe phase inclusions and segregation of primary Al_3Zr phase crystals (Figure 2d) that are absent in the EMC rod structure can be identified (Figure 2a,b) after slow solidification.

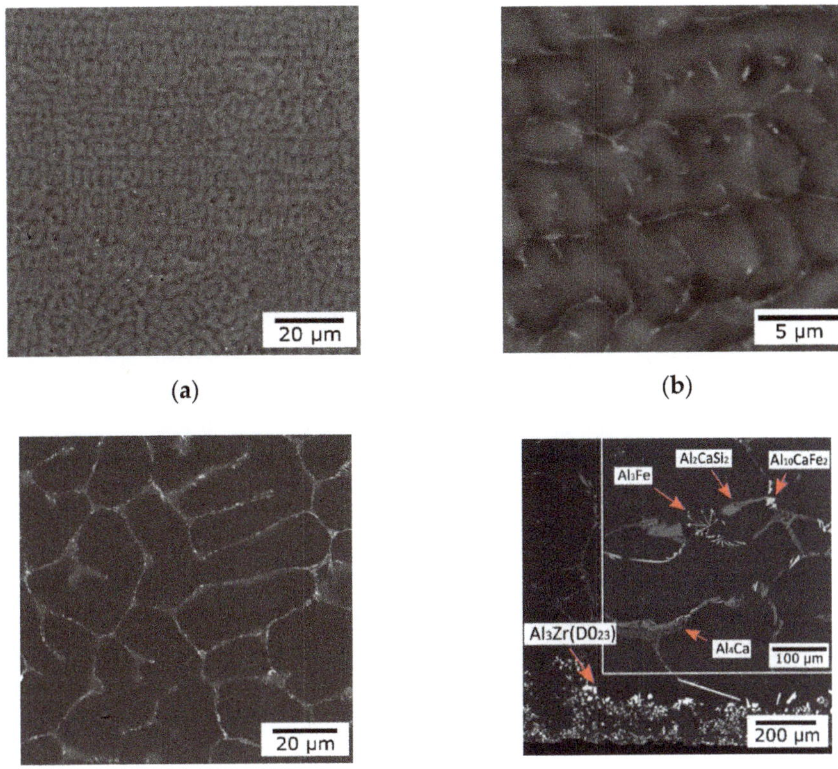

(a) (b) (c) (d)

Figure 2. As-cast microstructure of alloy ACZ after various cooling velocities, SEM: (**a**,**b**) ~10^4 (12 mm EMC rod, R in Table 2), (**c**) ~20 K/s (10 mm × 40 mm × 200 mm ingot), (**d**) ~0.1 K/s (cooling in furnace).

3.2. *Effect of Cold Deformation and Annealing on Structure, Hardness and Electrical Conductivity*

Due to the fine structure of the eutectic, the as-cast EMC rod possesses high ductility, even during cold forming. During both rolling and drawing, apart from the formation of a fibrous grain structure, fragmentation of Ca-containing eutectic particles can also be observed. Their size is not greater than 1 micron and they are uniformly distributed in the aluminum matrix (Figure 3a,b). Annealing at up to 450 °C inclusively does not lead to a significant change in the size of Ca-containing eutectic particles. However, the structure remains non-recrystallized. Coarsening of particles is observed at higher temperatures (Figure 3c,d).

As expected, cold deformation leads to a hardening of the ACZ alloy—the hardness increases up to 65HV for strip and up to 70 HV for wire (states R and W, respectively, see Table 3). A surprising result is a significant increase in the electrical conductivity—up to ~ 25 MS/m (both for strip and wire). In an earlier study for a Ca-free Al-0.6% Zr-0.4% Fe-0.4% Si alloy obtained by a similar process, this effect was not observed [22].

According to EMC rod hardness data obtained during annealing (Figure 4a), hardening reaches the highest level at 450 °C annealing stage temperature (the R450 state). Further increase in the annealing temperature leads to a significant decrease in HV, which is mainly due to the coarsening of the Al_3Zr precipitates [5–11,16]. Deformation hardening was retained upon strip annealing to 450 °C. At this temperature, the hardness of the EMC rod is the same as for the cold rolled strip (Figure 4a). Further increase in the annealing temperature leads to a significant softening due to the formation of a recrystallized structure. At the maximum annealing temperature used, 600 °C, they had approximately the same hardness (32–33 HV) due to the coarsening and transformation of the Al_3Zr precipitates to the equilibrium $D0_{23}$ phase [6,15].

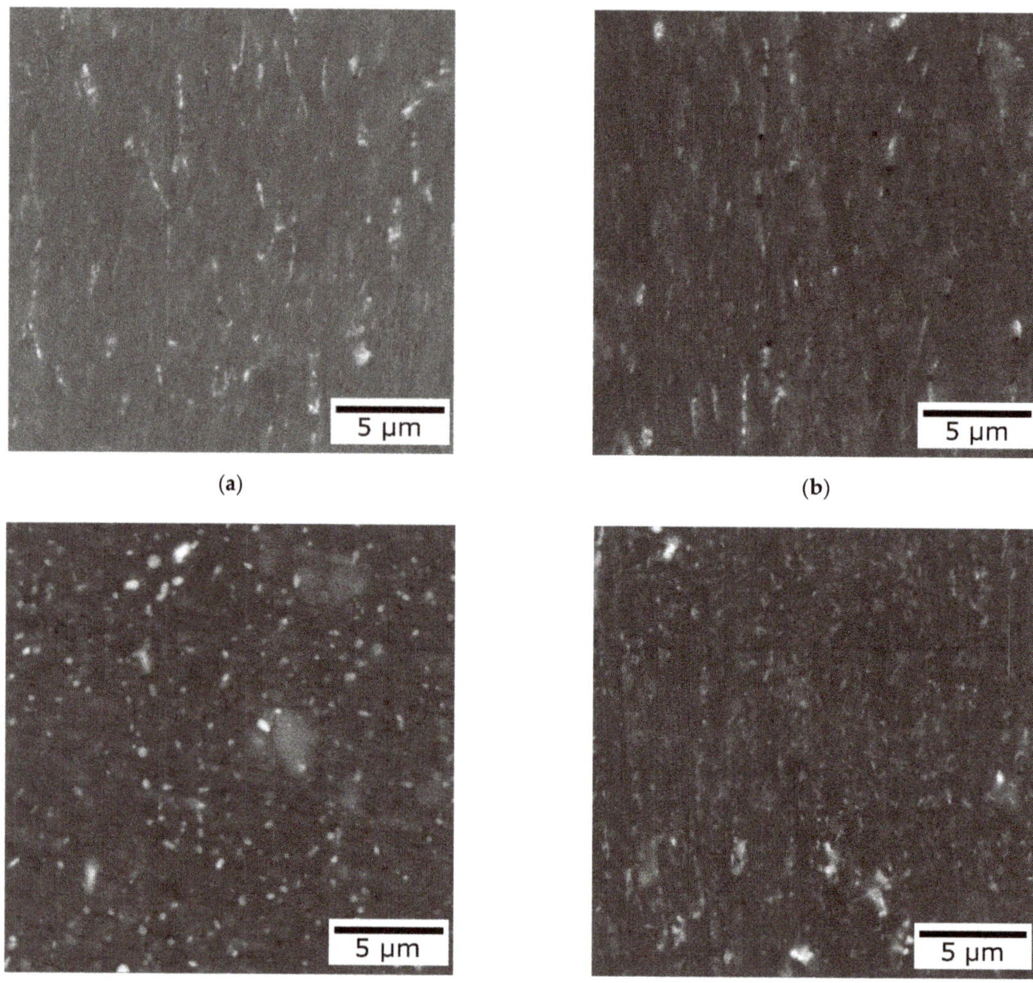

Figure 3. Microstructure of wire alloy ACZ in (**a**,**c**) 3 mm wire and (**b**,**d**) 2 mm strip: (**a**), as-drawn, (**b**) as-rolled, (**c**,**d**) annealed at 600 °C, SEM, (**a**) W, (**b**) S, (**c**) W600, (**d**) S600, designations see Table 2.

Table 3. Mechanical and electrical properties of ACZ wire alloy.

State	Mechanical Properties				Electrical Properties		
	YS, MPa	UTS, MPa	El, %	HV, MPa	ER, nΩm	EC, MS/m	EC, %IACS
W	246 ± 12	282 ± 7	4.1 ± 0.4	70 ± 3	40.26	24.8	42.9
W350	–	–	–	–	36.10	27.7	47.8
W400	197 ± 2	218 ± 1	8.9 ± 2.5	64 ± 3	34.24	29.2	50.4
W450	181 ± 6	202 ± 1	11.9 ± 1.3	60 ± 2	31.54	31.7	54.7

Decomposition of the aluminum solid solution with the formation of $L1_2$ (Al_3Zr) nanoparticles during annealing promotes the increase in electrical conductivity (EC), as shown in Figure 4b. At the same time, the difference between the EC values for the EMC rod and for the 2 mm strip remains approximately the same at all annealing temperatures up to 500 °C inclusively (2.5–3.0 MS/m). At this temperature, the maximum EC for a 2 mm strip is reached, but it corresponds to the softening stage (Figure 4a). Considering the maximum hardness and high electrical conductivity in the cold rolled strip after annealing at 450 °C, we used the same heat treatment for the prepared wire.

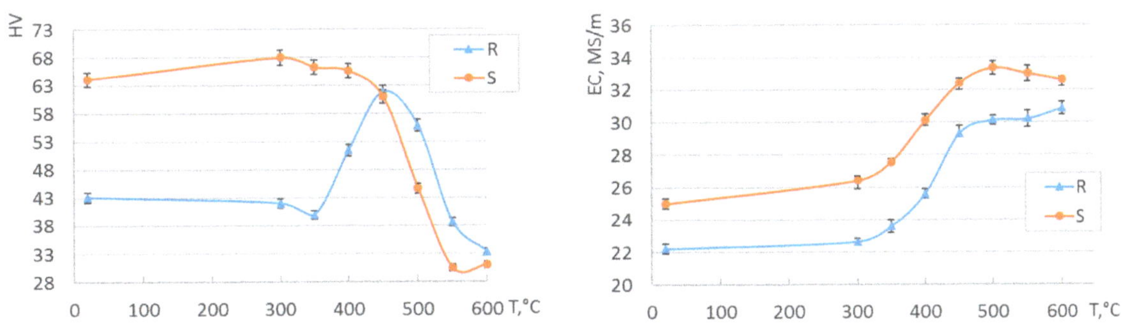

(**a**) (**b**)

Figure 4. (**a**) Hardness (HV) and (**b**) electrical conductivity (EC) curves for EMC rod (R) and cold rolled strip (S) subjected to stepwise annealing in accordance with the processing route shown in Table 2 (see R (temperature)/S (temperature) designation).

TEM microstructure of the ACZ alloy was examined for cold rolled strip in the S350 °C (Figure 5a,b) and S450 °C (Figure 5c,d) states to confirm the main structure changes described. According to obtained data for the both states, the fine individual particles of the eutectic Ca-containing intermetallics (dark in appearance) with a less than 1 μm size can be detected at sub-grain boundaries confirming their high pinning ability. However, detailed analysis revealed very few subtle particles in the S350 °C state (Figure 5b) which can be attributed to the initial stage of $L1_2$-Al_3Zr phase formation. This assumption meets well with the electrical conductivity data presented in Figure 4b. In contrast, for the S450 °C state an exceptionally high number density and uniform distribution of the $L1_2$-Al_3Zr phase nanoparticles are detected (Figure 5d,e). The result obtained suggests almost complete decomposition of the aluminum solid solution which converges well with data on the maximum electrical conductivity (Figure 4b) and hardness of the EMC rod (Figure 4a).

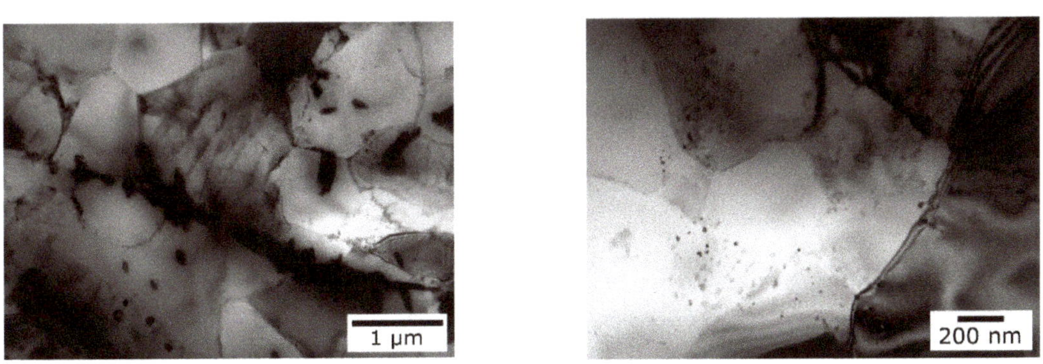

(**a**) (**b**)

Figure 5. *Cont.*

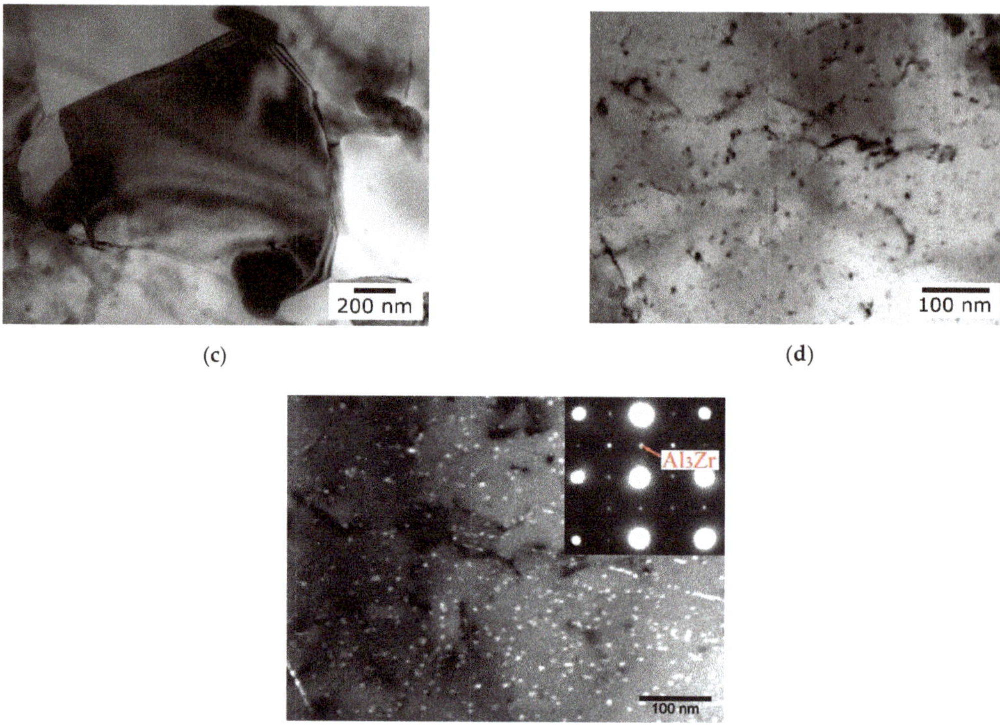

(c) (d)

(e)

Figure 5. TEM structure of wire alloy ACZ (strip) and after annealing according to (**a**,**b**) S350 °C and (**c**–**e**) S450 °C regimes, (**a**–**d**) bright field, (**e**) dark field and diffraction patterns.

3.3. Properties of Wire

Mechanical and electrical properties of the wire ACZ alloy in as-drawn and annealed states are given in Table 3. In the initial state, the alloy has a good combination of strength (UTS ~ 280 MPa and YS ~ 250 MPa) and ductility (El ~ 4%), however, the electrical conductivity is small (~43% IACS). Annealing of the 3 mm wire according to the modes given in Table 2 allows one to increase the EC value significantly. In this case, the EC values are approximately the same as for the cold rolled strip (Figure 4b). This suggests that the decomposition of (Al) proceeds in a similar way. As can be seen from Table 3, the experimental alloy in the W450 state has the best combination of strength (UTS ~ 200 MPa and YS ~ 180 MPa), elongation (El ~ 12%) and electrical conductivity (54.7% IACS).

Fractography of wire samples after a tensile test revealed a fine-dimpled ductile structure of the fracture surface (Figure 6). Calcium-bearing particles found inside the dimples (Figure 6a,b, BSE mode) are much smaller than the average diameter of the dimples. It should also be noted that no oxides or nonmetallic inclusions were observed which meets well with a previous study [22] confirming melt refining tendency when using EMC technology.

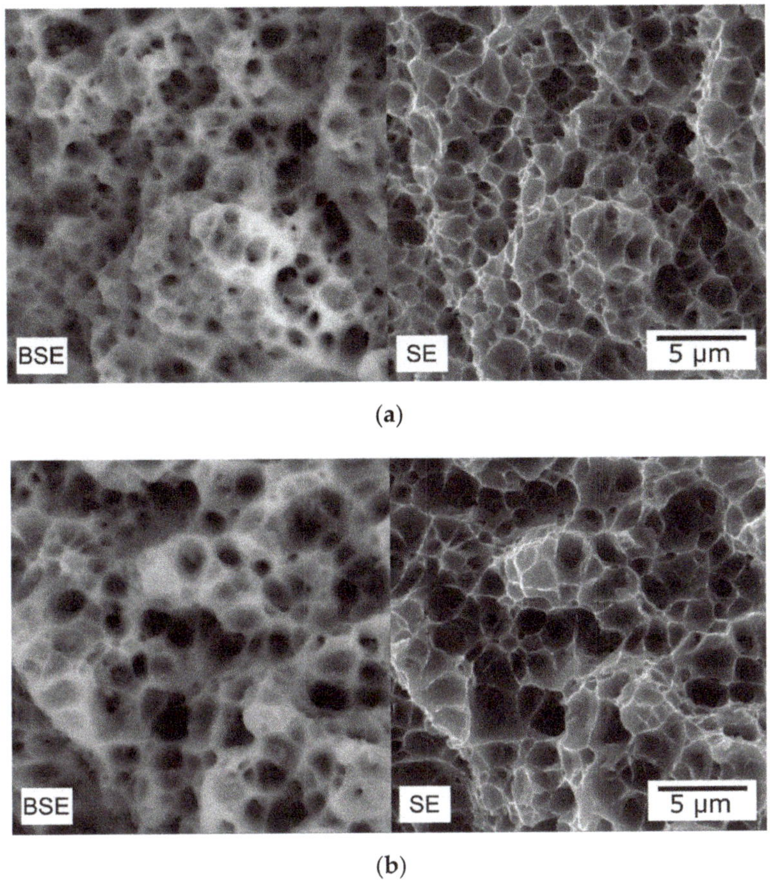

Figure 6. Fracture surfaces of experimental wire alloy (**a**) in initial (W) and (**b**) annealed (W450) states, SEM, **left**—backscattered electron image, **right**—secondary electron image, SEM.

4. Discussion

When comparing the basic characteristics of the new ACZ alloy and the previously studied Al-0.6% Zr-0.4% Fe-0.4% Si alloy [22] obtained under the same conditions, it can be seen that at close values of strength and electrical conductivity, the heat resistance of the Ca-containing alloy is significantly higher (450 vs. 400 °C) despite a smaller content of Zr (0.5 vs. 0.6 wt.%). Indeed, the hardness of the Ca-free alloy decreases down to 40 HV after annealing at 450 °C, which is significantly lower than that for the new alloy with calcium (Table 3). Obviously, the reason for the increase in the heat resistance is that the ACZ alloy contains calcium. Therefore, study of the effect of this element deserves special consideration. Considering that zirconium forms only the Al_3Zr phase (stable and metastable modifications), to understand the distribution of calcium, iron and silicon between phases, it is necessary to consider the Al–Ca–Fe–Si system. Earlier [26], the structure of this phase diagram in the range of Al-Ca alloys (concentration of Ca much higher than Fe and Si) was reviewed but other fields (with small Ca content) were not studied. Using the calculation in the Thermo-Calc software [31], as well as the results of previously published works [22,24–26,30,32] and additional experiments, we proposed structures of the Al–Ca–Fe–Si system in the aluminum corner including distribution of phases in the solid state (Figure 7a) and polythermal projection (Figure 7b). According to the proposed version, this system contains five four-phase regions: I—(Al) + Al_4Ca +

$Al_{10}CaFe_2$ + Al_2CaSi_2, II—(Al) + $Al_{10}CaFe_2$ + Al_2CaSi_2 + Al_3Fe, III—(Al) + Al_2CaSi_2 + Al_3Fe + Al_8Fe_2Si, IV—(Al) + Al_2CaSi_2 + Al_8Fe_2Si + Al_5FeSi, V—(Al) + Al_2CaSi_2 + Al_5FeSi + (Si). From the distribution shown in Figure 7a it follows that the Al_2CaSi_2 phase is present in all regions of this quaternary system. Considering the low solubility of Ca in (Al), this means that even at a small amount of calcium in Fe and Si containing alloys, the formation of this particular ternary compound is inevitable. Excess calcium should lead to the formation of the Al_4Ca and $Al_{10}CaFe_2$ phases. The influence of calcium on the equilibrium phase composition of the Al-0.5% Fe-0.25% Si alloy (i.e., with the same concentrations as in the ACZ alloy) is reflected in Table 4. The calculation results show that the phase composition of the base ternary alloy has a very high sensitivity to the calcium content. In this case, the ACZ alloy must certainly be in region I. However, for non-equilibrium solidification, the phase composition can differ greatly from the equilibrium one. This is largely due to the occurrence of incomplete peritectic reactions. As follows from the polythermal projection (Figure 7b), there are three invariant peritectic and three eutectic reactions in the aluminum corner of the Al–Ca–Fe–Si system. The compositions of the liquid phase and the temperatures of these reactions are given in Table 5.

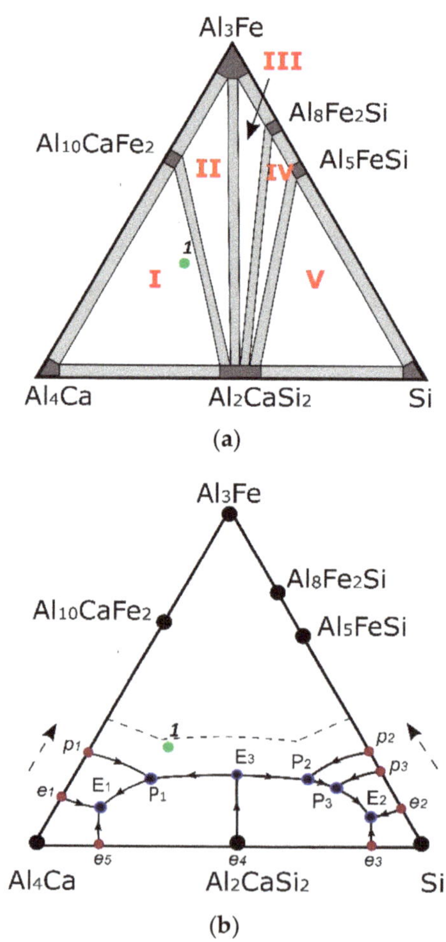

Figure 7. Phase diagram of Al–Ca–Fe–Si system in aluminum corner: (**a**) distribution of phases in in the solid state and (**b**) polythermal projection. Point **1** corresponds to the chemical composition of the ACZ alloy.

Table 4. Effect of Ca concentration on phase composition of alloy Al-0.5% Fe-0.25% Si (calculated at 200 °C).

Concentration of Ca, wt.%	Phase Field	Designation in Figure 7a
0–<0.01	(Al) + Al_8Fe_2Si + Al_5FeSi	–
<0.01–0.10	(Al) + Al_2CaSi_2 + Al_8Fe_2Si + Al_5FeSi	IV
~0.10	(Al) + Al_2CaSi_2 + Al_8Fe_2Si	IV/III
0.10–0.18	(Al) + Al_2CaSi_2 + Al_3Fe + Al_8Fe_2Si	III
~0.18	(Al) + Al_2CaSi_2 + Al_3Fe	III/II
0.18–0.38	(Al) + $Al_{10}CaFe_2$ + Al_2CaSi_2 + Al_3Fe	II
~0.38	(Al) + $Al_{10}CaFe_2$ + Al_2CaSi_2	II/I
>0.38	(Al) + Al_4Ca + $Al_{10}CaFe_2$ + Al_2CaSi_2	I

Table 5. Calculated parameters of invariant reactions in the Al–Ca–Fe–Si system.

Reaction	T, °C	Point in Figure 7b	Concentration in Liquid, wt.%			
			Ca	Fe	Si	Al
L→(Al) + Al_4Ca + $Al_{10}CaFe_2$ + Al_2CaSi_2	611	E_1	6.92	0.68	0.50	-
L→(Al) + Al_5FeSi + (Si) + Al_2CaSi_2	576	E_2	<0.01	0.84	12.48	86.68
L→(Al) + Al_3Fe + Al_2CaSi_2 [1]	632	E_3	1.52	1.61	2.24	94.63
L + Al_3Fe→(Al) + $Al_{10}CaFe_2$ + Al_2CaSi_2	~620	P_1	~6	~1	~1	~92
L + Al_3Fe→(Al) + Al_8Fe_2Si + Al_2CaSi_2	629	P_2	0.60	1.90	3.69	93.91
L + Al_8Fe_2Si→(Al) + Al_5FeSi + Al_2CaSi_2	612	P_3	0.07	1.60	7.10	91.23

[1] In quasi-ternary section.

As follows from the polythermal projection (Figure 7b), in a quaternary alloy containing 0.8% Ca, 0.5% Fe and 0.25% Si (point 1 in Figure 7b), after primary crystallization of (Al), eutectic reactions L→(Al)+Al_3Fe and L→(Al) + Al_3Fe + Al_2CaSi_2 (line E_3–P_1), and then peritectic L + Al_3Fe→(Al) + $Al_{10}CaFe_2$ + Al_2CaSi_2 (point P_1) one should proceed. The incompleteness of the latter explains the presence of needle-like particles in the microstructure of the slowly solidified ACZ alloy (Figure 2d). With an increase in the cooling rate, the phase boundaries shift towards higher iron content (dashed line in Figure 7b), therefore the Al_3Fe phase is not formed and it is absent in the as-cast microstructure of the EMC rod (Figure 2a,b). The solidification of this quaternary alloy (and hence the ACZ alloy) should end via eutectic reactions with the formation of three Ca-containing phases (point P_1). To further confirm the proposed structure of the Al–Ca–Fe–Si system, several quaternary alloys were annealed at 600 °C. As can be seen from Figure 8a, in the alloy containing 0.8% Ca, 0.5% Fe and 0.25% Si, only globular particles are detected, which, according to the distribution maps for Ca (Figure 8b), Fe (Figure 8c) and Si (Figure 8d) elements, can be identified as Al_4Ca, $Al_{10}CaFe_2$ and Al_2CaSi_2.

To confirm the effect of cold deformation on the electrical conductivity of the ACZ alloy mentioned above, additional studies on the effect of calcium on EC of both as-cast ingots and cold rolled sheets containing 0.5% Fe and 0.25% Si at varying calcium concentrations (Table 1) were carried out (Figure 9). As can be seen from Figure 9, the difference between the ingot and cold rolled sheet is small at small Ca content but at 0.75–1% Ca it reaches ~2 MS/m, i.e., similar to the ACZ alloy (Figure 4b). Taking into account the low solubility of Ca in (Al) and the invariability of the phase composition during deformation, this effect can probably be caused by the influence of the dislocation structure, vacancies and other defects of the crystal structure. This requires special study. A slight increase in EC should be noted with the addition of 0.1% Ca to the ternary alloy. This can be explained by a decrease in the concentration of Si in (Al) due to the formation of Al_2CaSi_2 compound.

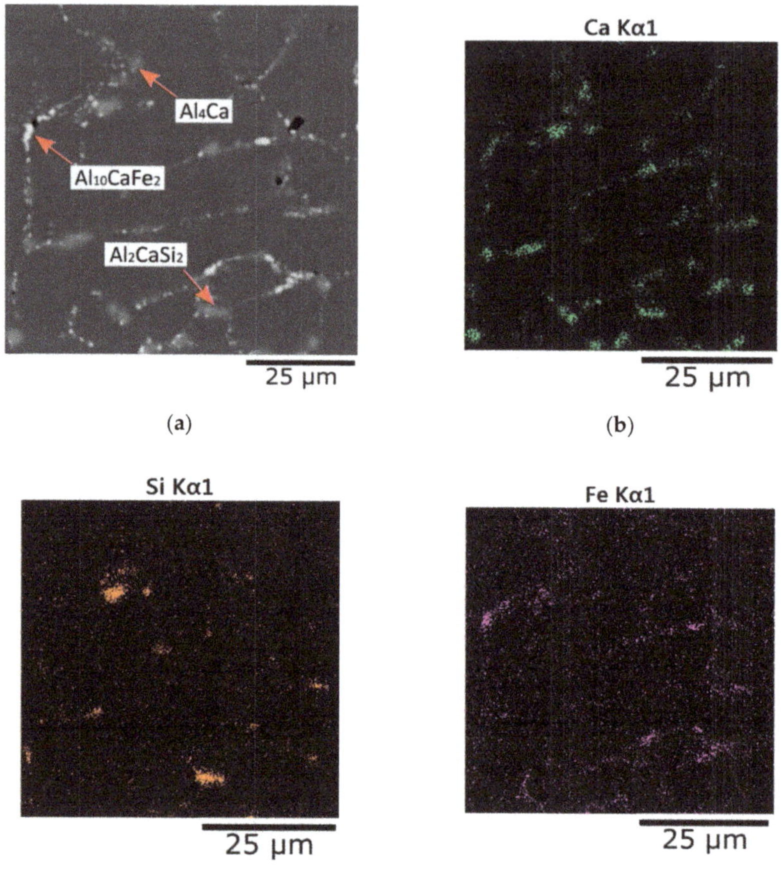

(**a**) (**b**) (**c**) (**d**)

Figure 8. Microstructure of alloy Al-1% Ca-0.5% Fe-0.25% Si (ingot 10 mm × 40 mm × 200 mm) after annealing at 600 °C: (**a**) SEM, (**b–d**) electron microprobe analysis (EMPA) mapping (**b**—Ca, **c**—Si, **d**—Fe).

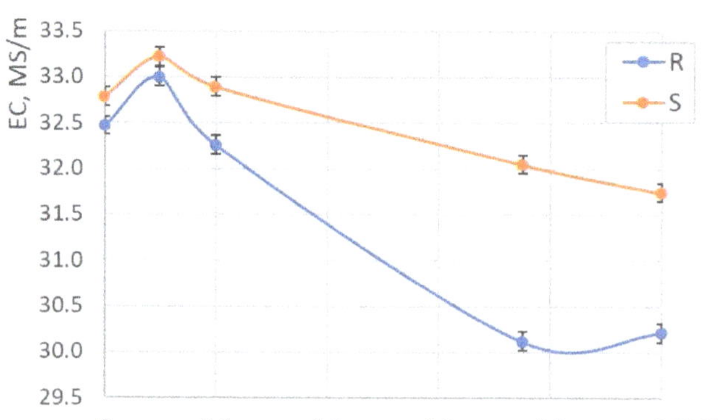

Figure 9. Electrical conductivity of Al–Ca–Fe–Si alloys containing 0.5% Fe and 0.25% Si vs. calcium content curves: R—EMC rod, S—cold rolled strip.

5. Summary

1. Experimental aluminum alloy containing 0.8% Ca, 0.5% Zr, 0.5% Fe and 0.25% Si (wt.%) in the form of a long-length rod 12 mm in diameter and ≈20 m in length was manufactured using an electromagnetic casting (EMC) technique. The extremely high cooling rate during alloy solidification (≈10^4 K/s) caused the formation of a favorable microstructure containing fine eutectic particles and full dissolution of Zr in Al solid solution. Due to the microstructure formed, the ingots possess high manufacturability for both cold drawing of a wire 3 mm in diameter and cold rolling of a strip in 2 mm in thickness.
2. EMC rod hardness reaches the highest value at 450 °C annealing temperature which is associated with the formation of the $L1_2$ type Al_3Zr phase precipitates with an average diameter of 10–20 nm and a high number density. Decomposition of aluminum solid solution during annealing also promotes the increase in the electrical conductivity (EC). Further increase in the annealing temperature leads to significant softening due to the coarsening of the Al_3Zr precipitation structure. Deformation hardening in the cold rolled strip can be maintained at up to 450 °C annealing, which is associated with the high pinning ability of zirconium and calcium containing phases: Al_4Ca, $Al_{10}CaFe_2$ and Al_2CaSi_2. Further increase in the annealing temperature leads to softening due to the recrystallization. The best combination of hardness and EC in the cold rolled strip was reached after annealing at 450 °C.
3. The general structure of Al–Ca–Fe–Si system in the Al corner (including distribution of phases in the solid state and polythermal projection) was proposed.
4. The experimental wire alloy had the best combination of strength (UTS = 200 MPa, YS = 180 MPa), electrical conductivity (54.8% IACS) and thermal stability (up to 450 °C) as compared with alloys on the basis of Al–Zr and Al–REM systems.
5. It was shown that the presence of calcium in the model alloy increases the electrical conductivity after cold forming operations (both drawing and rolling).

Author Contributions: Conceptualization, N.B.; investigation, N.K. and M.M.; methodology, V.T. and T.A.; supervision, N.B.; writing—original draft, N.B.; investigation, A.F., writing—review and editing, M.M., T.A, and N.K. All authors have read and agreed to the published version of the manuscript.

Funding: The authors gratefully acknowledge the financial support of the Ministry of Science and Higher Education of the Russian Federation in the framework of Increase Competitiveness Program of MISiS (No. P02-2017-2-10) (TEM, tensile tests, discussion) and the Russian Science Foundation (project No. 20-79-00239) (SEM, hardness and electrical conductivity tests, calculations).

Institutional Review Board Statement: Not applicable.

Informed Consent Statement: Not applicable.

Data Availability Statement: Data are available from the corresponding author on reasonable request.

Acknowledgments: The results were obtained by using the equipment of RPC Magnetic hydrodynamics LLC, Krasnoyarsk, Russia; Institute of Physics of Advanced Materials, Ufa State Aviation Technical University, Ufa, Russia; and Department of metal forming, National University of Science and Technology MISiS, Moscow, Russia.

Conflicts of Interest: The authors declare no conflict of interest.

References

1. Polmear, I.J. *Light Alloys. From Traditional Alloys to Nanocrystals*, 5th ed.; Butterworth-Heinemann: Oxford, UK, 2006; pp. 129–130. Available online: https://www.elsevier.com/books/light-alloys/polmear/978-0-08-099431-4 (accessed on 29 October 2020).
2. ASTM B941-16. *Standard Specification for Heat Resistant Aluminum-Zirconium Alloy Wire for Electrical Purposes*; ASTM International: West Conshohocken, PA, USA, 2016; pp. 1–4.
3. Brubak, J.P.; Eftestol, B.; Ladiszlaidesz, F. Aluminium Alloy, a Method of Making It and an Application of the Alloy. U.S. Patent 5,067,994, 26 November 1991. Available online: https://patents.google.com/patent/US5067994A/en?oq=5067994 (accessed on 4 November 2020).

4. Knych, T.; Jablonsky, M.; Smyrak, B. New aluminium alloys for electrical wires of fine diameter for automotive industry. *Arch. Metall. Mater.* **2009**, *54*, 671–676. Available online: https://www.researchgate.net/publication/263734063_New_aluminium_alloys_for_electrical_wires_of_fine_diameter_for_automotive_industry (accessed on 6 November 2020).
5. Orlova, T.S.; Mavlyutov, A.M.; Latynina, T.A.; Ubyivovk, E.V.; Murashkin, M.Y.; Schneider, R.; Gerthsen, D.; Valiev, R.Z. Influence of severe plastic deformation on microstructure strength and electrical conductivity of aged Al-0.4Zr (wt.%) alloy. *Rev. Adv. Mater. Sci.* **2018**, *55*, 92–101. [CrossRef]
6. Belov, N.A.; Alabin, A.N.; Matveeva, I.A.; Eskin, D.G. Effect of Zr additions and annealing temperature on electrical conductivity and hardness of hot rolled Al sheets. *Trans. Nonferrous Met. Soc. China* **2015**, *25*, 2817–2826. [CrossRef]
7. Fu, J.; Yang, Z.; Deng, Y.; Wu, Y.; Lu, J. Influence of Zr addition on precipitation evolution and performance of Al-Mg-Si alloy conductor. *Mater. Charact.* **2020**, *159*, 110021. [CrossRef]
8. Knipling, K.E.; Karnesky, R.A.; Lee, C.P.; Dunand, D.C.; Seidman, D.N. Precipitation evolution in Al–0.1Sc, Al–0.1Zr and Al–0.1Sc–0.1Zr (at.%) alloys during isochronal aging. *Acta Mater.* **2010**, *58*, 5184–5195. [CrossRef]
9. Deschamp, A.; Guyo, P. In situ small-angle scattering study of the precipitation kinetics in an Al–Zr–Sc alloy. *Acta Mater.* **2007**, *55*, 2775–2783. [CrossRef]
10. Lefebvre, W.; Danoix, F.; Hallem, H.; Forbord, B.; Bostel, A.; Marthinsen, K. Precipitation kinetic of Al_3(Sc,Zr) dispersoids in aluminium. *J. Alloys Compd.* **2009**, *470*, 107–110. [CrossRef]
11. Forbord, B.; Lefebvre, W.; Danoix, F.; Hallem, H.; Marthinsen, K. Three dimensional atom probe investigation on the formation of Al_3(Sc,Zr)-dispersoids in aluminium alloys. *Scr. Mater.* **2004**, *51*, 333–337. [CrossRef]
12. Clouet, E.; Barbu, A.; Lae, L.; Martin, G. Precipitation kinetics of Al_3Zr and Al_3Sc in aluminum alloys modeled with cluster dynamics. *Acta Mater.* **2005**, *53*, 2313–2325. [CrossRef]
13. Çadırl, E.; Tecer, H.; Sahin, M.; Yılmaz, E.; Kırındı, T.; Gündüz, M. Effect of heat treatments on the microhardness and tensile strength of Al–0.25 wt.% Zr alloy. *J. Alloys Compd.* **2015**, *632*, 229–237. [CrossRef]
14. Robson, J.D.; Prangnell, P.B. Dispersoid precipitation and process modelling in zirconium containing commercial aluminium alloys. *Acta Mater.* **2001**, *49*, 599–613. [CrossRef]
15. Gao, T.; Ceguerra, A.; Breen, A.; Liu, X.; Wu, Y.; Ringer, S. Precipitation behaviors of cubic and tetragonal Zr-rich phase in Al–Si–Zr alloys. *J. Alloys Compd.* **2016**, *674*, 125–130. [CrossRef]
16. Yea, J.; Guana, R.; Zhaoa, H.; Yinc, A. Effect of Zr content on the precipitation and dynamic softening behavior in Al–Fe–Zr alloys. *Mater. Charact.* **2020**, *162*, 110–181. [CrossRef]
17. Vlach, M.; Stulíková, I.; Smola, B.; Žaludová, N.; Černá, J. Phase transformations in isochronally annealed mould-cast and cold-rolled Al–Sc–Zr-based alloy. *J. Alloys Compd.* **2010**, *492*, 143–148. [CrossRef]
18. Belov, N.A.; Korotkova, N.O.; Alabin, A.N.; Mishurov, S.S. Influence of a silicon additive on resistivity and hardness of the Al–1% Fe–0.3% Zr alloy. *Russ. J. Non-Ferrous Met.* **2018**, *59*, 276–283. [CrossRef]
19. Knych, T.; Piwowarska, M.; Uliasz, P. Studies on the process of heat treatment of conductive AlZr alloys obtained in various productive processes. *Arch. Metal Mater.* **2011**, *56*, 687–692. [CrossRef]
20. Avdulov, A.A.; Usynina, G.P.; Sergeev, N.V.; Gudkov, I.S. Otlichitel'nyye osobennosti struktury I svoystv dlinnomernykh slitkov malogo secheniya iz alyuminiyevykh splavov, otlitykh v elektromagnitnyy kristallizator. [Distinctive features of the structure and properties of long ingots of small cross section from aluminum alloys cast in an electromagnetic mold]. *Tsvet. Met.* **2017**, *7*, 73–77. [CrossRef]
21. Dobatkin, V.I.; Elagin, V.I.; Fedorov, V.M. *Bystrozakristallizovannye Alyuminievye Splavy (Rapidly Solidified Aluminum Alloys)*; VILS: Moscow, Russia, 1995; pp. 43–59.
22. Belov, N.A.; Murashkin, M.Y.; Korotkova, N.O.; Akopyan, T.K.; Timofeev, V.N. Structure and properties of Al–0.6wt.%Zr(Fe,Si) wire alloy manufactured by direct drawing of electromagnetically cast wire rod. *Metals* **2020**, *10*, 769. [CrossRef]
23. Korotkova, N.O.; Belov, N.A.; Timofeev, V.N.; Motkov, M.M.; Cherkasov, S.O. Influence of heat treatment on the structure and properties of an Al–7% REM conductive aluminum alloy casted in an electromagnetic crystallizer. *Phys. Met. Metallogr.* **2020**, *121*, 173–179. [CrossRef]
24. Belov, N.A.; Akopyan, T.K.; Korotkova, N.O.; Naumova, E.A.; Pesin, A.M.; Letyagin, N.V. Structure and properties of Al-Ca(Fe, Si, Zr, Sc) wire alloy manufactured from as-cast billet. *JOM* **2020**, *72*, 3760–3768. [CrossRef]
25. Korotkova, N.O.; Belov, N.A.; Avxentieva, N.N.; Aksenov, A.A. Effect of calcium additives on the phase compositionand physicomechanical properties of a conductive alloy Al–0.5% Fe–0.2% Si–0.2% Zr–0.1% Sc. *Phys. Met. Metallogr.* **2020**, *121*, 95–101. [CrossRef]
26. Belov, N.A.; Naumova, E.A.; Akopyan, T.K.; Doroshenko, V.V. Phase diagram of the Al-Ca-Fe-Si system and its application for the design of aluminum matrix composites. *JOM* **2018**, *70*, 2710–2715. [CrossRef]
27. RPC Magnetic Hydrodynamics LLC. Available online: http://www.npcmgd.com (accessed on 10 November 2020).
28. IEC 60468:1974. *Method of Measurement of Resistivity of Metallic Materials*, 1st ed.; IEC: Geneva, Switzerland, 1974; pp. 1–31.
29. Bäckerud, L.; Chai, G.; Tamminen, J. *Solidification Characteristics of Aluminum Alloys. Vol. 1: Foundry Alloys*; Skanaluminium: Oslo, Norway, 1986; pp. 9–26.
30. Glazoff, M.V.; Khvan, A.V.; Zolotorevsky, V.S.; Belov, N.A.; Dinsdale, A.T. *Casting Aluminum Alloys. Their Physical and Mechanical Metallurgy*; Elsevier: Oxford, UK, 2019; pp. 180–192.

31. Thermo-Calc Software—Computational Materials Engineering. Available online: http://www.thermocalc.com (accessed on 1 December 2020).
32. Belov, N.A.; Eskin, D.G.; Aksenov, A.A. *Multicomponent Phase Diagrams: Applications for Commercial Aluminum Alloys*; Elsevier: Amsterdam, The Netherlands, 2005; pp. 19–31.

metals MDPI

Article

Effect of La Addition on Solidification Behavior and Phase Composition of Cast Al-Mg-Si Alloy

Vladislav Deev [1,2], Evgeny Prusov [3,*], Pavel Shurkin [2], Ernst Ri [4], Svetlana Smetanyuk [5], Xizhang Chen [6,7] and Sergey Konovalov [6,7]

1 School of Mechanical Engineering and Automation, Wuhan Textile University, Wuhan 430073, China; deev.vb@mail.ru
2 Department of Metal Forming, National University of Science and Technology MISiS, 119049 Moscow, Russia; p.shurkin@misis.ru
3 Department of Functional and Constructional Materials Technology, Vladimir State University Named after Alexander and Nikolay Stoletovs, 600000 Vladimir, Russia
4 Department of Foundry Engineering and Metal Technology, Pacific National University, 680042 Khabarovsk, Russia; erikri999@mail.ru
5 Department of Foundry, Siberian State University, 660041 Krasnoyarsk, Russia; smetanyuk.sv@mail.ru
6 College of Mechanical and Electrical Engineering, Wenzhou University, Wenzhou 325035, China; chenxizhang@wzu.edu.cn (X.C.); ksv@ssau.ru (S.K.)
7 Department of Metals Technology and Aviation Materials, Samara National Research University, 443086 Samara, Russia
* Correspondence: eprusov@mail.ru; Tel.: +7-4922-479-821

Received: 3 November 2020; Accepted: 11 December 2020; Published: 14 December 2020

Abstract: The current study focusses on the phase composition, solidification path, and microstructure evaluation of gravity cast Al-4Mg-0.5Si-xLa aluminum alloy, where x = 0, 0.1, 0.25, 0.5, 0.75, and 1 wt.% La. A computational CalPhaD approach implemented in Thermo-Calc software and scanning electron microscopy technique equipped with electron microprobe analysis (EMPA) was employed to assess its above-mentioned characteristics. The thermodynamic analysis showed that the equilibrium solidification path of La-containing Al-Mg-Si alloys consists of only binary phases $LaSi_2$ and Mg_2Si precipitation along with α-Al from the liquid and further solid-state transformation of this mixture into α-Al + $Al_{11}La_3$ + Mg_2Si + Al_3Mg_2 composition. Scheil–Gulliver simulation showed a similar solidification pathway but was accompanied by an increase in the solidification range (from ~55 °C to 210 °C). Furthermore, microstructural observations were congruent with the calculated fraction of phases at 560 °C and related to α-Al + $LaSi_2$ + Mg_2Si three-phase region in terms of formation of La-rich phase having both eliminating effect on the eutectic Mg_2Si phase. Quantitative EMPA analysis and elemental mapping revealed that the La-rich phase included Al, La, and Si and may be described as Al_2LaSi_2 phase. This phase shows a visible modifying effect on the eutectic Mg_2Si phase, likely due to absorbing on the liquid/solid interface.

Keywords: cast aluminum alloy; Al-Mg-Si-La system; lanthanum; Mg_2Si; eutectic modification; microstructure; solidification; rare earths; intermetallics

1. Introduction

Recently, many studies showed that micro-alloying with rare earth (RE) elements may have remarkable effects related to refining of the as-cast structure, thus making them efficient modification agents for aluminum alloys [1–5]. Among the Light Rare-Earth Elements (LREE), also known as the cerium group (Sc, La, Ce, Pr, Nd, Pm, Sm, Eu, and Gd), lanthanum and cerium are the most abundant in the Earth's crust (31 ppm and 63 ppm, respectively) and the least expensive [6]. According to binary Al-La and Al-Ce systems [7], these elements both provide a formation of eutectic reactions,

L→α-Al + $Al_{11}La_3$ (11.7 wt.% La, 640 °C) and L→α-Al + $Al_{11}Ce_3$ (12.2 wt.% Ce, 621 °C). In this respect, several previous studies related to the design of alternative cast alloys for high-temperature applications are published [8–10]. Moreover, both La and Ce may show de-gasifying and de-slagging performance, thereby improving the quality of cast products [11]. Despite affinities between La and Ce, the latter caused some contradictory results in terms of modifying effects on the structure. For example, the work [12] on 0.12 wt.% Ce-modified Al-Zn-Mg-Cu alloy reported that nucleation growth of α-Al on one of the crystal faces of the $Al_{11}Ce_3$ in the melt is very efficient upon solidification, also supported by a recent review [11]. On the other hand, on comparing the effect of 0.1–0.2 wt.% of La and Ce on the structure of the 6xxx alloy [13], La showed far better modifying ability compared to Ce, which was inefficient in this respect. Additionally, Ce showed a more detrimental effect in porosity formation because of greater oxidation tendency when compared to La [14]. Meanwhile, La is ubiquitously used as an addition for achieving favorably fine α-Al [15,16], eutectic silicon [1], Fe-bearing phases [5,13], as well as Mg_2Si phase [17,18], one of the main structural components in 511 type cast aluminum alloys studied in this paper.

When considering grain refining function, 0.03–0.2 wt.% La suppresses the growth of α-Al grains via enriching upon their solidification front. Moreover, it was observed that La may react with other modifying elements such as Ti and V, thus bonding with $Al_{20}(Ti,V)_2La$ inter-metallic and acting as heterogeneous nucleation sites [19]. However, the research data on the modifying effect of La on the eutectic Mg_2Si phase is very limited. Most papers are focused on the primary Mg_2Si phase playing a reinforcing function in aluminum matrix in-situ composites [18,20–22]. Research on Mg-5Si alloy [17] has reported the segregation of La on the growth front of the primary Mg_2Si phase, thus changing its surface energy by lattice distortion, poisoning the growth steps, and suppressing the directional growth of the primary phase. In Al-based alloys, not only the surface activity of La was confirmed [18], but also it was shown that La atom clusters may act as effective nuclei due to the similarity in the crystal structure of La and Mg_2Si phase.

The above-stated research shows that RE La may serve as a multi-modifying agent for α-Al and intermetallics. However, regarding commercial cast Mg-rich Al-Mg-Si alloys, the available research does not give an ambiguous explanation of its effect on the solidification path and morphology of the eutectic structure. Due to these limitations, the present work plans to discover microstructure evolution and solidification behavior of the gravity-cast Al-4%Mg-0.5%Si aluminum alloy (511 type) after the addition of the different amounts of La (0, 0.1, 0.25, 0.5, 0.75, and 1 wt.%).

2. Materials and Methods

In terms of chemical composition, the experimental alloys studied in this work corresponded to the standard cast aluminum alloy of 511 grade (Aluminum Association, Arlington, VA, USA) with various additions of lanthanum. Alloys of a nominal composition Al-4Mg-0.5Si-xLa, where $x = 0$, 0.1, 0.25, 0.5, 0.75, and 1 wt.% La, were prepared from pure materials Al (99.99%), Mg (99.9%), Si (99%), and La (99.95%). Melting was carried out in a 500 g capacity alundum crucible, using a vertical electric resistance furnace in an air atmosphere without the addition of protective gas. The melt temperature was kept at 750 °C for each alloy. After the melting of the base Al-Mg-Si alloy, foil-wrapped lanthanum was added and mixed using a graphite stick down to its completed dissolution. Then, the molten metal was held for 10 min for its homogenization, skimmed, stirred, and poured at the temperature of 720–740 °C into a metal mold of Ø20 mm × 100 mm in size. The cooling conditions provided a dendrite cell size (d) of approximately 20 μm and, hence, the cooling rate (V_c) of about 10^2 °C/s, as it was estimated by well-known dependency $V_c = (A/d)^{1/n}$ [23]. The chemical composition determined by spectral analysis and the calculated phase composition of the experimental alloys is shown in Table 1.

Table 1. Chemical and phase composition (at 20 °C) of the experimental alloys.

Designation	Concentrations, wt. %							
	Elements [1,2]				Phases [3]			
	Mg	Si	Fe	La	α-Al	$Al_{11}La_3$	Mg_2Si	Al_3Mg_2
L0	4 (3.85)	0.5 (0.58)	0.12	0	91.03	0	1.36	7.60
L1	4 (4.11)	0.5 (0.49)	0.09	0.10 (0.12)	90.86	0.17	1.36	7.60
L2	4 (4.21)	0.5 (0.39)	0.06	0.25 (0.32)	90.59	0.44	1.36	7.60
L3	4 (4.14)	0.5 (0.45)	0.04	0.50 (0.57)	90.17	0.86	1.36	7.61
L4	4 (3.91)	0.5 (0.41)	0.12	0.75 (0.81)	89.72	1.30	1.36	7.61
L5	4 (3.89)	0.5 (0.43)	0.11	1.00 (1.06)	89.30	1.71	1.36	7.62

[1] Nominal and actual (in brackets) composition; [2] Al—balance; [3] calculated based on the nominal composition.

The solidification paths and phase compositions of the experimental alloys were investigated using the Thermo-Calc software (Version 3.1, TCAl4 Al-based alloy database, Thermo-Calc Software AB, Stockholm, Sweden) [24]. Single point equilibrium, phase diagram, property diagram, and Scheil–Gulliver solidification simulation options were used.

The microstructure was examined by optical microscopy (OM, Axio Observer MAT, Carl Zeiss Microscopy GmbH, Oberkochen, Germany), scanning electron microscopy (SEM, TESCAN VEGA3, Tescan Orsay Holding, Brno, Czech Republic) with an electron microprobe analysis system (EMPA, Oxford Instruments plc, Abingdon, UK), and the Aztec software (Version 3.0, Oxford Instruments plc, Abingdon, UK). The metallographic samples were ground with SiC abrasive paper and polished with 1 μm diamond suspension. A 1% hydrogen fluoride (HF) water solution was used for etching.

3. Results and Discussion

3.1. Thermodynamic Prediction

The polythermal section displayed in Figure 1a shows the solidification behavior of the experimental alloys. As can be seen, La had no significant influence on the equilibrium phase composition and transformation temperatures. In this respect, despite the high melting point of the element, alloying with La did not aggravate industrial compatibility of the base alloy since the liquidus temperature, the key factor for energy consumption in melting, was relatively low (635–637 °C), thus starting the solidification from the α-Al phase. Moreover, as can be seen from the magnified section (Figure 1b), the liquidus line goes down slightly along with the area related to the formation of the α-Al phase, likely due to nearness to the eutectic point (4.5 wt.% La, 625 °C) adjoining to the area for $LaSi_2$ primary phase appearance. The liquidus projection shown in Figure 1c reveals that the region for undesirable primary intermetallic crystallization was highly beyond the experimental concentrations of La and Si.

The equilibrium solidus decreased with an increase in La concentration but remained relatively high (578–583 °C). It should be noted that the TCAl4 database does not consider solubility of Al in the $LaSi_2$ phase, i.e., the existence of the ternary Al_2LaSi_2 found in the study [25], though a wide composition range of this phase was reported previously and showed a wide homogeneity in Al content [26]. Since this work does not aim to discover new intermetallics, possible La- and Si-rich phases will be referred to as AlLaSi phase. This may bring some limitations for reliable predicting, but some important conclusions can be deduced due to the depletion of the aluminum melt with Si ultimately being considered. The experimental alloys complete their solidification in the three-phase region α-Al + $LaSi_2$ + Mg_2Si. However, the calculation shows that further transformations may proceed in the solid-state with the consistent formation of $Al_{11}La_3$ phase instead of the $LaSi_2$ phase and Al_3Mg_2 phase. Solid-state diffusion of La in α-Al is negligible, and this transformation can probably be suppressed, which will lead to the formation of as-cast structure α-Al + $LaSi_2$ + Mg_2Si (near-equilibrium solidus region).

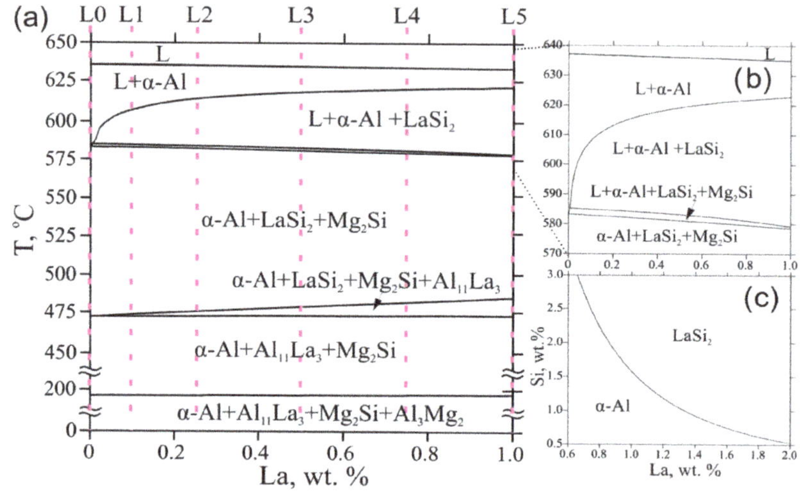

Figure 1. (**a**) Polythermal section of the Al-Mg-Si-La system at 4 wt.% Mg and 0.5 wt.% Si; (**b**) magnified polythermal section showing the change in liquidus and solidus temperatures; (**c**) liquidus projection of the Al-Mg-Si-La system at 4 wt.% Mg.

The equilibrium phase mass fractions in the L1 (0.1 wt.% La) and L5 (1 wt.% La) alloys are plotted in Figure 2a. Generally, in the temperature range 0 to 485 °C, La does not influence the amount of Mg_2Si and Al_3Mg_2 phases (1.36 and 7.6 wt.% respectively) but sufficiently changes the amount of the $Al_{11}La_3$ phase (from 0.17 wt.% to 1.71 wt.%, respectively). It can be assumed that the fractions of phases in the as-cast state may correspond to the three-phase region α-Al + $LaSi_2$ + Mg_2Si due to low solubility of La in α-Al during the suppressed transformation of α-Al + $LaSi_2$ + Mg_2Si into α-Al + $Al_{11}La_3$ + Mg_2Si. It is profoundly visible from Figure 2b that in comparison to the plot at 20 °C, at 560 °C (corresponded to α-Al + $LaSi_2$ + Mg_2Si region near-equilibrium solidus) the higher La content leads to the higher fraction of the $LaSi_2$ phase and the lower fraction of the Mg_2Si phase. This relationship achieves equality (0.5 wt.%) at about 0.48 wt.% La after which the ternary phase is the dominating phase (e.g., 1.2 wt.% $LaSi_2$ vs. 0.1 wt.% Mg_2Si in L5 alloy).

The considered equilibrium solidification analysis is very helpful for predicting of phase composition of the alloys. However, the actual solidification always occurs in non-equilibrium conditions, implying suppressed diffusion and termination after precipitation of the lowest melt-point phase. The non-equilibrium solidification path was plotted according to the Scheil–Gulliver model ($D_{Liquid} = \infty$, $D_{Solid} = 0$, where D is a diffusion coefficient). The non-equilibrium solidification terminates in a phase region α-Al + $Al_{11}La_3$ + Mg_2Si + Al_3Mg_2, where the breakup of the $LaSi_2$ phase is also considered (Figure 3). It has already been stated that La slightly decreases liquidus temperature, but it was also observed to have a quite significant influence on the non-equilibrium solidus as well. In comparison to the equilibrium solidification range, the non-equilibrium one for all alloys was approximately fourfold higher. Since the formation of the $LaSi_2$ phase in alloys L1–L5 promotes a reduction in Mg_2Si and increase in free Mg, it bonds with Al into Al_3Mg_2 phase, which solidifies at a relatively low temperature (450 °C for the base L0 alloy vs. 424 °C for other ones, see in Table 2). Accordingly, this may aggravate casting properties due to the wider solidification range. Moreover, the curves show a reduction in a region for primary α-Al formation as a result of La addition (down to halvation at 1 wt.% La), as well as strong domination of La-containing phases. On the one hand, this result seems to be promising due to the increase in eutectic fraction, possibly providing better casting properties and possible heterogeneous nucleation of α-Al on the eutectic particles' surfaces. Nevertheless, on the other hand, this may cause inadequate growth of the excessive brittle intermetallics, especially the $LaSi_2$ phase nucleating before other ones, adversing ductility, and fracture toughness. To gain further insight into the La effect on the structure formation, the as-cast state samples will be studied in the next sections.

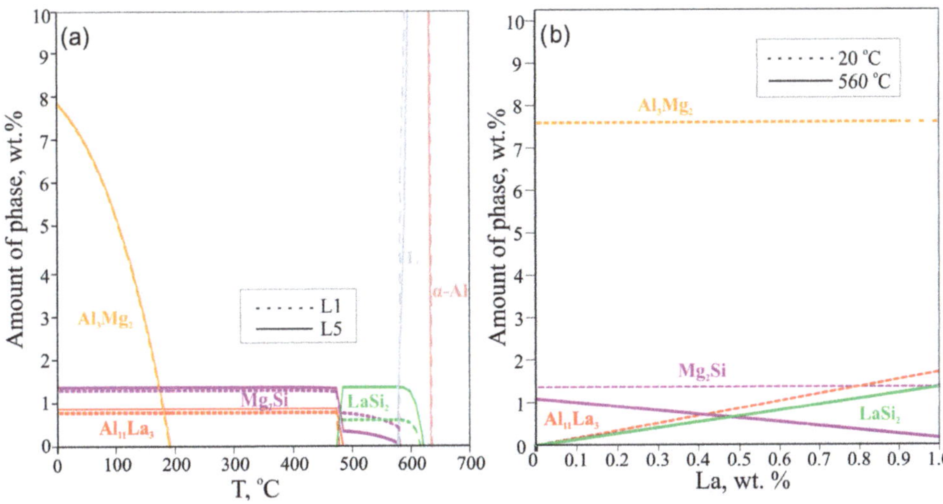

Figure 2. Calculated phase fractions for L1 and L5 alloys as a function of temperature (**a**); calculated phase fractions in Al—4 wt.% Mg—0.5 wt.% Si alloy for 20 °C and 560 °C as a function of La content (**b**).

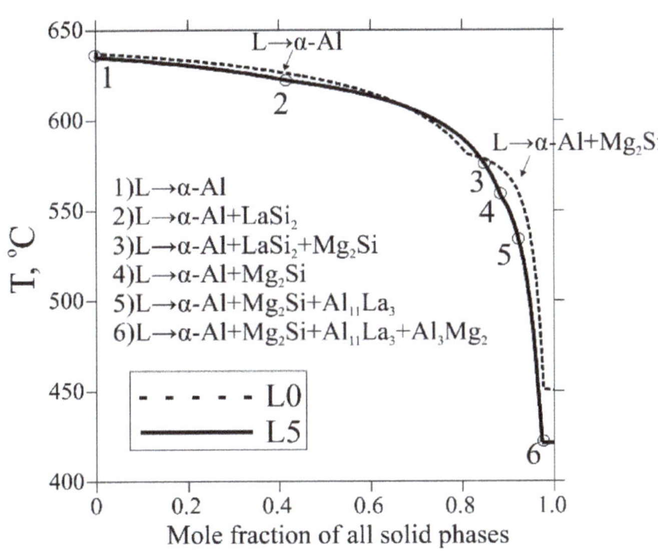

Figure 3. Calculated non-equilibrium solidification curves of the L0 and L5 alloys according to Scheil–Gulliver simulation.

Table 2. Calculated critical solidification temperatures of the experimental alloys.

Temperature, °C	Alloy					
	L0	**L1**	**L2**	**L3**	**L4**	**L5**
Liquidus	637.5	637.1	636.5	635.6	635.1	634.6
Solidus	583.3	582.9	582.1	581.4	579.5	578.3
Solidification range	54.2	54.2	54.4	54.2	55.6	56.3
Non-equilibrium solidus	450	424	424	424	424	424
Non-equilibrium solidification range	187.5	213.1	212.5	211.6	211.1	210.6

3.2. *Microstructural Evolution*

To understand the microstructure and phases of the Al-Mg-Si-xLa alloys the backscattered SEM images are presented in Figure 4. Overall, the microstructures include gray α-Al dendrite arms and intermetallic particles of various shapes and colors in the surroundings. The Mg_2Si phase, appeared in black color and was quite visible in all the images. Other phases mostly contained La and some traces of Fe impurity, elements with a high atomic number, which were presented as bright inclusions. The composition of different phase-referred areas, determined by EMPA, are summarized in Table 3. As is shown, there was no appearance of Al_3Mg_2 in La-rich alloys due to dissolution of Mg in α-Al.

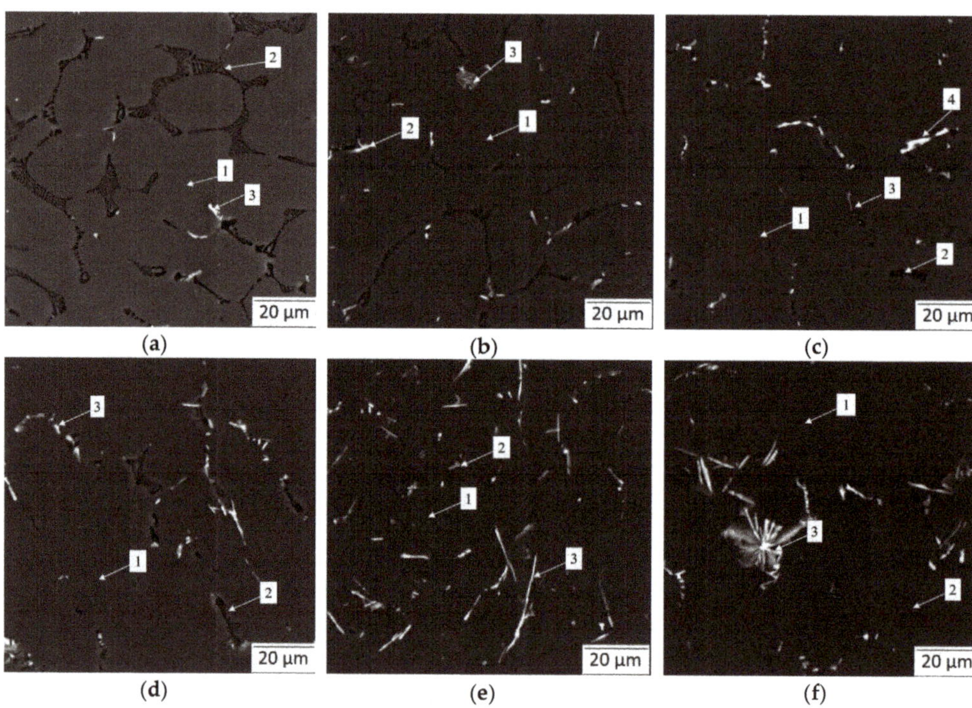

Figure 4. Backscattered SEM micrographs showing as-cast microstructural evolution of Al—4 wt.% Mg—0.5 wt.% Si alloy with increasing in La content: (**a**) L0 alloy; (**b**) L1 alloy; (**c**) L2 alloy; (**d**) L3 alloy; (**e**) L4 alloy; (**f**) L5 alloy.

Table 3. Chemical composition of the microstructural components in experimental alloys according to electron microprobe analysis (EMPA) [1].

Referenced Alloy	Referenced Specter	Concentrations, wt.%				Phase
		Mg	Si	Fe	La	
L0	1	1.83	0.06	0.04	0	α-Al
	2	10.37	6.12	0.04	0	α-Al + Mg_2Si
	3	3.55	0.31	11.34	0	α-Al + $Al_{13}Fe_4$
L1	1	2.88	0.02	0	0	α-Al
	2	3.43	0.20	18.21	0	α-Al + $Al_{13}Fe_4$
	3	6.63	0.35	0.06	15.43	α-Al + AlLaSi
L2	1	3.69	0	0.02	0.04	α-Al
	2	10.88	11.83	0	1.36	α-Al + Mg_2Si
	3	10.22	0.17	9.74	2.11	α-Al + $Al_{13}Fe_4$
	4	5.97	1.53	0.06	33.04	α-Al + AlLaSi
L3	1	3.70	0	0	0	α-Al
	2	4.79	1.98	0.10	0.13	α-Al + Mg_2Si
	3	2.46	9.67	0	42.63	α-Al + AlLaSi
L4	1	5.36	0.07	0	0.21	α-Al
	2	9.66	0.03	9.86	0.74	α-Al + $Al_{13}Fe_4$
	3	3.20	8.09	0.15	35.79	α-Al + AlLaSi
L5	1	2.32	0.01	0.07	0	α-Al
	2	10.57	4.62	0.09	0.04	α-Al + Mg_2Si
	3	2.89	4.33	0.02	28.49	α-Al + AlLaSi

[1] See referenced alloys microstructure images and referenced specters in Figure 4.

Following the base L0 alloy as a reference (Figure 4a), a relatively high volume of lamellar and needle-like Mg_2Si phase, and bright small inclusions of Fe-containing phase (likely, $Al_{13}Fe_4$ phase, according to calculation) were recorded. The latter was very fine and of particular low proportion compared to the Mg-rich phase. Apart from the formation of the Fe-rich phase, it is remarkable how precisely the microstructure agrees with Scheil–Gulliver prediction.

Observations of the La-containing alloys show significant changes in intermetallic shapes. The addition of 0.1 wt.% La resulted in thinning of the Mg_2Si phase branches, though they retained lamellar morphology. Meanwhile, some traces of La-rich phase resembled the AlLaSi compound due to the prominent presence of Si and some amount of Mg possibly captured during spectral emission (Table 3). Besides, having exhibited the $Al_{13}Fe_4$ phase, it was complicated to distinguish two binary intermetallics (Fe- and La-rich phases). However, the AlLaSi phase was far more differentiated, while the $Al_{13}Fe_4$ phase was mostly of a small blocky shape. The appearance of the AlLaSi phase becomes more evident with an increase in La due to thicker branches of intermetallics and less capturing of Mg-rich α-Al phase coupled by more evident Si content. Strikingly, the addition of 0.25 wt.% La almost eliminated the presence of the Mg_2Si phase. As is seen from the microstructure of the L2 alloy (Figure 4c), the AlLaSi phase is located in the vicinity of a very fine eutectic Mg_2Si phase. This partially confirms the absorbing effect of the La-rich phase. This effect may have suppressed the anisotropic growth of the Mg_2Si phase inhibiting its shape factor. On the other hand, since the amount of the Mg-rich phase is visibly low, the thermodynamically predicted relationship between the volume of AlLaSi and Mg_2Si is confirmed by the presence of a higher amount of a AlLaSi phase and a lower amount of the Mg_2Si phase. On further investigation of different alloys, the incredible growth of the AlLaSi phase and evolution of the Mg_2Si phase down to submicron in thickness, accompanied by an increase in Mg solubility in α-Al (e.g., from 1.83 wt.% in the L0 alloy vs. 5.36 wt.% in the L4 alloy) was noted. Under the further increase in La content, the shape of the AlLaSi phase progresses dramatically from acicular in the L3 alloy (Figure 4d) up to needle-like the L4 alloy (Figure 4e) and to coarse star-like shape in L5 alloy intrinsic for $Al_{13}Fe_4$ phase in Fe-rich aluminum alloys. This advancement is accompanied by a prominent change in morphology of the Mg_2Si phase, which became both slender and vermicular in the L5 alloy microstructure.

In addition, the microstructural analysis revealed that almost all the microstructures included Fe-rich intermetallics ($Al_{13}Fe_4$). Generally, they must have a commonly adverse needle-like morphology in most alloys, even at a low concentration of Fe. It is of particular note that in the experimental alloys, $Al_{13}Fe_4$ shows a short flake-like shape at a relatively high Fe content (up to 0.12 wt.%). Indeed, it was reported that La may hamper the orientation growth of the $Al_{13}Fe_4$ phase in a similar manner as of Mg_2Si phase [5]. However, La content of more than 0.25 wt.% brings degradation and formation of very coarse acicular intermetallics related to the AlLaSi phase, which may act as stress raisers, thus being detrimental to mechanical properties.

It should be taken into account that quantitative EPMA analysis produces a misleading backlighting effect due to the capturing elemental emissions from nearby areas. However, an EMPA mapping of the representative L4 alloy (Figure 5) is strongly consistent with previous thermodynamic predictions and microstructural investigations. The existence of three types of microstructural components was noted, and the corresponding elemental correlations provide a reliable qualitative explanation of their chemical composition. It was found that Mg (red-color in the maps) is distributed mostly in the region related to the α-Al phase, advocating its dissolution. Its presence in regions adjacent to La-bearing phases (turquoise-color in the maps) may also be observed, which is probably related to Mg micro-segregation enriching to the edges of the α-Al phase dendrite cells because of non-equilibrium solidification conditions [23]. Besides, Mg along with Si (deep-green-color in the maps) are incorporated into the Mg_2Si phase of worm-like shape. Ultimately, while Fe is bonded into unevenly distributed fine particles of less than 3 μm in size, the most dominating microstructural component is a very coarse AlLaSi phase revealed by correlation among Al, Si, and La enriched regions.

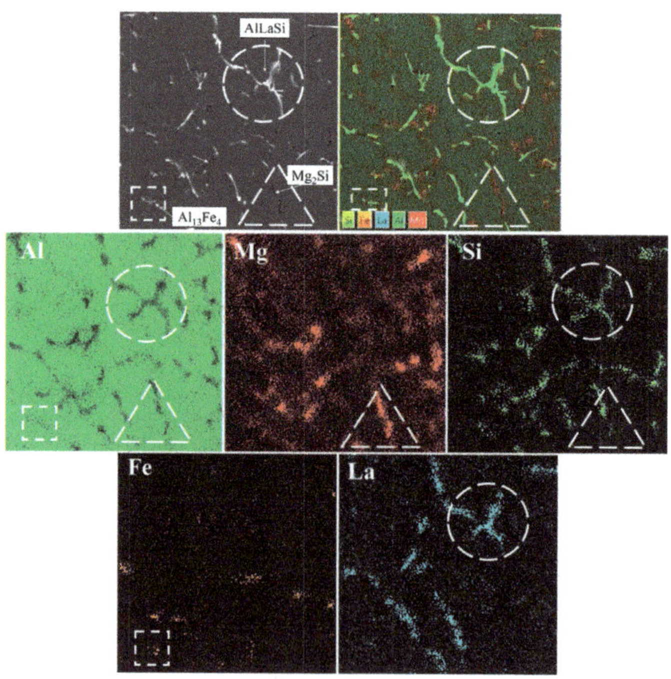

Figure 5. EMPA mapping (Al, Mg, Si, Fe, La) of the phases presented in the microstructure of the L4 alloy.

4. Discussion

Firstly, for explaining the absence of thermodynamically predicted formation of the Al_3Mg_2 phase and uncompromised formation of the Mg_2Si phase, the solidification behavior of ternary Al-Mg-Si alloys will be considered. According to the binary Al–Mg phase diagram, the solubility of Mg in solid aluminum is about 17%. According to [27], the binary alloys with less than 3.5 wt.%Mg exhibit a single-phase microstructure containing only α-Al. However, the presence of the Al_8Mg_5 phase (also known and notified in this study as Al_3Mg_2) may be inhibited at a Mg content of more than 5 wt.% as a result of an increase in cooling rate due to deceleration of Mg enrichment in the melt during the final solidification stage [28]. Furthermore, the addition of silicon greatly reduced the solid solubility of Mg in α-Al. Conversely, the solid solubility of Si is reduced under the appearance of Mg (e.g., 1.17 wt.% Mg and 0.68 wt.% Si at the quasibinary line in Al-Mg-Si diagram [29]). Under these conditions, they both segregate to the front of solid/liquid interface, concentrating together in the final solidification zone, thus promoting the formation of the Mg_2Si phase without alternative.

Secondly, for considering La-containing alloys, it was substantiated that they complete solidification in the α-Al + AlLaSi + Mg_2Si region that, in turn, contradicts with a thermodynamic calculation showing the formation of only binary phases. Whilst the inhibition of the Al_3Mg_2 phase seems to be clear, the formation of the AlLaSi phase is under discussion. However, the latter has received a lot of attention recently by the discovery of the new phases ($AlSi_2La$, Al_2Si_2La, and $(Al_{1-x}Si_x)_2La$ ($0.075 \leq x \leq 0.18$) in Al-Si-La system [26], the results are still not implemented in describing the solidification of commercial Al-Mg-Si alloys modified with La. Actually, the TCAl4 database, implemented into Thermo-Calc software, likely uses thermodynamic data regarding Al-La, La-Si, and Mg-La due to the lack of thermodynamic optimization of the Al-Mg-Si-La quaternary diagram even for the novel advanced TCAl7 database [30]. The formation of aluminides was observed in this study. The interaction between Mg and La in La-modified Al-4%Mg-0.5%Si alloys may not be anticipated. Not only the binary LaMg phase but also the ternary $(Al,Mg)_3La$ was observed in Mg-Al-La alloys [31]. However, since we showed a substantial Mg-enrichment of α-Al out of solidification and the clear presence of Mg_2Si phases at up to 1% La, the Mg content is negligible for providing the formation of Mg-La phases. Studies also supported this and revealed such phases only in Mg alloys or Al alloys with more than 30% Mg content [32]. Besides, the study [33] reported an influence of isomorphic La and Ce on Al-3%Mg alloy and found only $Al_{11}La_3$ and $Al_{11}Ce_3$ phases.

For an explanation of intermetallic formation, we discovered very extensive recent efforts on Si-alternative eutectic-forming elements (Ce, Ni, Ca, La, etc.). Meanwhile, there are similar compounds, Al_2CeSi_2 [1–3] and Al_2CaSi_2 [34], observed in Al-Si alloys that show a needle-like shape similar to the AlLaSi phase in the present work. Based on this, it is most likely that the composition of the observed AlLaSi phase is the same as of Al_2LaSi_2, corresponding to the same space group P3m1 as the ternary Al_2CeSi_2 and Al_2CaSi_2 [26]. In regards to RE Ce, current investigations on phase equilibria are focused on Si-rich Al-Mg-Si alloys solidifying under quaternary eutectic reaction L→α-Al + (Si) + Al_2LaSi_2 + Mg_2Si [35]. Either way, there is no appropriate research on the solidification of Mg-rich Al-Mg-Si with REM. In contrast, the efforts [36] toward the quaternary Al-Ca-Mg-Si diagram show a very similar phase composition at room temperature (Al_3Mg_2 + Al_4Ca + Mg_2Si) but also consider a formation of hexagonal Al_2CaSi_2 phase. It was also reported that peritectic reaction L + Al_2CaSi_2→α-Al + Al_4Ca + Mg_2Si is suppressed. In this respect, RE La may act as an alternative to Si because it is a common eutectic-forming element. It promotes the reduction of both in equilibrium liquidus and solidus, whilst the non-equilibrium solidus remains constant under an increase in La content. Therefore, we assumed that equilibrium invariant peritectic reaction L + AlLaSi→α-Al + Al_{11}La3 + Mg_2Si likely exists supported by Al depletion of the melt due to its solubility in AlLaSi phase and negligible in fraction (see point 3 in Figure 3) transition area L→α-Al + $LaSi_2$(or AlLaSi) + Mg_2Si. This reaction is likely to be suppressed, leading to the formation of the AlLaSi in the as-cast structure along with phases α-Al and Mg_2Si.

Finally, a substantial effect of La on the eutectic Mg_2Si phase was shown, notwithstanding the adverse degradation of the La-rich intermetallics at more than 0.5 wt.% La. Generally, the modifying effect may be described by heterogeneous nucleation and surface absorption. A study [18] reported the modification of the primary Mg_2Si phase, referring to small (3.72%) lattice misfit between the La and Mg_2Si phase causing heterogeneous nucleation of the Mg_2Si particles on the La clusters during the early stage of the solidification. However, in our study, the Mg_2Si phase precipitates according to the eutectic reaction along with AlLaSi (or Al_2LaSi_2) phase. The difference between two phases Mg_2Si and Al_2LaSi_2 in lattice parameters and lattice-type is enormous (cubic lattice, Fm3m space group, 12 atoms/cell, a = 6.35 Å for Mg_2Si [36] and hexagonal lattice, P-3m1 space group, a = 4.23 Å for Al_2LaSi_2 [26]). The possibility of the AlLaSi compound being a nucleus is minor based on the before mentioned observation. On the other hand, as it is shown in the section related to microstructural observation, the Mg_2Si phase precipitates in the vicinity of the AlLaSi phase, showing the latter increased around and may act as a surfactant for the Mg_2Si eutectic phase. This means that after the formation of the α-Al phase, the AlLaSi compound is further segregated at the solid-liquid interface of the Mg_2Si phase. A quite similar effect is shown by La in modifying eutectic Si particles, which, as shown in [37], have a great similarity with the Mg_2Si phase in terms of properties and solidification behavior.

The presented results are promising steps toward the development of the currently static research area on modification of eutectic Mg_2Si in cast 5xx aluminum alloys. While the behavior of the intermetallics formation has been comprehended, the detailed mechanism of the modification process of eutectic Mg_2Si with La remains debatable and additional investigations to find out will be performed in further works.

5. Conclusions

In summary, the influence of La at concentrations up to 1 wt.% on the phase composition and intermetallics formation during solidification of the nominal Al-4Mg-0.5Si aluminum alloy (AA 511 grade) has been investigated. Based on the results obtained, the main conclusion are as follows:

(1) Thermodynamic prediction using the CalPhaD approach is limited due to the lack of optimized databases resulting in the formation of only binary Mg_2Si, $LaSi_2$, $Al_{11}La_3$, and Al_3Mg_2 intermetallics of eutectic origin. However, the tendency toward change in intermetallics fraction and critical temperature may be considered as reliable. The addition of La causes an increase in $LaSi_2$ along with a sharp decrease in the Mg_2Si phase at the temperature under solidus. Moreover, the calculated equilibrium solidification ranges of the experimental alloys are near 55 °C that is increased fourfold in non-equilibrium conditions based on calculations using the Scheil–Gulliver model.

(2) In contradiction to calculation, the microstructures of the La-containing alloys exhibited dominating ternary AlLaSi phase that is likely to be Al_2LaSi_2 compound. An increase in La content leads to the growth of the ternary phase. This growth is particularly enormous at over 0.5 wt.% La when the intermetallics evolved as slender needles (at 0.75 wt.% La) and stars (at 1 wt.% La).

(3) A remarkable modifying effect of La on the eutectic Mg_2Si phase was observed. The latter's shape changed slightly after the addition of 0.1 wt.% La but evolved incredibly at a higher concentration from the lamellar down to slender flakes. This effect was referred to as adsorption of AlLaSi (Al_2LaSi_2) phase in the solid/liquid interface inhibiting the growth of the Mg_2Si.

Author Contributions: Conceptualization, V.D., E.P. and P.S.; methodology, X.C.; formal analysis, V.D. and S.K.; investigation, V.D., E.P., S.S. and P.S.; writing—original draft preparation, V.D. and P.S.; writing—review and editing, E.P., E.R. and S.K.; project administration, S.K.; funding acquisition, V.D. All authors have read and agreed to the published version of the manuscript.

Funding: This research was funded by the Russian Science Foundation (project No. 20-19-00687).

Acknowledgments: The authors would like to thank Natalia Korotkova for providing SEM investigations.

Conflicts of Interest: The authors declare no conflict of interest.

References

1. Gursoy, O.; Timelli, G. Lanthanides: A focused review of eutectic modification in hypoeutectic Al–Si alloys. *J. Mater. Res. Tech.* **2020**, *9*, 8652–8666. [CrossRef]
2. De Giovanni, M.; Kaduk, J.A.; Srirangam, P. Modification of Al–Si Alloys by Ce or Ce with Sr. *JOM* **2019**, *71*, 426–434. [CrossRef]
3. Abdelaziz, M.H.; Samuel, A.M.; Doty, H.W.; Valtierra, S.; Samuel, F.H. Effect of additives on the microstructure and tensile properties of Al–Si alloys. *J. Mater. Res. Technol.* **2019**, *8*, 2255–2268. [CrossRef]
4. Lin, G.; Li, K.; Feng, D.; Feng, Y.; Song, W.; Xiao, M. Effects of La–Ce addition on microstructure and mechanical properties of Al–18Si–4Cu–0.5Mg alloy. *Trans. Nonferr. Met. Soc. China* **2019**, *29*, 1592–1600. [CrossRef]
5. Jiang, H.; Li, S.; Zheng, Q.; Zhang, L.; He, J.; Song, Y.; Deng, C.; Zhao, J. Effect of minor lanthanum on the microstructures, tensile and electrical properties of Al-Fe alloys. *Mater. Des.* **2020**, *195*, 108991. [CrossRef]
6. Balaram, V. Rare earth elements: A review of applications, occurrence, exploration, analysis, recycling, and environmental impact. *Geosci. Front.* **2019**, *10*, 1285–1303. [CrossRef]
7. Cao, Z.; Kong, G.; Che, C.; Wang, Y.; Peng, H. Experimental investigation of eutectic point in Al-rich Al-La, Al-Ce, Al-Pr and Al-Nd systems. *J. Rare Earths* **2017**, *35*, 1022–1028. [CrossRef]
8. Akopyan, T.K.; Belov, N.A.; Naumova, E.A.; Letyagin, N.V. New in-situ Al matrix composites based on Al-Ni-La eutectic. *Mater. Lett.* **2019**, *245*, 110–113. [CrossRef]
9. Akopyan, T.K.; Letyagin, N.V.; Sviridova, T.A.; Korotkova, N.O.; Prosviryakov, A.S. New casting alloys based on the Al + Al_4 (Ca,La) eutectic. *JOM* **2020**. [CrossRef]
10. Weiss, D. Improved High-Temperature Aluminum Alloys Containing Cerium. *J. Mater. Eng. Perform.* **2019**, *28*, 1903–1908. [CrossRef]
11. Czerwinski, F. Cerium in aluminum alloys. *J. Mater. Sci.* **2020**, *55*, 24–72. [CrossRef]
12. Yu, X.; Sun, J.; Li, Z.; Dai, H.; Fang, H.; Zhao, J.; Yin, D. Solidification behaviour and elimination of undissolved Al_2CuMg phase during homogenization in Ce-modified Al–Zn–Mg–Cu alloy. *Rare Met.* **2020**, *39*, 1279–1287. [CrossRef]
13. Hosseinifar, M.; Malakhov, D.V. Effect of Ce and La on microstructure and properties of a 6xxx series type aluminum alloy. *J. Mater. Sci.* **2008**, *43*, 7157–7164. [CrossRef]
14. Mahmoud, M.G.; Elgallad, E.M.; Ibrahim, M.F.; Samuel, F.H. Effect of rare earth metals on porosity formation in A356 alloy. *Int. J. Metalcast.* **2018**, *12*, 251–265. Available online: http://dx.doi.org/10.1007/s40962-017-0156-5 (accessed on 3 November 2020). [CrossRef]
15. Chen, Y.; Pan, Y.; Lu, T.; Tao, S.; Wu, J. Effects of combinative addition of lanthanum and boron on grain refinement of Al–Si casting alloys. *Mater. Des.* **2014**, *64*, 423–426. [CrossRef]
16. Yao, D.; Xia, Y.; Qiu, F.; Jiang, Q. Effects of La addition on the elevated temperature properties of the casting Al–Cu alloy. *Mater. Sci. Eng. A* **2011**, *528*, 1463–1466. [CrossRef]
17. Liping, W.; Erjun, G.; Baoxia, M. Modification effect of lanthanum on primary phase Mg_2Si in Mg-Si alloys. *J. Rare Earths* **2008**, *26*, 105–109. [CrossRef]
18. Bai, G.; Liu, Z.; Lin, J.; Yu, Z.; Hue, Y.; Wen, C. Effects of the addition of lanthanum and ultrasonic stirring on the microstructure and mechanical properties of the in situ Mg_2Si/Al composites. *Mater. Des.* **2016**, *90*, 424–432. [CrossRef]
19. Yao, D.; Qiu, F.; Jiang, Q.; Li, Y.; Arnberg, L. Effect of Lanthanum on grain refinement of casting aluminum-copper alloy. *Int. J. Metalcast.* **2013**, *7*, 49–54. [CrossRef]
20. Pramod, S.L.; Bakshi, S.R.; Murty, B.S. Aluminum-Based Cast In Situ Composites: A Review. *J. Mater. Eng. Perform.* **2015**, *24*, 2185–2207. [CrossRef]
21. Sun, Y.; Ahlatci, H. Mechanical and wear behaviors of Al-12Si-XMg composites reinforced with in situ Mg_2Si particles. *Mater. Des.* **2011**, *32*, 2983–2987. [CrossRef]
22. Prusov, E.; Deev, V.; Rakhuba, E. Aluminum Matrix In-Situ Composites Reinforced with Mg_2Si and Al_3Ti. *Mater. Today Proc.* **2019**, *11*, 386–391. [CrossRef]
23. Glazoff, M.; Khvan, A.; Zolotorevsky, V.; Belov, N.; Dinsdale, A. *Casting Aluminium Alloys: Their Physical and Mechanical Metallurgy*, 2nd ed.; Elsevier: Amsterdam, The Netherlands, 2018. [CrossRef]

24. *Thermo-Calc Software TCAL4—TCS Al-Based Alloy Database*, version 4.0; Thermo-Calc Software AB: Stockholm, Sweden, 2015; Available online: http://www.thermocalc.com (accessed on 15 October 2020).
25. Hosseinifar, M.; Malakhov, D.V. The sequence of intermetallics formation during the solidification of an Al-Mg-Si alloy containing La. *Metall. Mater. Trans. A* **2011**, *42*, 825–833. [CrossRef]
26. Du, J.; Tu, H.; Peng, H.; Liu, Y.; Wu, C.; Wang, J.; Su, X. Phase equilibria of the Al-Si-La system between 0 and 50 at.% La at 600 and 800 °C. *JALCOM* **2018**, *765*, 608–615. [CrossRef]
27. Liu, Y.L.; Kang, S.B. Solidification and segregation of Al-Mg alloys and influence of alloy composition and cooling rate. *Mater. Sci. Technol.* **1997**, *13*, 331–336. [CrossRef]
28. Liu, Y.; Luo, L.; Han, C.; Ou, L.; Wang, J.; Liu, C. Effect of Fe, Si and cooling rate on the formation of Fe- and Mn-rich intermetallics in Al–5Mg–0.8Mn alloy. *J. Mater. Sci. Technol.* **2016**, *32*, 305–312. [CrossRef]
29. Liu, Y.L.; Kang, S.B. The solidification process of Al–Mg–Si alloys. *J. Mater. Sci.* **1997**, *32*, 1443–1447. [CrossRef]
30. *Thermo-Calc Software TCAL7—TCS Al-Based Alloy Database*, version 4.0; Thermo-Calc Software AB: Stockholm, Sweden, 2020; Available online: http://www.thermocalc.com (accessed on 15 October 2020).
31. Wong, C.; Styles, M.J.; Zhu, S.; Qiu, D.; McDonald, S.D.; Zhu, Y.; Gibson, M.A.; Abbott, T.B.; Easton, M.A. $(Al,Mg)_3La$: A new phase in the Mg–Al–La system. *Acta Cryst.* **2018**, *B74*, 370–375. [CrossRef]
32. Gröbner, J.; Kevorkov, D.; Schmid-Fetzer, R. Thermodynamic modelling of Al–Ce–Mg phase equilibria coupled with key experiments. *Intermetallics* **2002**, *10*, 415–422. [CrossRef]
33. Zhang, X.; Wang, Z.; Zhou, Z.; Xu, J. Influence of rare earth (Ce and La) addition on the performance of Al-3.0 wt%Mg alloy. *J. Wuhan Univ. Technol.-Mater. Sci. Ed.* **2017**, *32*, 611–618. [CrossRef]
34. Wang, M.; Xu, W.; Han, Q. Study of Refinement and Morphology Change of AlFeSi Phase in A380 Alloy due to Addition of Ca, Sr/Ca, Mn and Mn, Sr. *Mater. Trans.* **2016**, *57*, 1509–1513. [CrossRef]
35. Wu, X.; Zhang, H.; Jiang, H.; Mi, Z.; Zhang, H. Multi-Refinement Effect of Rare Earth Lanthanum on α-Al and Eutectic Si Phase in Hypoeutectic Al-7Si Alloy. *Metals* **2020**, *10*, 621. [CrossRef]
36. Belov, N.A.; Naumova, E.A.; Akopyan, T.K.; Doroshenko, V.V. Phase Diagram of Al-Ca-Mg-Si System and Its Application for the Design of Aluminum Alloys with High Magnesium Content. *Metals* **2017**, *7*, 429. [CrossRef]
37. Zhang, J.; Fan, Z.; Wang, Y.; Zhou, B. Microstructural development of Al–15wt.% Mg_2Si in situ composite with mischmetal addition. *Mater. Sci. Eng. A* **2000**, *281*, 104–112. [CrossRef]

metals

MDPI

Review

Asymmetric (Hot, Warm, Cold, Cryo) Rolling of Light Alloys: A Review

Denis Pustovoytov [1], Alexander Pesin [1,*] and Puneet Tandon [2]

1 Laboratory of Mechanics of Gradient Nanomaterials, Nosov Magnitogorsk State Technical University, 455000 Magnitogorsk, Russia; pustovoitov_den@mail.ru
2 deLOGIC Lab, PDPM Indian Institute of Information Technology, Design and Manufacturing, Jabalpur 482005, India; ptandon@iiitdmj.ac.in
* Correspondence: pesin@bk.ru; Tel.: +7-3519-063056

Abstract: Asymmetric sheet rolling is a process used when there are differences in any technological parameters in the horizontal plane across the width of the deformation zone or in the vertical plane between the top and bottom surfaces of the deformation zone. Asymmetry can either have random causes, or it can be created purposefully to reduce rolling force, improve sheet flatness, minimize the ski effect, obtain thinner sheets and for grain refinement and improvement of texture and mechanical properties of sheet metals and alloys. The purpose of this review is to analyze and summarize the most relevant information regarding the asymmetric (hot, warm, cold, cryo) rolling processes in terms of the effect of purposefully created asymmetry on grain size and mechanical properties of pure Mg, Al, Ti and their alloys. The classification and fundamentals of mechanics of the asymmetric rolling process are presented. Based on the analysis of publications related to asymmetric rolling, it was found that a superior balance of strength and ductility in pure Mg, Al, Ti and their alloys could be achieved due to this processing. It is shown that asymmetric rolling in comparison with conventional severe plastic deformation methods have an undeniable advantage in terms of the possibility of the production of large-scale sheets.

Keywords: asymmetric rolling; differential speed rolling; shear strain; severe plastic deformation; microstructure; mechanical behavior; Mg alloys; Al alloys; Ti alloys

Citation: Pustovoytov, D.; Pesin, A.; Tandon, P. Asymmetric (Hot, Warm, Cold, Cryo) Rolling of Light Alloys: A Review. *Metals* **2021**, *11*, 956. https://doi.org/10.3390/met11060956

Academic Editor: Nikolay A. Belov

Received: 10 May 2021
Accepted: 10 June 2021
Published: 13 June 2021

1. Introduction

Asymmetric sheet rolling is a process used when there are differences in any technological parameters in the horizontal plane across the width of the deformation zone or in the vertical plane between the top and bottom surfaces of the deformation zone. Asymmetric rolling due to purposefully created differences in the circumferential speeds of the work rolls is also called "differential speed rolling". For such a process, a degree of asymmetry is defined by a ratio of circumferential speeds V_1 and V_2 of the work rolls according to Equation (1):

$$S_R = \frac{V_1}{V_2} \tag{1}$$

where S_R is the "speed ratio" and $V_1 > V_2$. The first theoretical description of the asymmetric rolling process was proposed in 1941 by Siebel [1]. Experimentation has been used to quantify rolling force and torque. In 1947, Sachs and Klinger [2] first identified the region of cross shear, due to the fact that friction forces act in opposite directions in the deformation zone. The first invention of a rolling process with high ratio of circumferential speeds of work rolls ($S_R = 3$) was announced in 1940 in the USSR by Razuvaev [3]. He was the first to propose a rolling process at a circumferential speed ratio equal to the ratio of sheet thicknesses before and after rolling according to Equation (2):

$$\frac{V_1}{V_2} = \frac{h_0}{h_1} \tag{2}$$

where h_0 is the thickness of the sheet before the rolling pass, h_1 is the thickness of the sheet after the rolling pass. This rolling process was proposed to reduce rolling force, obtain thinner sheets and reduce the number of rolling passes by increasing the reduction in each pass.

In the 1950s and 1960s, the first laboratory and industrial experiments on asymmetric rolling were performed and technological results were published [4,5]. Between the 1970s and 1990s, many experiments and theoretical calculations were carried out and the development of asymmetric rolling processes was focused mainly on the improvement of sheet geometry and technological aspects of the hot and cold rolling processes [6–13]. Asymmetry was used to reduce rolling force, improve sheet flatness, minimize the ski effect and obtain thinner sheets. Thus, from the 1940s to the end of the 1990s, the development of asymmetric rolling processes took place in research, which can be named as the "geometry approach" (Figure 1). The introduced asymmetry was rather low. Usually, the S_R value did not exceed 1.5. Since the end of the 1990s, a new direction in the research on the asymmetric rolling processes has occurred. The development of asymmetric rolling processes has taken place in research, which can be named as the "material approach". The introduced asymmetry is rather high. Usually, the S_R value is more than or equal to 2.0.

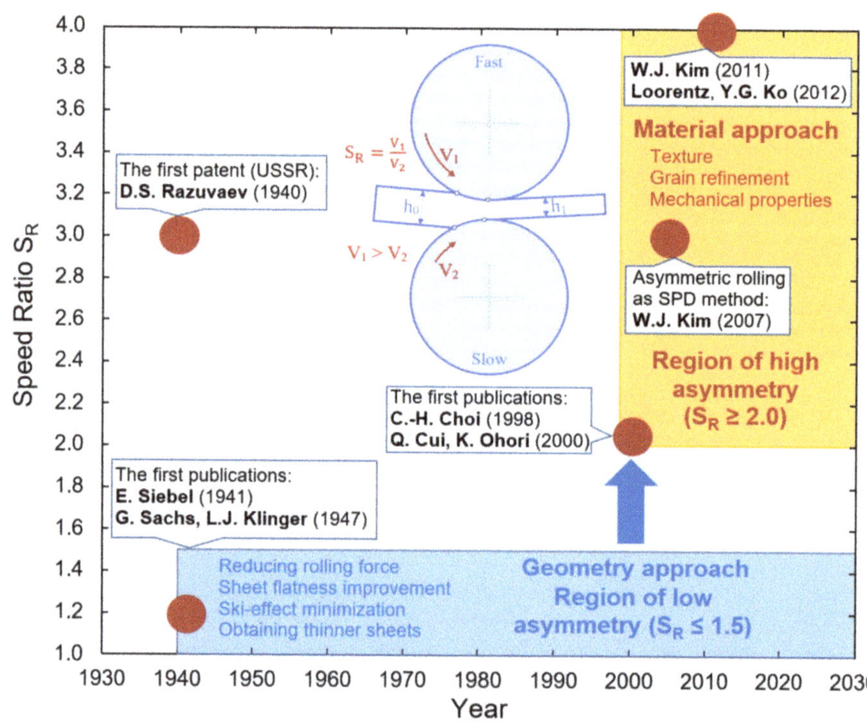

Figure 1. Different research lines for asymmetrical rolling processes: "geometry approach" and "material approach".

The first work about using asymmetric rolling with a high speed ratio (S_R = 2.0) to control the texture and mechanical properties of Al alloys was published in 1998 by Choi et al. [14]. It was shown that AA1100 and AA3005 sheets composed of only shear texture with high {111} components could be obtained by asymmetric rolling. A remarkable improvement in the average Lankford value was also obtained. In 2000, Cui and Ohori [15] demonstrated that the presence of shear strain is more important than the maximum level of thickness reduction, when considering grain refinement by asymmetric rolling. They achieved grain refinement (~2 μm) in high-purity Al by asymmetric cold rolling. Fine grains evolved during asymmetric rolling and were stable at temperatures below 423 K.

In 2007, Kim et al. [16] introduced and developed an asymmetric rolling process known as the severe plastic deformation (SPD) method. Differential speed rolling with a high speed ratio ($S_R = 3.0$) between the top and bottom work rolls was applied to AZ31 sheets (2 mm thick and 140 mm wide). The diameters of the top and bottom work rolls were identical and equal to 300 mm. From a starting thickness of 2 mm, sheets were reduced to 0.6 mm (70% reduction) by a single pass without lubrication. The effective strain accumulated during asymmetric rolling was 3.53 on average [16], which was comparable to that obtained by ECAP on a rod after three pressings. A very fine grain size of 1.4 μm and a high yield stress of over 300 MPa were obtained after asymmetric rolling at 413 K. Significant grain refinement was achieved due to the introduction of high shear strain. The best compromise between strength and ductility was achieved at 433 K. This result indicated that SPD can be imposed by differential speed rolling with a high speed ratio. This process was called "high-ratio differential speed rolling" (HRDSR) [17–20]. In 2009, it was applied to obtain high-strength sheets from Al alloy 6061 [19], and in 2010, to obtain high-strength sheets from pure Ti [20]. Kim et al. [21] in 2011, Loorentz and Ko [22] in 2012 and Polkowski [23] in 2013 introduced differential speed rolling with speed ratio $S_R = 4.0$ for improvement of the microstructure and texture in Ti [21], Al [22] and Cu [23] alloys.

Thus, from 1998 to the present, asymmetric rolling with the "material approach" has been applied to light alloys of Mg, Al and Ti, which are the materials selected for this review. These materials have very attractive features such as light weight and high specific strength. The development of asymmetric rolling technology, such as the severe plastic deformation method, for industrial production of large-scale sheets with ultrafine grain structure and enhanced mechanical properties has been the major research line for the last 20 years. Depending on temperature conditions, four asymmetric rolling processes can be implemented: (1) Asymmetric hot rolling, when the temperature of the sheet is in the range (0.6...0.8) T_{melt}, where T_{melt} is a melting temperature in K; (2) asymmetric warm rolling, when the temperature of the sheet is in the range (0.3...0.6)T_{melt}; (3) asymmetric cold rolling, when the temperature of the sheet is below 0.3T_{melt}; (4) asymmetric cryorolling, when the sheet is cooled in liquid nitrogen until the temperature reaches −153–−196 °C. Asymmetric cryorolling was first proposed in 2012 by Yu [24] as a new technique that combines the features of asymmetric rolling and cryorolling [25,26].

Although some reviews of asymmetric rolling processes have already been presented [27,28], the purpose of this review is to analyze and summarize the most relevant information regarding the asymmetric (hot, warm, cold, cryo) rolling processes in terms of the effect of purposefully created asymmetry on grain size and mechanical properties of pure Mg, Al, Ti and their alloys. This work is divided into six sections. First, the classification of asymmetric rolling processes is presented. Secondly, the fundamentals of the mechanics of the asymmetric rolling process due to purposefully created differences in the peripheral speeds of the work rolls are presented. This is followed by an analysis of asymmetric (hot, warm, cold, cryo) rolling of Mg alloys, Al alloys and Ti alloys. The review ends with conclusions and future prospects for these technologies.

2. Classification of Asymmetric Rolling Processes

Asymmetry can either have random causes, or it can be created purposefully (Figure 2). Asymmetry in the horizontal plane along the width of the deformation zone due to random causes can be caused by differences in thickness, temperatures, friction, displacement of the sheet along the work roll length, skewed work rolls or incorrect setting of the right and left hydraulic pressure devices.

Asymmetry in the vertical plane between the top and bottom surfaces of the deformation zone due to random causes can be caused by inclined entry of the sheet into the roll gap, differences in temperatures, friction, differences in the work roll diameters at the same angular speeds or differences in the work roll angular speeds with the same diameters. In any case, asymmetric rolling caused by random causes is undesirable. For example, a negative consequence of asymmetric rolling is the bending of the plate, usually in hot rolling mills, more often known as the "ski effect" [29,30]. The negative effect of such asymmetry is prevented by improving equipment, stabilizing technology and using automatic regulators. However, asymmetry can also be purposefully introduced to improve the rolling process, for example, reducing rolling force, improving sheet flatness, minimizing the ski effect, obtaining thinner sheets, grain refinement and improvement of the texture and mechanical properties of sheet metals and alloys.

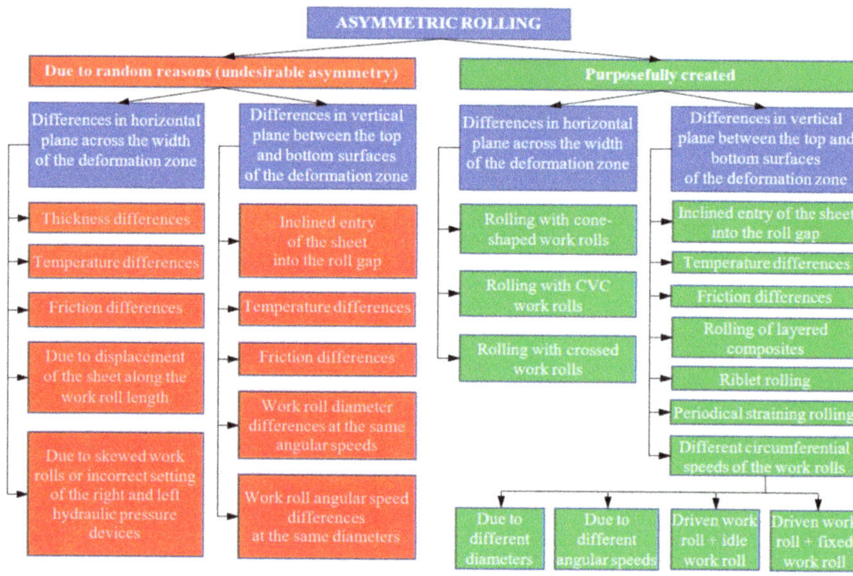

Figure 2. Classification of asymmetric rolling processes.

Asymmetry in the horizontal plane across the width of the deformation zone can be purposefully created by rolling with cone-shaped work rolls [31]. A schematic illustration of this process is shown in Figure 3a. When the ratio of the largest diameter of the roll to the smallest one D/d = 1.5, shear strain increases by 1.5–2.0 times in comparison with conventional rolling [32].

A little asymmetry in the horizontal plane across the width of the deformation zone can be an effective tool to improve sheet shape. The continuous variable crown (CVC) system (Figure 3b) was developed by Schloemann-Siemag (SMS) as a powerful tool for sheet shape control. The work rolls have an "S" shape and are arranged asymmetrically [33]. Through the axial shifting of two work rolls towards each other, a continuously adjustable rolling gap contour can be obtained. The CVC system is employed in many cold rolling and hot rolling mills worldwide. A little asymmetry in the horizontal plane can also be obtained by crossed work rolls (Figure 3c). The first hot rolling mill with asymmetry due to crossed work rolls was used at Nippon Steel's Kimitsu Works in 1991 [34]. In the work roll crossed system, the axes of the top and bottom work rolls are crossed as shown in Figure 3c. Work roll crossing is designed to control the roll gap profile. This action provides control over the shape and profile of the sheet. Thus, a little asymmetry across the width of the deformation zone provided by CVC or crossed work rolls is used in industrial rolling mills to improve the shape and flatness of sheets.

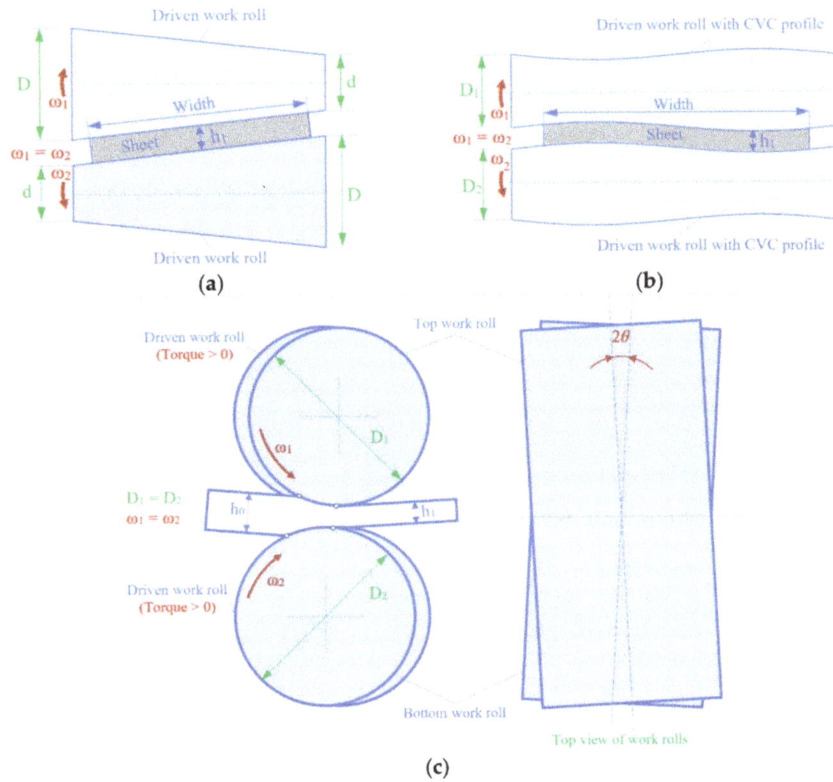

Figure 3. Schematic illustration of asymmetric rolling processes with purposefully created asymmetry in the horizontal plane along the width of the deformation zone: (**a**) Rolling with cone-shaped work rolls; (**b**) rolling with CVC work rolls; (**c**) rolling with crossed work rolls.

Asymmetry in the vertical plane between the top and bottom surfaces of the deformation zone can be purposefully created by inclined entry of the sheet into the roll gap [35] (Figure 4a) or due to differences in temperatures, when one work roll is purposefully heated and the other one is cold (Figure 4b).

Asymmetry during rolling can be purposefully created due to friction differences [36,37]. This rolling process is called "differential friction rolling" [37]. To change the friction coefficients on the contact surfaces of the deformation zone, the lubricant can be supplied only from one side (Figure 5a). Another way to create friction differences is rolling with different roughnesses of the top and bottom work rolls (Figure 5b). Utsunomiya et al. [37] proposed a differential friction rolling process for improvement in the r-value of aluminum sheets. A two-high rolling mill with TiC-coated rolls 130 mm in diameter was used for rolling of commercial aluminum AA1050 sheets 3 mm thick and 30 mm wide. A single-pass thickness reduction of 50% was conducted at 473 K. Low friction was achieved by applying a thin polytetrafluoroethylene (PTFE) film on the sheet. Teflon (PTFE) was sprayed by aerosol on the upper surface of the aluminum sheet and dried. However, the lower surface remained unlubricated. Both the top and bottom rolls revolved at the same speed of 2 m/min. The shear strain introduced by differential friction rolling was comparable to that introduced by differential speed rolling [37]. The annealed sheet showed a higher r-value than the symmetrically rolled sheet.

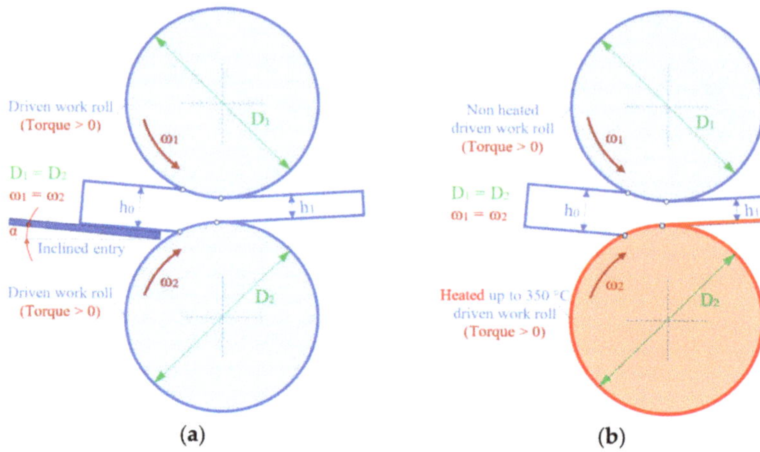

Figure 4. Schematic illustration of asymmetric rolling processes: (**a**) Rolling with inclined entry of the sheet into the roll gap; (**b**) rolling with temperature differences.

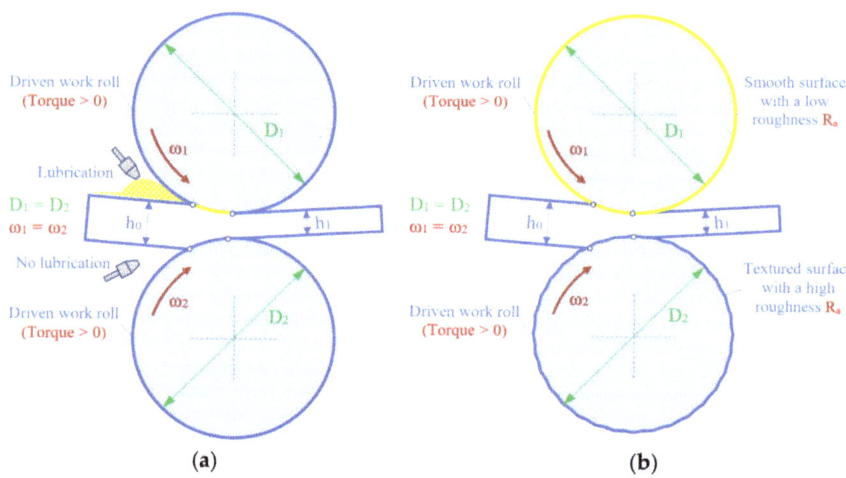

Figure 5. Schematic illustration of asymmetric rolling processes due to friction differences: (**a**) Rolling with different lubrications for top and bottom work rolls; (**b**) rolling with different roughnesses of top and bottom work rolls.

Asymmetry during rolling can be purposefully created due to differences in the circumferential speeds of the work rolls. Asymmetric rolling can be implemented due to disconnecting one work roll to make another work roll the only driven work roll (Figure 6a). This process called "single-roll-driven rolling" [38]. Sakai et al. [39] proposed two-pass single-roll-driven unidirectional shear rolling. Shear strain greater than 1.0 was introduced throughout the thickness of the AA5052 sheet. A shear texture was developed throughout the thickness of the sheets. The average r-value equaled unity. These values resulted in a superior drawability of the unidirectionally shear rolled and annealed sheets to that of sheets fabricated by the conventional process.

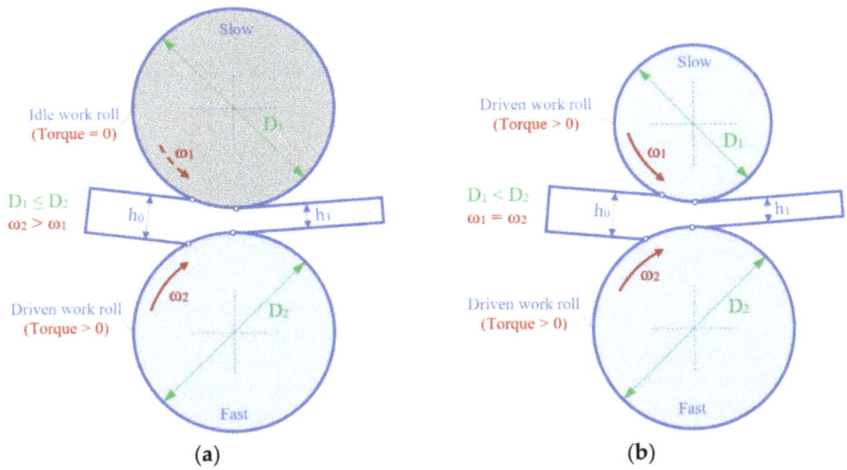

Figure 6. Schematic illustration of asymmetric rolling processes due to different circumferential speeds of the work rolls: (**a**) Rolling with idle work roll ("Single roll driven rolling"); (**b**) rolling with different diameters of top and bottom work rolls.

Asymmetry can also be achieved by using work rolls with different diameters while rotating at the same angular speed (Figure 6a). Choi et al. [14] used asymmetric rolling with different diameters of 248 mm and 126 mm of the top and bottom work rolls to control the texture and mechanical properties of Al alloys. Pesin [40] developed an improved process of asymmetric rolling by using different diameters of work rolls in combination with a bending roll (Figure 7a). This process is called the "combined process of asymmetric rolling and plastic bending". This technology is used to produce segments of large cylinders (Figure 7b) as curved plates with an angular size of 45 ... 60 degrees, plate thickness of 40 ... 220 mm, a width of up to 4300 mm, a length of up to 5000 mm and a radius of curvature from 1850 to 5000 mm. Industry testing of the combined process of asymmetric rolling and plastic bending is shown in Figure 8.

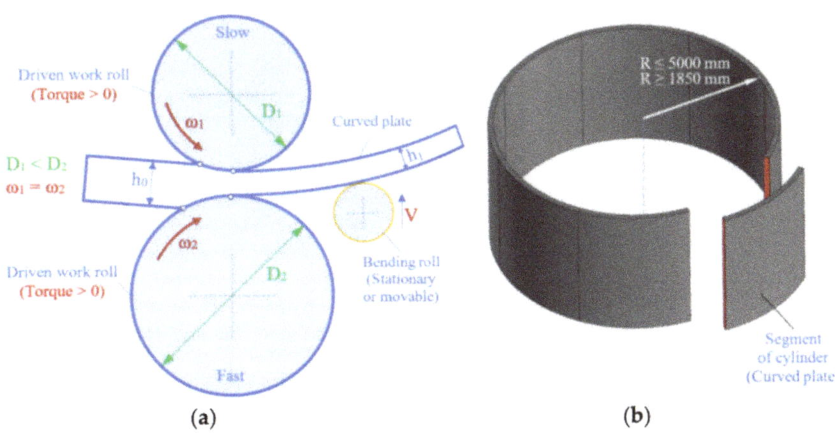

Figure 7. Schematic illustration of combined process of asymmetric rolling and plastic bending (**a**) and a curved plate as a segment of a cylinder (**b**).

(**a**) (**b**) (**c**)

Figure 8. Industry testing on hot plate mill 4500 at OJSC Magnitogorsk Iron and Steel Works: (**a**,**b**) Combined process of asymmetric rolling and plastic bending; (**c**) finished curved plate.

Asymmetric rolling can be achieved by using a single driven work roll, when one of the rolls is kept stationary. In the late 1970s, Potapkin and Fedorinov [41] proposed an asymmetric rolling process called "deformation by a stationary and driven work roll" (Figure 9a). The process is carried out at a sufficiently high front tension of the strip. The diameter of the stationary work roll is 3–10 times smaller than the diameter of the driven work roll. This provides a significant reduction in rolling force due to a decrease in the contact area. The technological capabilities and advantages of this process increase with a decrease in the diameter of the stationary roll. However, this decrease is limited by a decrease in its strength and rigidity and complications of fastening (the possibility of work roll twisting). In the late 1980s, Pesin [42] proposed to use, instead of a stationary cylindrical work roll, a prism-shaped tool with a cylindrical part with a very small radius up to 5 mm (Figure 9b). This solves the problem of the strength and rigidity of the deforming tool in combination with ensuring a small radius of the contact surface of this tool.

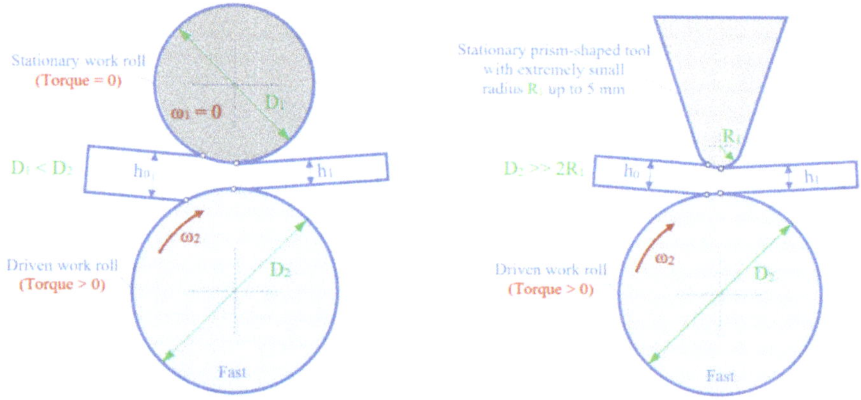

(**a**) (**b**)

Figure 9. Schematic illustration of asymmetric rolling processes: (**a**) Deformation by a stationary and driven work roll; (**b**) deformation by a stationary prism-shaped tool and driven work roll.

Asymmetric rolling can be carried out due to different angular speeds of the work rolls with the same diameters (both rolls are driven) (Figure 10a). The ratio of these speeds is one of the most important parameters of the process, since it will influence the velocity field along the sheet thickness (together with other parameters, such as contact friction and thickness reduction per pass). The asymmetric rolling due to purposefully created differences in the circumferential speeds of the work rolls is also called "differential speed rolling".

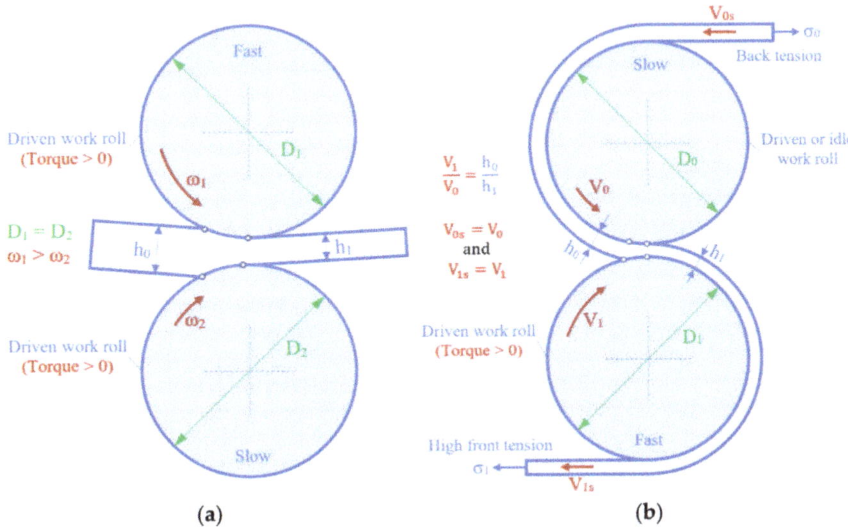

Figure 10. Schematic illustration of asymmetric rolling processes due to different circumferential speeds of the work rolls: (**a**) Rolling with different angular speeds of the work rolls; (**b**) "rolling–drawing" process.

In the 1960s and 1970s, the energy theory of the asymmetric rolling process was developed by Vydrin [43]. As a result, a new asymmetric rolling process called "rolling–drawing" (Figure 10b) was proposed in 1971 [44]. This process is carried out under the following kinematic conditions:

$$\frac{V_1}{V_0} = \frac{h_0}{h_1},\ V_{0s} = V_0 \text{ and } V_{1s} = V_1 \quad (3)$$

where V_{0s}, V_{1s} are the circumferential speeds of the strip before and after the deformation zone, respectively; V_0, V_1 are the circumferential speeds of the work rolls; h_0 is the thickness of the strip before the rolling pass; h_1 is the thickness of the strip after the rolling pass. The rolling–drawing process is carried out at a sufficiently high front tension of the strip. Such kinematic conditions provide one-zone and counter-directional friction forces in the deformation zone. Since the rolling–drawing process requires a high front tension of the strip, the thickness reduction per pass should not exceed 10–40%. This limits the possibility of using this process.

Thus, different circumferential speeds of the work rolls can be created in the following ways: (1) due to different diameters of the work rolls at the same angular speeds (both rolls are driven); (2) when one work roll is driven and another work roll is idle; (3) when one work roll is driven and another work roll or tool is stationary; (4) due to different angular speeds of the work rolls with the same diameters (both rolls are driven). The main disadvantage of the first three ways is the low technological flexibility of these processes due to the impossibility to control the speed ratio S_R of the work rolls when creating shear strain, which is necessary for grain refinement and improving the texture and mechanical properties of various metals and alloys. From an industrial point of view, rolling with different angular speeds of the work rolls with the same diameters (both rolls are driven) is the most suitable way to implement asymmetric rolling.

Asymmetry during rolling can be purposefully created due to differences in the surface geometry of the top and bottom work rolls. In 2014, Shimoyama et al. [45] proposed the new rolling process called "periodical straining rolling" that enables control of the microstructure and texture of Mg alloys sheets. The periodical straining rolling process consists of two deformation stages. At the first pass, the sheet is rolled with a grooved work roll with, for example, a sine profile of the surface, and a flat work roll (Figure 11a). As a result, small rack-like dimples are formed periodically on the sheet surface. At a subsequent rolling pass, the rack-like sheet is flattened by conventional flat work rolls. Finally, the smooth sheet surface is obtained. The objective of the proposed rolling process is to introduce localized strain, to cause microstructure and texture changes in the rolled sheet. Periodical straining rolling can be easily adapted in existing strip rolling mills by replacing the ordinary work roll profiles.

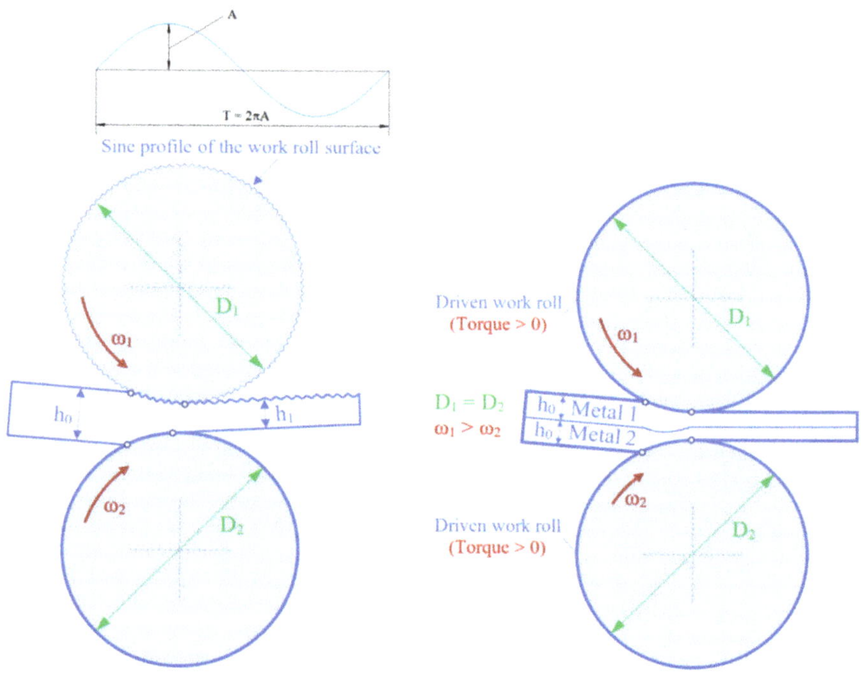

Figure 11. Schematic illustration of periodical straining rolling (**a**) and asymmetric roll bonding of bimetal layered composites (**b**).

Hot or cold rolling is widely used for the manufacturing of bimetal layered composites [46]. As a result of the difference between the flow stresses of the sheets to be bonded, this process belongs essentially to the category of asymmetric rolling. However, this process is usually simply called "hot or cold roll bonding". Additional purposefully created asymmetry due to different circumferential speeds of the work rolls can be introduced to the rolling process of bimetal layered composites [47]. In this case, the rolling process is called "asymmetric hot or cold roll bonding" (Figure 11b). The harder metal contacts the fast work roll and the softer metal contacts the slow work roll. The deformation zone of asymmetric rolling forms a cross shear zone, which increases the bonding strength.

In 2019, Wang [48–50] proposed the new asymmetric rolling process called "corrugated rolling and flat rolling" (Figure 12) to achieve strong interface bonding and substantial grain refinement in Mg/Al and Cu/Al layered composites. The surface of a corrugated work roll was designed as a sine curve with an amplitude of 0.5 mm and a period of 0.06 radians. The diameters of the corrugated and flat work rolls were both 150 mm [48].

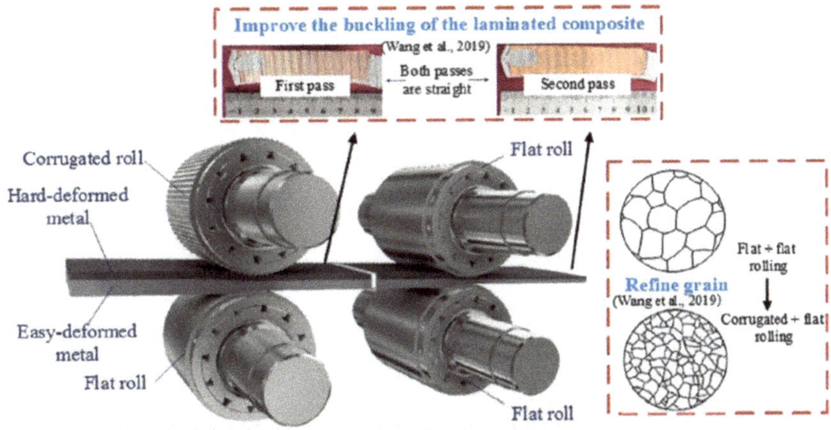

Figure 12. Schematic illustration of corrugated rolling and flat rolling, reproduced from [50], with permission from Elsevier, 2021.

During the first corrugated rolling, compressive stress of various orientations is formed in the corrugated Mg/Al laminated composite [49]. This means that asymmetric deformation occurs, as shown in Figure 13 [49]. The upper corrugated roll results in an inhomogeneous local stress distribution at different positions. Thus, the grain structure of AZ31B might not be uniform and the microstructure might change with the curve of the corrugated roll. On the contrary, the lower conventional flat roll corresponds to 5052 aluminum and stresses on the aluminum sheet near the flat roll appear to be homogeneous. During the second flattened rolling process, the AZ31B metal at the peak position prepared by the first pass flows downward and sideways, enduring severe plastic deformation under the rolling force. The original coarse grain might be broken and recrystallize during the rolling process and subsequent heat treatment. Considering the matching of surface ripples between the corrugated Mg/Al composite and corrugated roll, the rolling force appears to be uniform along the width direction, which is different from the force along the rolling direction. Therefore, for the flattened as-rolled Mg/Al laminated composite, the longitudinal and transverse interfacial microstructures are significantly different. The Mg/Al laminated composites prepared by corrugated rolling and flat rolling exhibited outstanding tensile properties along both RD and TD, which can be attributed to the microstructure refinement induced by the severe shear strain. The tensile properties along TD were higher than that along RD in both the as-rolled and heat-treated state, and this significant anisotropy of the tensile property was mainly due to the microstructure spatial distribution and the interfacial intermetallic compound layer along the different directions [49].

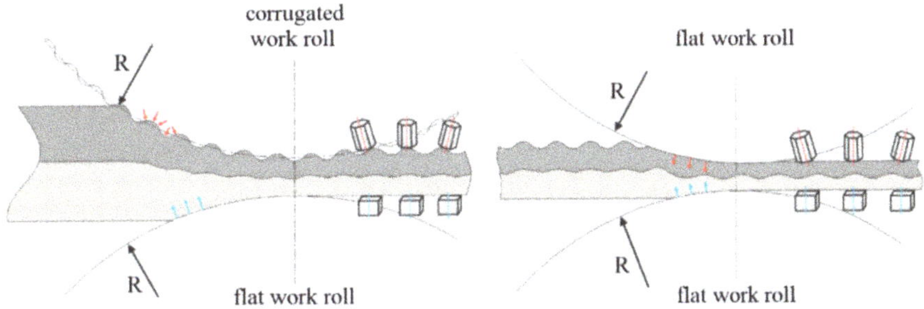

Figure 13. Schematic illustration showing the shear stress during the corrugated rolling and flat rolling process [49].

Pesin et al. [51] proposed asymmetric rolling based on the corrugated rolling and flat rolling process for the manufacturing of Al–steel layered composites with improved bonding strength between the stacked layers. It was numerically shown [51] that mechanical clinching and elevated plastic strain at the interface due to the creation of wave-like contact surfaces between the Al and steel, with the mutual penetration of hard material into soft material when the shear strain is activated, can provide superior bonding strength.

In 2007, Hirt and Thome [52,53] proposed the new asymmetric rolling process called "riblet rolling", which is used to obtain specific functional surface structures. Before rolling, a fine high-strength steel wire is tightly wound around the upper roll to provide it with a negative riblet imprint. Afterwards, the riblet structure is formed on the sheet by riblet rolling (Figure 14). The aim is to approach the production of semi-circular shaped riblet structures with a spacing s of 90–100 μm. In the ideal case, the ratio of riblet height h to riblet spacing s should be about 0.5. Riblet spacing s is defined by the diameter d_w of the used steel wire. This surface is characterized by almost perfect negative semi-circular riblet structures with very sharp ground radii. The accuracy of the winding process—i.e., a constant pitch p of the steel wire on the roll—must be very high.

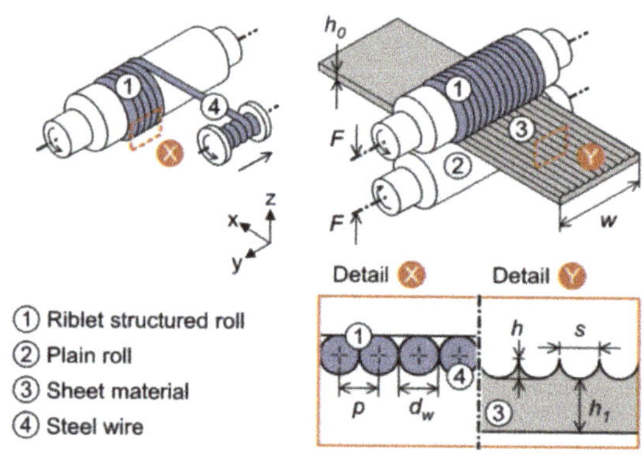

Figure 14. Schematic illustration of riblet rolling, reproduced from [52], with permission from Elsevier, 2008.

Rolling of functional riblet structures allowing for significant drag reduction on surfaces in fluid dynamic applications like airplanes and trains can be applied for the manufacturing of aluminum alloy sheets. Polished cross-sections of structured clad and bare AA 2024-T351 sheets are shown in Figure 15 [54]. Rolling of uniformly shaped riblets with a lateral distance of less than 100 μm is assumed to be possible on large Al sheets.

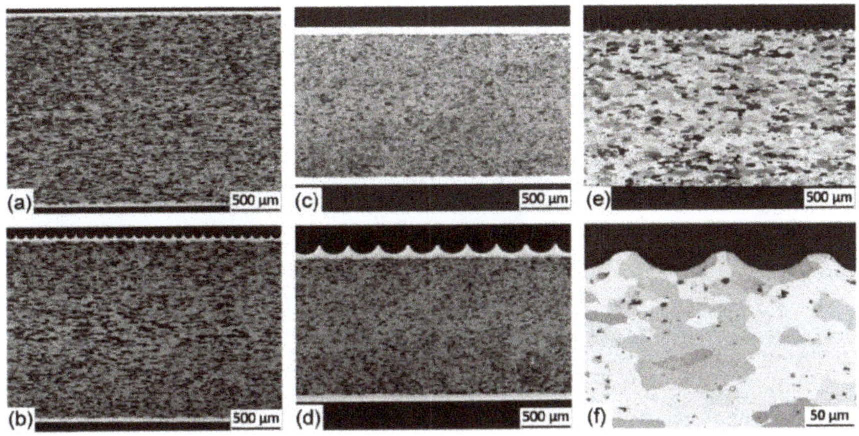

Figure 15. Polished cross-sections. (**a**) A 2 mm thick AA 2024-T351 sheet combined with a clad layer of commercially pure Al (Alclad 2024-T351). (**b**) Structured sheet of Alclad 2024-T351 with 80 µm riblet diameter. (**c**) A 1.6 mm thick Alclad 2024-T351 sheet. (**d**) Structured sheet of Alclad 2024-T351 with 300 µm riblet diameter. (**e**,**f**) A 2 mm thick AA2024-T351/T4 bare sheet with 100 µm riblet diameter, reproduced from [54], with permission from Elsevier, 2014.

3. Fundamentals of Differential Speed Rolling

3.1. Deformation Zone

The sheet is deformed in the deformation zone between the entrance and the exit, where the initial thickness h_0 is reduced to the final thickness h_1, as shown in Figure 16. During symmetric rolling, as shown in Figure 16a, the circumferential speed of the bottom work roll V_1 equals a circumferential speed of the top work roll V_2, and the speed of the bottom surface of the sheet V_{1s} equals the speed of the top surface of the sheet V_{2s}. The neutral points on the top and bottom contact surfaces are at the same locations in the direction of rolling. Neutral points divide the deformation zone into two zones: a backward slip zone and a forward slip zone. In the backward slip zone, the contact friction forces are active, and in the forward slip zone, they are reactive. If pressure p or the shear strength k is known at the contact surface, the frictional stress τ can be expressed as μp in accordance with the Coulomb friction law or mk in accordance with the shear friction law. Here, μ is the Coulomb friction coefficient and m is the shear friction factor. During differential speed rolling when V_1 is greater than V_2, the neutral point associated with the slow work roll is shifted toward the entrance of the roll gap, while the neutral point associated with the fast work roll is moved toward the exit of the roll gap (Figure 16b). As a result, along with the backward slip zone and the forward slip zone, a third zone appears—the cross-shear zone—in which the forces of contact friction are oppositely directed.

Depending on length of the backward and forward slip zones from the side of the top and bottom work rolls, the following cases of differential speed rolling may occur:

(1) General case, when the deformation zone is divided into three zones: backward slip zone, forward slip zone and cross-shear zone (Figure 16b). There are two neutral points in the deformation zone.

(2) Semi-limiting case, when the deformation zone is divided into two zones: backward slip zone or forward slip zone and cross-shear zone. In this case, one neutral point is between the entrance and exit of the deformation zone, and the other neutral point is at the entrance or at the exit of the deformation zone. This neutral point may also be completely absent. This case corresponds to the asymmetric rolling due to deformation by a stationary and driven work roll (Figure 9a) or due to deformation by a stationary prism-shaped tool and driven work roll (Figure 9b).

(3) Limiting case, when only the cross-shear zone occupies the entire deformation zone. The first neutral point is at the entrance of the deformation zone, and the second neutral point is at the exit of the deformation zone. This case corresponds to the rolling–drawing process which requires a high front tension, as mentioned above (Figure 10b).

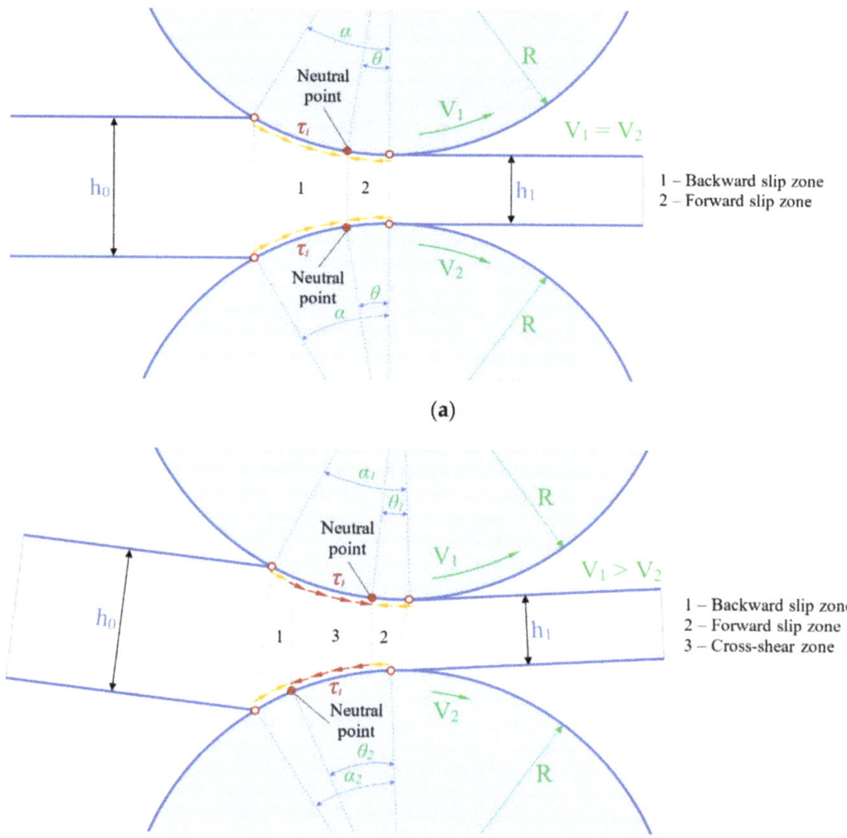

Figure 16. Schematic illustration of deformation zone during symmetric (**a**) and differential speed rolling (**b**).

In symmetric rolling, the pressure distribution at the work roll/sheet interface has a typical "friction hill". The rolling pressure increases gradually from the entrance of the deformation zone, reaches a maximum value corresponding to the neutral point and then gradually falls towards the exit of the deformation zone. In asymmetric rolling, the "friction hill" is cut off due to the cross-shear zone. As a result, rolling force is decreased. This is one of the main advantages of asymmetric rolling. Figure 17 shows pressure distribution along the deformation zone during symmetric and asymmetric rolling. Zuo et al. [55] reported that in the case of hot asymmetric rolling of aluminum alloy 5182, the rolling force was 5–30% lower than that of symmetric hot rolling. It was also confirmed by Tian et al. [56] that the extent of the cross-shear zone increases with the increase in the circumferential speed ratio, whereas the rolling force decreases.

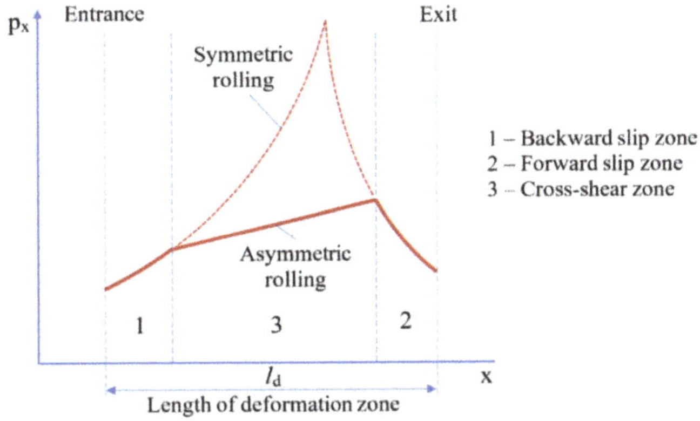

Figure 17. Schematic illustration of pressure distribution along the deformation zone during symmetric and asymmetric rolling.

3.2. Dissipation of Power and Speed Profiles in Deformation Zone

The power provided by the faster work roll is represented in Figure 18 by the area of a-e-e′-d-a minus the area of e-b″-c-e′-e multiplied by the frictional stress, while that by the slower work roll is the area of a″-f-f′-d-a″ minus the area of f-b′-c-f′-f multiplied by the frictional stress [57]. The power used for deformation by the faster work roll is the area of a′-g-e-e′-d-a′ minus the area of e-b-c-e′-e multiplied by the frictional stress, while that by the slower work roll is the area of a′-f-f′-d-a′ minus the area of f-g′-b-c-f′-f multiplied by the frictional stress. The power wasted by friction is the power provided by work rolls minus the power used for deformation. In particular, the power used for shear deformation due to the differential speed is represented by the area of a′-g-b-g′-a′ multiplied by the frictional stress [57].

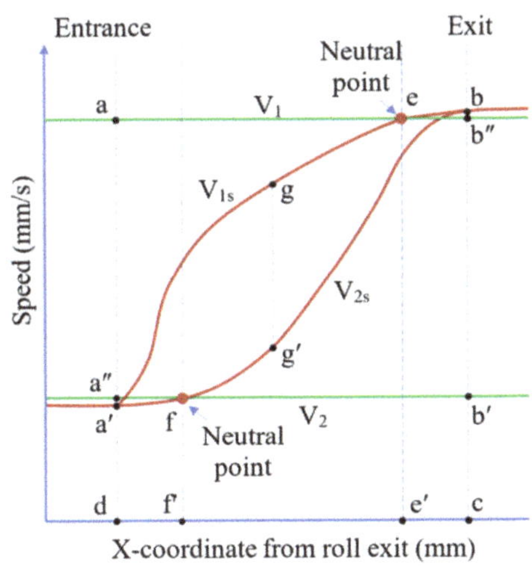

Figure 18. Schematic illustration of speed profiles in deformation zone during differential speed rolling, reproduced from [57], with permission from Elsevier, 2015.

Park [57] suggests Relation (4) that restricts the speed ratio to be less than the thickness ratio. If this relation is not satisfied, a portion of the power provided by the rolls becomes unnecessarily wasted as frictional heat, rather than used for deformation of the sheet.

Additionally, an excessive tensile stress could be developed in the sheet, resulting in fracture if the material of the sheet is as brittle as a magnesium alloy.

$$\frac{V_1}{V_0} < \frac{h_0}{h_1} \tag{4}$$

where V_0, V_1 are the circumferential speeds of the work rolls; h_0, h_1 are the thickness of the sheet before and after rolling.

3.3. Shear Strain and Equivalent Strain

The level of a shear strain and equivalent strain plays a key role in terms of the possibility of using the asymmetric rolling process as a method of SPD. Two types of shear strain are known: pure shear and simple shear. Pure shear corresponds to symmetric rolling. During pure shear, the square is converted to a rectangle and the circle is converted to an ellipse (Figure 19). The thickness "AB" is compressed, and the length "AD" is elongated. The axes of the strain ellipsoid do not rotate (Figure 20).

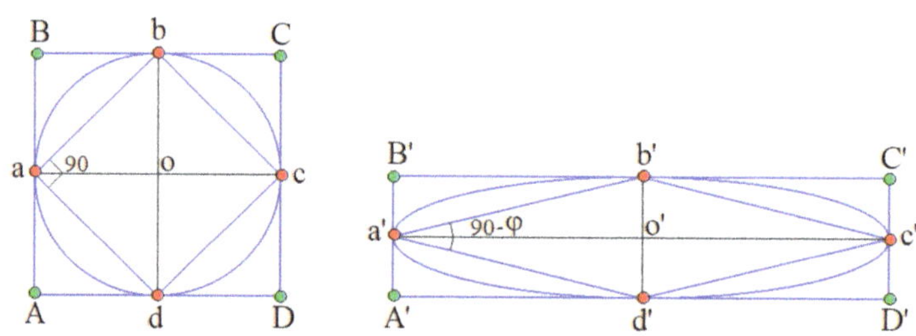

Figure 19. Schematic illustration of pure shear during symmetric rolling.

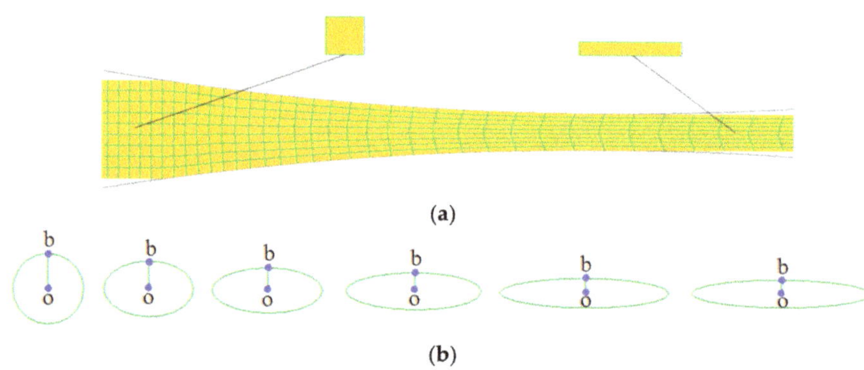

Figure 20. FEM simulation (this work) of pure shear during symmetric rolling (**a**) and transformation of strain ellipsoid (in the middle layer) when passing through the deformation zone (**b**).

Simple shear corresponds to equal channel angular pressing (ECAP) (Figure 21). During simple shear, the square is converted to a parallelogram and the circle is converted to an ellipse (Figure 22). The thickness "AB" and the length "AD" are not changed. Simple shear involves rotation of the axes of the strain ellipsoid (Figure 23). Exact rotation leads to grain refinement and the formation of high angle boundaries during ECAP.

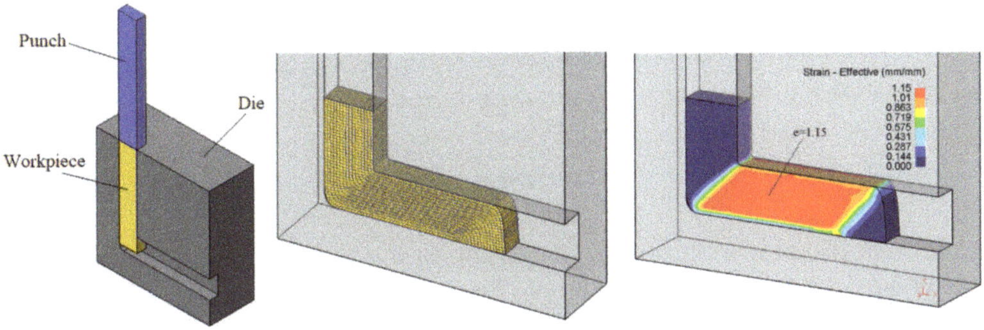

Figure 21. Schematic illustration of ECAP.

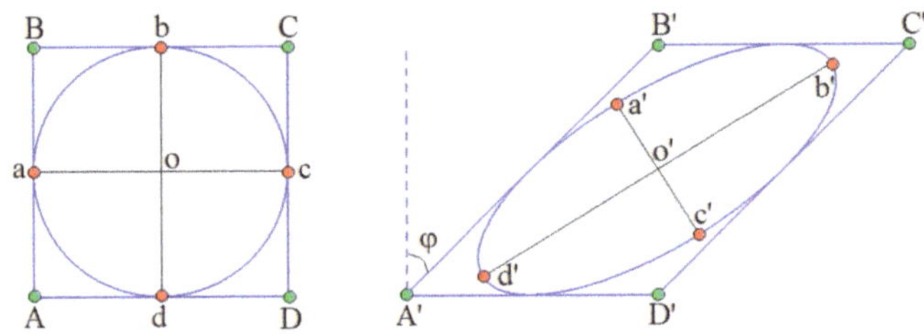

Figure 22. Schematic illustration of simple shear during ECAP.

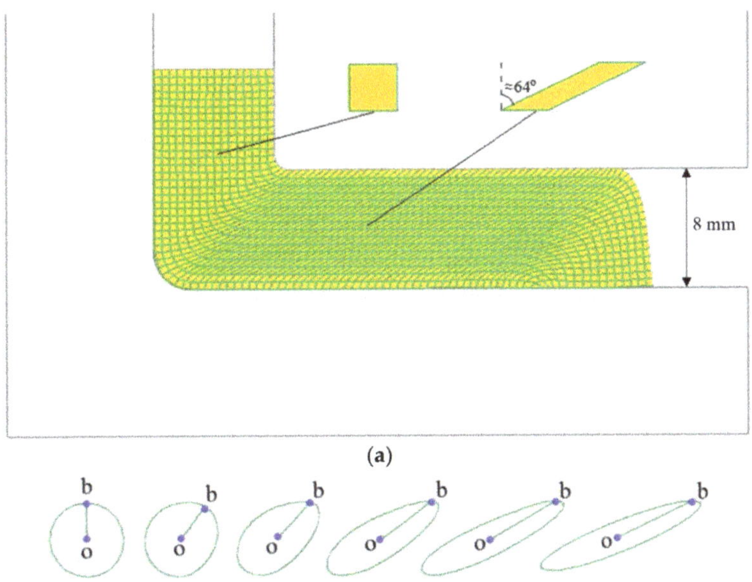

Figure 23. FEM simulation (this work) of simple shear during ECAP (**a**) and rotation of the axes of the strain ellipsoid when passing through the deformation zone (**b**).

Simultaneous pure and simple shear combine the advantages of simple shear (rotation of the material as with ECAP) and the advantages of pure shear (compression and elongation of the material as with symmetric rolling). Simultaneous pure and simple shear (Figure 24) corresponds to asymmetric rolling. The square is converted to a parallelogram and the circle is converted to an ellipse. The thickness "AB" is compressed, and the length "AD" is elongated. The ellipse is rotated (Figure 25) [58].

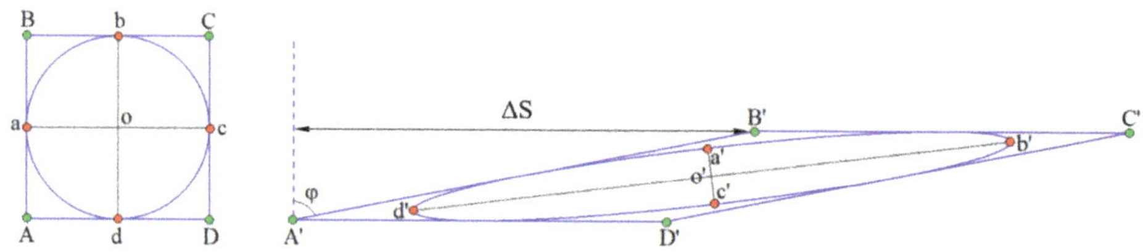

Figure 24. Schematic illustration of simultaneous pure and simple shear during asymmetric rolling.

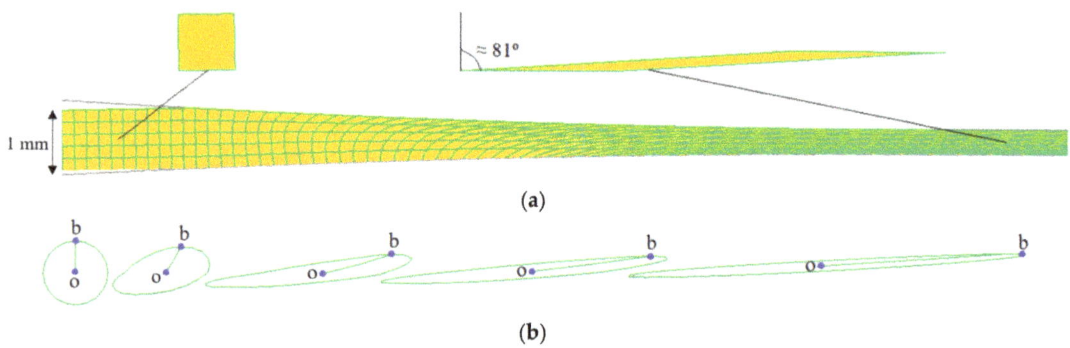

Figure 25. Results of FEM simulation (this work) of simultaneous pure and simple shear during asymmetric rolling (**a**) and rotation of the axes of the strain ellipsoid when passing through the deformation zone (**b**) (simulation conditions: room temperature, material AA5083, initial sheet thickness 1.0 mm, thickness reduction 60%, diameters of work rolls 500 mm, Coulomb friction coefficient 0.4, speed of the bottom work roll 43 mm/s, speed of the top work roll 100 mm/s, FEM code DEFORM 2D™).

The equivalent strain $\bar{\varepsilon}$ for asymmetric rolling process can be defined as von Mises strain [23,59–62]:

$$\bar{\varepsilon} = \sqrt{\frac{2}{3}e_{ij}e_{ij}} \tag{5}$$

$$\bar{\varepsilon} = \sqrt{\frac{2}{9}\left[(\varepsilon_{11}-\varepsilon_{22})^2 + (\varepsilon_{22}-\varepsilon_{33})^2 + (\varepsilon_{33}-\varepsilon_{11})^2 + 6\left(\varepsilon_{12}^2 + \varepsilon_{23}^2 + \varepsilon_{31}^2\right)\right]} \tag{6}$$

As, $\varepsilon_{11} = \varepsilon_{xx}$, $\varepsilon_{22} = \varepsilon_{yy}$, $\varepsilon_{33} = \varepsilon_{zz}$, $\varepsilon_{12} = \varepsilon_{xy}$, $\varepsilon_{23} = \varepsilon_{yz}$, $\varepsilon_{31} = \varepsilon_{zx}$, then

$$\bar{\varepsilon} = \sqrt{\frac{2}{9}\left[\left(\varepsilon_{xx}-\varepsilon_{yy}\right)^2 + \left(\varepsilon_{yy}-\varepsilon_{zz}\right)^2 + (\varepsilon_{zz}-\varepsilon_{xx})^2 + 6\left(\varepsilon_{xy}^2 + \varepsilon_{yz}^2 + \varepsilon_{zx}^2\right)\right]} \tag{7}$$

The deformation process during asymmetric rolling can be approximated by a two-dimensional strain state of compressive strain along the normal direction z (ε_{zz} and $\varepsilon_{zz} = -\varepsilon_{xx}$) together with simple shear strain along the rolling direction x (ε_{xz}):

$$\begin{bmatrix} \varepsilon_{xx} & 0 & 0 \\ 0 & 0 & 0 \\ 0 & 0 & \varepsilon_{zz} = -\varepsilon_{xx} \end{bmatrix} + \begin{bmatrix} 0 & 0 & \varepsilon_{xz} \\ 0 & 0 & 0 \\ 0 & 0 & 0 \end{bmatrix} = \begin{bmatrix} \varepsilon_{xx} & 0 & \varepsilon_{xz} \\ 0 & 0 & 0 \\ 0 & 0 & \varepsilon_{zz} = -\varepsilon_{xx} \end{bmatrix} \tag{8}$$

Then, Equation (7) can be refined as:

$$\bar{\varepsilon} = \frac{2}{\sqrt{3}}\sqrt{\varepsilon_{xx}^2 + \varepsilon_{xz}^2} \tag{9}$$

The compressive strain component and simple shear strain component can be calculated, respectively, as follows: $\varepsilon_{xx} = ln\frac{h_0}{h_1}$, $\varepsilon_{xz} = \frac{\gamma_{xz}}{2} = \frac{\gamma}{2}$, where h_0 is the initial sheet thickness, h_1 is the final sheet thickness. Then, Equation (9) can be refined as:

$$\bar{\varepsilon} = \frac{2}{\sqrt{3}}\sqrt{\left(ln\frac{h_0}{h_1}\right)^2 + \left(\frac{\gamma}{2}\right)^2} \tag{10}$$

If shear strain $\gamma = 0$, then Equation (10) transforms into the well-known equivalent strain $\bar{\varepsilon}$ for the symmetric rolling process:

$$\bar{\varepsilon} = \frac{2}{\sqrt{3}}ln\frac{h_0}{h_1} \tag{11}$$

If compressive strain $\varepsilon_{xx} = 0$, when $h_0 = h_1$, then Equation (10) transforms into the well-known equivalent strain $\bar{\varepsilon}$ for simple shear:

$$\bar{\varepsilon} = \frac{\gamma}{\sqrt{3}} \tag{12}$$

$$\gamma = tg\varphi \tag{13}$$

where γ is the simple shear strain, φ is the shear angle (see Figure 22). Fueled by advances in severe plastic deformation, there has been much discussion about computing the equivalent strain $\bar{\varepsilon}$ in simple shear [63]. Since the increment in equivalent strain can be defined via the increment in plastic work, the authors of [63,64] argue for using the von Mises strain (11). On the other hand, Onaka [65] argues that one must use the Hencky relation [66]. The justification given by Onaka is the need to exclude rotations from the deformation gradient tensor in large simple shear deformations. Beygelzimer [67] showed that the equivalent strain of simple shear must be linear in γ. This approach gives evidence in favor of using the von Mises strain (12) for computing the equivalent strain in simple shear.

Saito et al. (1986) [68], Sakai et al. (1988) [69], Cui and Ohori (2000) [15] and Angella et al. (2013) [70] used the following equations for calculation of the equivalent strain $\bar{\varepsilon}$ and shear strain γ in asymmetric rolling:

$$\bar{\varepsilon} = \frac{2}{\sqrt{3}}\phi \ln\frac{1}{1-r} \tag{14}$$

$$r = 1 - \frac{t_1}{t_0},\ \phi = \sqrt{1 + \left\{\frac{(1-r)^2}{r(2-r)}\tan\theta\right\}^2} \tag{15}$$

$$\gamma = 2\sqrt{\phi^2 - 1}\ln\frac{1}{1-r} \tag{16}$$

where t_0, t_1 are the thickness of the sheet before and after rolling; γ is the shear strain; θ is the shear angle.

In 2005, Kang et al. [71] reported that the shear strain component γ imposed by asymmetric rolling with different diameters of the top and bottom work rolls could be calculated by the following equation:

$$\gamma = \frac{1}{\langle D\rangle}\left[R_1 cos^{-1}\left(\frac{R_1 - \Delta D/2}{R_1}\right) - R_2 cos^{-1}\left(\frac{R_2 - \Delta D/2}{R_2}\right)\right] \tag{17}$$

where $\langle D\rangle = \frac{1}{2}(D_1 + D_2)$, $\Delta D = D_1 - D_2$; D_1 is the initial sheet thickness; D_2 is the final sheet thickness; R_1 is the larger radius of the work roll; R_2 is the smaller radius of the work roll.

In 2008, Sidor et al. [60], by using the assumed shear strain γ during asymmetric rolling in accordance with Equation (17), reported that the equivalent von Mises strain could be calculated by the following equation:

$$\bar{\varepsilon} = \frac{\sqrt{2}}{3}\left[6\left[ln\left(1-\frac{(h_i-h_f)}{h_i}\right)\right]^2 + \frac{6}{(h_i+h_f)^2}\left[R_1cos^{-1}\left(\frac{2R_1-(h_i-h_f)}{2R_1}\right) - R_2cos^{-1}\left(\frac{2R_2-(h_i-h_f)}{2R_2}\right)\right]^2\right]^{1/2} \quad (18)$$

where h_i is the initial sheet thickness, h_f is the final sheet thickness, R_1 is the larger radius of the work roll, R_2 is the smaller radius of the work roll.

In 2016, Ma et al. [72] reported that the shear strain component γ imposed by asymmetric rolling with the same diameters of work rolls but different angular speeds could be calculated by the following equation:

$$\gamma = \frac{1}{h_i + h_f} R \cos^{-1}\left(1 - \frac{h_i - h_f}{2R}\right)\left(1 - \frac{v_2}{v_1}\right) \quad (19)$$

where h_i is the initial sheet thickness, h_f is the final sheet thickness, R is the radius of the work roll, v_1 is the speed of the slower work roll, v_2 is the speed of the faster work roll.

In 2015, Park J.-J. [57] reported that the shear strain component γ imposed by asymmetric rolling with different circumferential speeds could be calculated by the following equation:

$$\gamma = \frac{\Delta s}{t_{ave}} = \frac{(V_{us} - V_{ls})\Delta t}{t_{ave}} = \frac{C(V_{ur} - V_{lr})\Delta t'}{t_{ave}} = \frac{C'L}{t_{ave}} \quad (20)$$

where Δs is the displacement (see Figure 24); $t_{ave} = \frac{h_0+h_1}{2}$ is the average thickness of the sheet; h_0 is the initial sheet thickness; h_1 is the final sheet thickness; Δt is the the time required for a material point on the fast surface of the sheet to pass through the deformation zone; V_{us}, V_{ls} are the tangential speeds of the upper (faster) and lower (slower) surfaces of the sheet, respectively; V_{ur}, V_{ls} are the tangential speeds of the upper (faster) and lower (slower) work rolls, respectively; $\Delta t'$ is the time required for a material point on the surface of the fast work roll to pass through the deformation zone; L is the length of the deformation zone; C varies from 0 to 0.7 and C' varies from 0 and 0.5 [57].

Substituting Equation (13) in Equation (10), the equivalent strain $\bar{\varepsilon}$ depending on the shear angle φ can be found for the asymmetric rolling process. Severe plastic deformation usually corresponds to equivalent strain $\bar{\varepsilon} \geq 3 \ldots 4$. This requires at least three or four passes of ECAP with a simple shear. This level can be obtained by single-pass asymmetric rolling with simultaneous pure and simple shear, when shear angle φ is no less than 80° (Figure 26) [73].

Taking C' = 0.5 in Equation (20) and simplifying that $L = \sqrt{R(h_0 - h_1)}$, and equating Equations (13) and (20), the limiting shear angle φ_{max} can be found:

$$\varphi_{max} = arctg\frac{\sqrt{R(h_0 - h_1)}}{h_0 + h_1} \quad (21)$$

The influence of initial thickness h_0 of the sheet and thickness reduction per pass ε on limiting shear angle φ during asymmetric rolling can be found with Equation (21) (Figure 27). From the graph, it follows that to create a shear angle φ of at least 80 degrees, it is necessary to make the thickness reduction per pass $\varepsilon \geq 50\%$ when initial sheet thickness $h_0 = 1$ mm, or $\varepsilon \geq 70\%$ when $h_0 = 2$ mm (in all cases, the diameter of work rolls is 300 mm).

The diameters of work rolls (cold rolling mills) at laboratory and industrial scales vary from 50 mm to 600 mm (Figure 28). The influence of diameter D of work rolls and thickness reduction per pass ε on limiting shear angle φ during asymmetric rolling is shown in Figure 29. From the graph, it follows that to create a shear angle φ of at least 80 degrees, it is necessary to make the thickness reduction per pass $\varepsilon \geq 50\%$ when the diameter of work rolls D = 600 mm, or $\varepsilon \geq 70\%$ when D = 300 mm (in all cases, the initial thickness of the sheet $h_0 = 2$ mm).

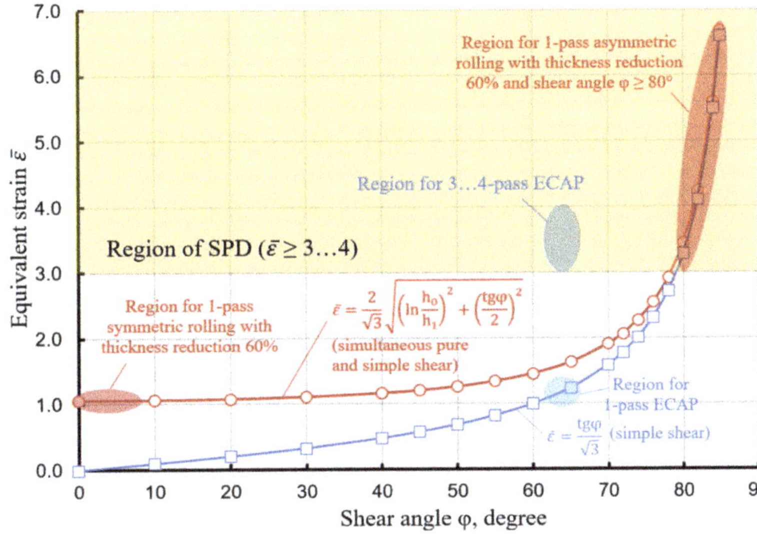

Figure 26. Influence of the shear angle φ on the equivalent strain $\bar{\varepsilon}$ during symmetric rolling, asymmetric rolling and ECAP.

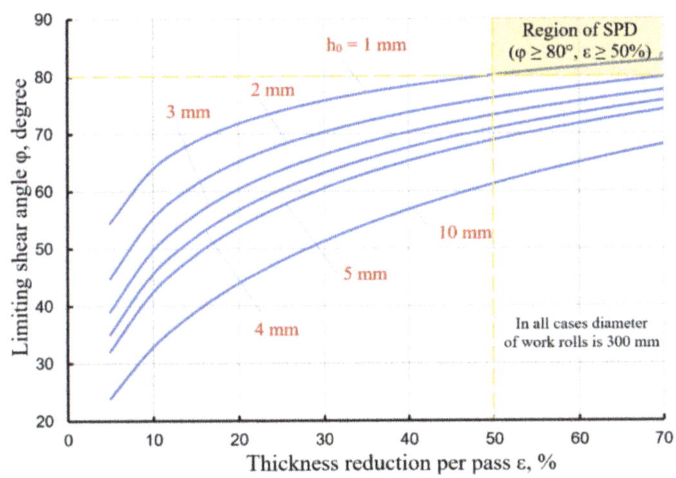

Figure 27. Influence of initial thickness h_0 of the sheet and thickness reduction per pass ε on limiting shear angle φ during asymmetric rolling.

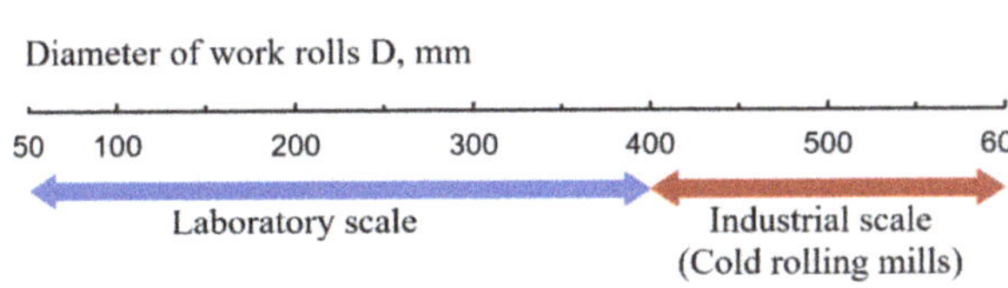

Figure 28. Diameter D (mm) of work rolls (cold rolling mills) at laboratory and industrial scales.

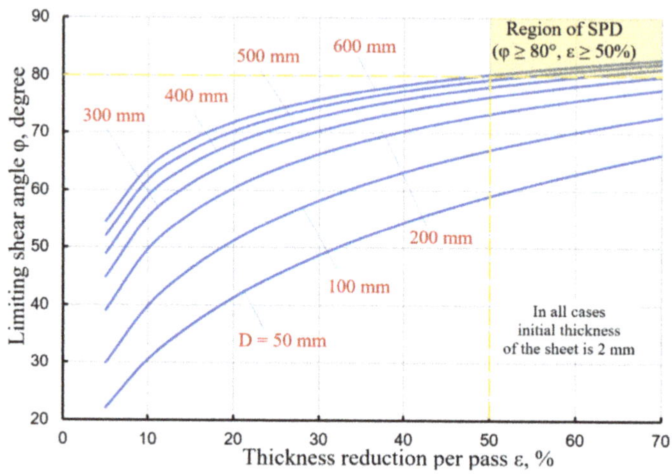

Figure 29. Influence of diameter D of work rolls and thickness reduction per pass ε on limiting shear angle φ during asymmetric rolling.

Hence, a very large equivalent strain $\bar{\varepsilon} \geq 3 \ldots 4$ during asymmetric rolling can be introduced with a large roll diameter (D ≥ 300 mm), high thickness reduction per pass (ε ≥ 50%) and thinner initial thickness of the sheet ($h_0 \leq 2$ mm). In this case, asymmetric rolling should be carried out on dry work rolls without lubrication (with high contact friction) and a high speed ratio of the work rolls ($S_R = 2 \ldots 4$).

3.4. Temperature Rise

Since asymmetric rolling is performed with high contact friction and high thickness reduction per pass, the temperature increase in the deformation zone can be significant [74–76]. The temperature increase may affect the additional microstructural changes which are related to thermal softening deteriorating the mechanical properties. The model to predict the temperature increase of the sheet during differential speed rolling was proposed in [74]. The temperature increase might be affected by several factors (Table 1). The first factor is the total stored energy which is related to external deformation work. The second factor is the internal energy determined by the mechanical capability of materials, which is defined as the total area beneath the stress–strain curve. The other factors are the energy input caused by frictional heat and the energy loss by non-isothermal conditions. Considering the fact that 90% of the deformation work is converted into heat and 50% of the frictional heat is transferred to the sheet, the energy-based state equation during differential speed rolling is expressed as follows [74]:

$$\rho C V \Delta T = 0.9 \sigma \varepsilon V + 0.5 m \tau v A \Delta t - A h \Delta T \Delta t \quad (22)$$

where ρ is the material density; C is the heat capacity; V is the main deformation volume; ΔT is the temperature increase in the main deformation volume; σ is the stress; ε is the strain; m is the friction factor; τ is the shear stress; v is the speed of the faster work roll; A is the contact area between the sheet and the faster work roll; Δt is the deformation time; h is the heat transfer coefficient.

Table 1. Factors and their equations influencing the temperature increase of the sheet during differential speed rolling (based upon the energy-based state equation), reproduced from [74], with permission from Elsevier, 2014.

Factors	Equations
Total stored energy by plastic work	$\rho CV\Delta T$
Internal energy generation by material capability	$0.9\sigma\varepsilon V$
Energy generation by frictional effect	$m\tau vA\Delta t$
Energy loss by non-isothermal conditions	$Ah\Delta T\Delta t$

The temperature increase in the sheet during differential speed rolling can be represented as [74]:

$$\Delta T = \frac{0.9\sigma\varepsilon + 0.5m\tau v\frac{A}{V}\Delta t}{\rho C + \frac{A}{V}h\Delta t} \quad (23)$$

where $V = Lw\left(\frac{h_0+h_1}{2}\right)$; $A = Lw$; $\Delta t = \frac{L}{v}$; h_0 is the initial sheet thickness; h_1 is the final sheet thickness; L is the length of the deformation zone; w is the width of the sheet.

Since severe plastic deformation should be performed at relatively low temperatures $T \leq (0.3 \ldots 0.4)T_{melt}$, the temperature increase in the sheet during differential speed rolling, especially for Mg and Al alloys, must be taken into account.

3.5. Deformation Routes

Symmetric rolling can be performed without any kind of rotation of the sheet between each pass. However, in asymmetric rolling, the purpose of imposing high levels of shear strains leads to the idea of rotating the sheet between each rolling pass, taking advantage of the difference in work roll speeds [77]. Several rotation routes can be performed, as shown in Figure 30 [78–81]. By the deformation route UD, the sheet is not rotated before and after each rolling pass, while in the other three cases it is rotated by 180° degrees. By the route RD, the rotation axis is the rolling direction, by the route TD it is the transverse direction and by the route ND it is the normal direction. From the observation of the possible reversal routes, it becomes clear that the UD and TD routes do not impose any reversal of strains on the sheet's surface, whereas RD and ND impose strain reversal. In the literature, the TD route is one of the most commonly used, but when studying the asymmetric rolling process, S.H. Lee and D.N. Lee [78] observed textures closer to the ideal "shear" components on specimens processed by the RD and ND reversing routes. Using FEM simulations, Kim and Lee [82] studied the influence of shear strain evolution during symmetric and asymmetric rolling. They found that in symmetric rolling, the sheet undergoes positive and negative shear strains, before and after the neutral point, respectively [82,83]. However, during asymmetric rolling, shear strains are positive. By using the shear strain history, determined for symmetric and asymmetric rolling, Kim and Lee calculated the resulting crystallographic textures. They found that the ideal shear textures were closely related to the shear strain reversal. In order to understand the influence of the deformation route, it is necessary to examine the shear patterns. These patterns are illustrated schematically in the form of the dominant directions of shear for routes UD, TD, RD and ND. It is apparent from Figure 30 that the shear patterns are significantly different between UD, TD and RD, ND deformation routes [84]. Routes UD and TD repeat shear on the same plane. By contrast, routes RD and ND have two shear planes intersecting at the shear angle. Y.G. Ko and K. Hamad [85], for AZ31 Mg alloy, experimentally showed that the RD route is the most effective in grain refinement, yielding equiaxed grains with an average size of 1.2 μm, while routes UD and TD yielded elongated grains with average sizes of 1.4 and 2.1 μm, respectively (Figure 31). Route ND, in turn, led to equiaxed fine grains (1.3 μm), but the trough thickness homogeneity achieved by the ND route was less than that achieved by the RD. The coarser-grained microstructures obtained for the sheets deformed by the UD and TD routes led to lower yield strengths compared to those deformed by the RD and ND routes [85].

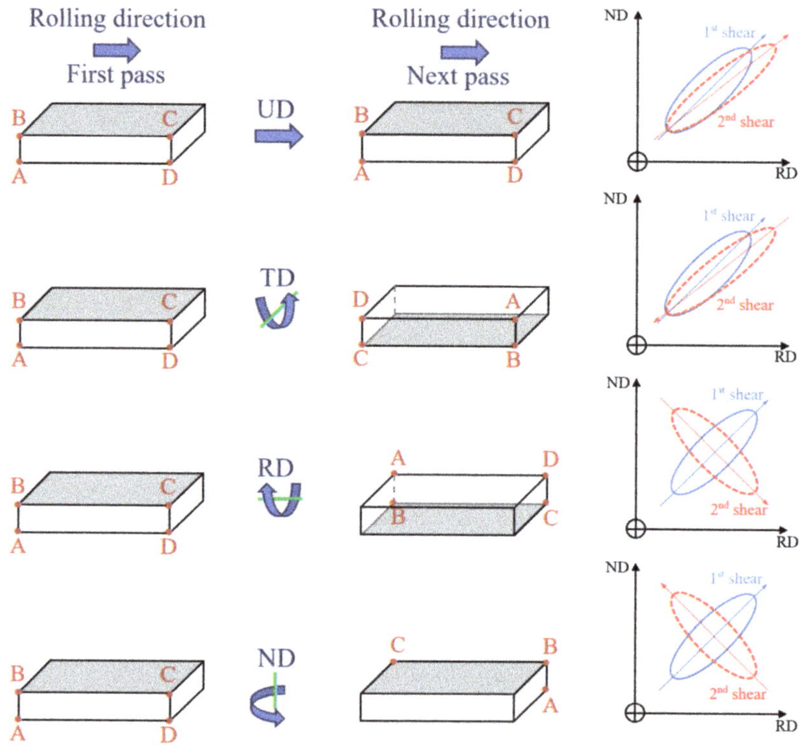

Figure 30. Schematic illustration of deformation routes and shear directions during asymmetric rolling. In UD rolling, rolling direction is not changed, whereas in TD, RD and ND rolling, the sheet is rotated through 180° about TD, RD and ND axes, respectively, for each pass. Reprinted from [84].

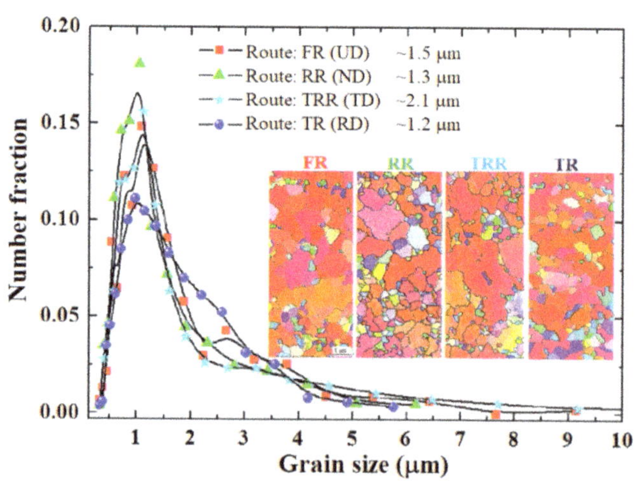

Figure 31. Grain size distributions and average grain sizes of the AZ31 Mg alloy samples deformed by differential speed rolling using the different deformation routes (thickness reduction from 4 mm to 1 mm by two passes, speed ratio $S_R = 4.0$), reproduced from [85], with permission from Elsevier, 2018.

Asymmetric rolling can give rise to severe plastic shear strains and in turn shear deformation textures through the sheet thickness. The ideal shear deformation texture of FCC metals can be approximated by the <111>//ND and {001}<110> orientations, among which the former improves the deep drawability. The ideal shear texture could not be obtained by using unidirectional asymmetric rolling, but only by using RD or ND reversing deformation routes [82,83].

One of the most important parameters in asymmetric rolling is the thickness reduction per pass [77]. When combined with a difference in work roll speeds, an increase in thickness reduction per pass imposes higher levels of shear strain on the sheet surface and, under the appropriate conditions, it is essential to achieve the total propagation of those shear strains along the sheet's thickness. Kim and Lee [82] studied the influence of the reduction per pass on the shear texture formation. They found that high values of this parameter were necessary to achieve shear textures. Low thickness reductions per pass led to plane strain textures, similar to those found with conventional rolling. At the same time, the thickness reduction per pass has a nonlinear effect on grain refinement. Lee and Lee [79,86], from the experimental results, concluded that a 5% or 50% reduction per pass gave rise to finer grains with a higher fraction of high angle boundaries than reductions of 10, 20 and 30% per pass (Figure 32). Figure 33a shows EBSD mappings of AA1050 sheets asymmetrically rolled by 93%. Uniformly distributed high-angle boundary structures developed in the sheets rolled with 5% and 50% reductions per pass. It can be seen that the 5% and 50% sheets have higher fractions of high-angle boundaries over 15° and finer grain sizes, as shown in Figure 32. The grain size decreased from the initial size of about 100 μm to 1.4–1.8 μm [79].

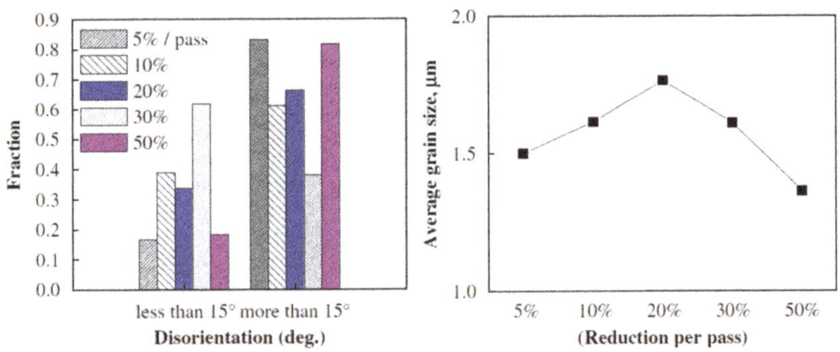

Figure 32. Disorientation distributions and average grain sizes in AA1050 sheets asymmetrically rolled by 93% after annealing at 195 °C for 60 min (thickness reduction from 4.2 mm to 0.3 mm, speed ratio $S_R \approx 2.0$ due to different diameters of work rolls of 248 mm and 128 mm), reproduced from [79], with permission from Elsevier, 2008.

The equivalent strain $\bar{\varepsilon}$ in the surface layer of the rolled sheet is a function of friction coefficient μ and the geometric shape factor L/t_{ave} (where $L = \sqrt{R(h_0 - h_1)}$ is the length of the deformation zone; $t_{ave} = (h_0 + h_1)/2$ is the average thickness of the sheet; h_0 is the initial sheet thickness; h_1 is the final sheet thickness) [79]. During rolling with high contact friction, the equivalent strain $\bar{\varepsilon}$ is the highest when the geometric shape factor L/t_{ave} is very small or, on the contrary, very large (Figure 33b) [79]. For a given thickness reduction, the equivalent strain $\bar{\varepsilon}$ is proportional to shear strains because the normal strains are the same, as can be seen in Equation (6). Therefore, the minimum equivalent strain $\bar{\varepsilon}$ in Figure 33b corresponds to the minimum shear strain at some value of geometric shape factor L/t_{ave}. The geometric shape factors L/t_{ave} at 5% and 50% thickness reductions per pass in [79,86] could be on the left and right sides of the minimum equivalent strain $\bar{\varepsilon}$ or the minimum shear strain, respectively. Thus, the reason why a 5% or 50% thickness reduction per pass gave rise to finer grains with a higher fraction of high-angle boundaries than thickness reductions of 10, 20 and 30% per pass can be explained.

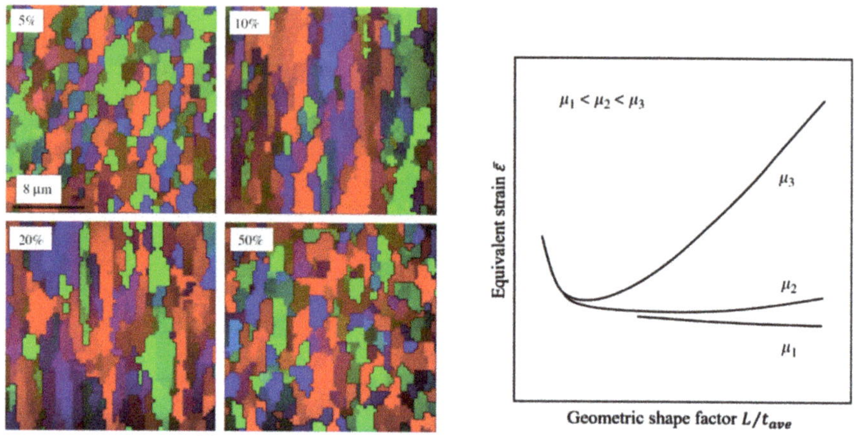

(a) (b)

Figure 33. (**a**) EBSD mapping along rolling plane of AA1050 sheets asymmetrically rolled by 93% after annealing at 195 °C for 60 min (thickness reduction from 4.2 mm to 0.3 mm, speed ratio $S_R \approx 2.0$ due to different diameters of work rolls of 248 mm and 128 mm). (**b**) Schematic representation of equivalent strain $\bar{\varepsilon}$ in the surface layer against geometric shape factor L/t_{ave} at different friction coefficients μ, reproduced from [79], with permission from Elsevier, 2008.

4. Asymmetric (Hot, Warm, Cold) Rolling of Mg Alloys

Magnesium alloys (especially these containing Al and Zn additions—so-called AZ series) are in many fields superior to aluminum alloys (e.g., they possess a lower density and thus a better specific strength). However, their problematic formability requires consideration [62]. Over the last 20 years, a number of scientific works have been devoted to a fabrication of Mg alloy sheets with a good drawability. It was proposed that the main reason for the very poor cold formability and high mechanical anisotropy is an induction of a strong {0 0 0 1} basal texture in conventional plastic-forming processing due to a limited number of slip systems in the hexagonal close-packed (hcp) crystal structure [62]. The results of an extensive study on various Mg alloys (Tables 2–4 and Figure 34) showed that the (hot, warm, cold) differential speed rolling has a great impact on the intensity of the basal texture, grain size and plasticity of these materials. It was established that increasing the shear strain by raising the roll speed ratio S_R leads to weakening of the basal texture through facilitating the activation of prismatic slip during deformation. The basal texture weakening effect at high speed ratios is attributed to extensive tension twinning that occurs in the basal-oriented matrix. Consequently, the asymmetrically rolled Mg alloy sheets are characterized not only by more isotropic properties but also by the enhanced plasticity combined with exceptionally high strength that is related to the simultaneous structure refinement. Therefore, the differential speed rolling process is considered to be one of the most efficient techniques for processing these materials.

Table 2. Experimental data related to asymmetric hot rolling of Mg alloys.

Material	Initial/Final Sheet Thickness, mm	Top/Bottom Work Roll Diameters, mm	Speed Ratio S_R	Total Reduction/Number of Passes	T, °C	Final Grain Size, µm	YS, MPa	UTS, MPa	EL, %	Ref.
Pure Mg	10/2.5	165/110	1.5	75%/2	350	6.74	-	-	-	[87]
Pure Mg	10/2.5	-	1.5	75%/2	350	6.7	-	-	-	[88]
Mg-3%Ca	2.0/0.7	-	2.0	65%/2	300	5.81	162.1	-	2.4	[89]
ZK60	1.0/0.55	-	1.5	45%/4	300	-	252	322	7.7	[90]
AZ31	6.0/2.0	700/600	1.25	67%/2	300	-	188	272	21	[91]
AZ31	4.3/1.8	220/220	3	58%/2	300	15	185	265	17.5	[92]
	4.3/1.8	220/220	3	58%/2	350	16	180	265	19	
AZ31	6.0/1.0	-	1.2	83.3%/6	350	10.3	145	240	28.5	[93]
AZ31	1.0/0.85	-	1.1	15%/1	300	-	241	289	13.6	[94]
AZ91	3.5/2.98	150/150	Single roll drive	15%/1	400	86.7	-	255.5	3.7	[95]
AZ31	4.0/1.5	-	1.167	63%/1	300	11	-	-	-	[96]
	4.0/1.5	-	1.167	63%/2	300	9	-	-	-	
	4.0/1.5	-	1.167	63%/4	300	10	-	-	-	
	4.0/1.5	-	1.167	63%/6	300	10	-	-	-	
	4.0/1.5	-	1.167	63%/8	300	12	-	-	-	
	4.0/1.5	-	1.167	63%/10	300	12	-	-	-	
ZK60	1.0/0.55	-	1.2	45%/4	300	-	246	331	6.1	[97]
	1.0/0.55	-	1.5	45%/4	300	-	253	323	7.7	
AZ31B	4.0/2.0	-	1.167	50%/6	430	-	-	-	-	[98–100]
	2.0/1.0	-	1.167	50%/8	300	-	145.5	259.2	25.5	
AM31	1.0/0.5	-	1.2	50%/2	300	5.6	195	265	17.5	[101]
	1.0/0.5	-	1.5	50%/2	300	4.9	181	257	16.9	
AZ31	10/1.5	-	1.05	85%/8	350	5.0				[102]
AZ80	5.0/1.0	-	1.36	80%/4	430	7.0	182	330	24.6	[103]
AZ31	2.0/0.6	-	3.0	70%/1	350	3.8	158	260	18	[16]
AM31	1.0/0.55	-	1.2	45%/2	300	6.9	222	283	18.3	[104]
	1.0/0.55	-	1.5	45%/2	300	4.9	217	259	17.6	
AZ91	2.0/0.5	-	2.0	75%/2	400	1~2	310	-	7 830 *	[105]
AZ31B	4.0/2.0	160/160	1.167	50%/6	430	-	-	-	-	[106]
	2.0/1.0	160/160	1.167	50%/8	300	12.3	-	-	-	
Mg-9.25Zn-1.66Y	2.0/0.61	-	2	69.5%/2	400	1~2	342	-	4.2	[107]
	2.0/0.63	-	3	68.5%/2	400	0.5~1.0	337	-	10.7	
Mg-4Y-3RE	5.0/1.0	-	1.36	80%/4	550	-	-	-	-	[108]
Mg-13Zn-1.55Y	2.0/0.6	-	3.0	70%/2	400	1~2			1021 **	[109]
AZ31	5.3/2.25	190/128	1.5	57.5%/1	500	-	-	-	-	[110]
	6.5/1.0	190/128	1.5	84.6%/9	500	-	-	-	-	

* Superplasticity at 300 °C and 1×10^{-3} s^{-1}; ** superplasticity at 250 °C and 1×10^{-3} s^{-1}.

Table 3. Experimental data related to asymmetric warm rolling of Mg alloys.

Material	Initial/Final Sheet Thickness, mm	Top/Bottom Work Roll Diameters, mm	Speed Ratio S_R	Total Reduction/Number of Passes	T, °C	Final Grain Size, µm	YS, MPa	UTS, MPa	EL, %	Ref.
Pure Mg	10/2.5	165/110	1.5	75%/2	200	3.95	-	-	-	[87]
AZ31	2.0/0.6	300/300	3.0	70%/1	200	3.0	-	-	-	[111]
Pure Mg	10/2.5	-	1.5	75%/2	200	4.0	-	-	-	[88]
AZ31B	2.0/1.6	280/280	1.5	20%/1	200	-	-	-	-	[112]
AZ31	1.6/0.4	70/60	1.17	75%/12	150	1.47	-	-	-	[113]
Mg-1Gd	1.6/0.4	70/60	1.17	75%/12	150	1.7	-	-	-	[113]
AZ31	4.3/2.8	220/220	3	35%/1	200	9.2	240	280	14	[92]
	4.3/1.8	220/220	3	58%/2	200	5.4	265	320	8.5	
	4.3/2.8	220/220	3	35%/1	250	12.1	180	265	17.5	
	4.3/1.8	220/220	3	58%/2	250	9.1	185	265	15.5	
AZ31	2.0/1.7	-	3.0	15%/1	150	-	-	-	-	[114]
AZ31	2.0/1.6	400/400	1.1	20%/1	240	6.3	236.9	295.5	12.2	[115]
	2.0/1.6	400/400	1.2	20%/1	240	4.3	241.5	296.3	13.8	
	2.0/1.6	400/400	1.5	20%/1	240	3.5	233.8	288.5	6.8	
	2.0/1.6	400/400	2	20%/1	240	6.1	209.6	281.8	16	
	2.0/1.6	400/400	3	20%/1	240	5.9	201	281.8	16.7	
AZ31	1.0/0.85	-	1.1	15%/1	200	-	242	295	18.5	[94]
	1.0/0.85	-	1.1	15%/1	250	-	237	290	17.3	
AZ31	2.0/0.68	300/300	2.0	66%/2	150	0.6	382	401	6.8	[116]
AZ31	2.0/0.51	400/400	2.0	74.5%/3	200	2.3	275.5	309.7	22.1	[117]
AZ31	2.0/0.6	300/300	3.0	70%/1	150	1.4	-	-	-	[17]
AZ31	2.0/0.9	90/90	3.0	55%/1	150	1.0	-	-	-	[118]
AZ31	2.0/0.6	-	3	70%/1	140	1.4	294	317	5	[16]
	2.0/0.6	-	3	70%/1	160	1.5	275	320	24	
	2.0/0.6	-	3	70%/1	200	2.9	188	274	23	
	2.0/0.6	-	3	70%/1	250	3	180	268	18	
AZ31	4.0/2.0	220/220	3.0	50%/2	180	3.3	198	228	7.1	[119]
AZ61	2.0/0.6	-	3.0	70%/1	200	0.3~0.5	-	-	850 *	[120]
AZ31	4.0/2.0	220/220	4.0	50%/2	150	1.2	295	-	21	[85]
ZK60	2.0/0.6	-	3.0	70%/1	200	1~2	286	338	30 1000 **	[18]

* Superplasticity at 250 °C and 3×10^{-4} s^{-1}; ** superplasticity at 280 °C and 1×10^{-3} s^{-1}.

Table 4. Experimental data related to asymmetric cold rolling of Mg alloys.

Material	Initial/Final Sheet Thickness, mm	Top/Bottom Work Roll Diameters, mm	Speed Ratio S_R	Total Reduction/Number of Passes	T, °C	Final Grain Size, µm	YS, MPa	UTS, MPa	EL, %	Ref.
Pure Mg	10/2.5	165/110	1.5	75%/2	20	2.11	-	-	-	[87]
Pure Mg	10/2.5	-	1.5	75%/2	20	2.3	-	-	-	[88]
AZ91	3.5/2.98	-	Single roll drive	15%/1	20	98	-	-	-	[121]
ZK60	2.0/0.64	400/400	3.0	68%/1	20 *	1.0	-	-	-	[122]

* Cold sheet was fed into hot (200 °C) rolls.

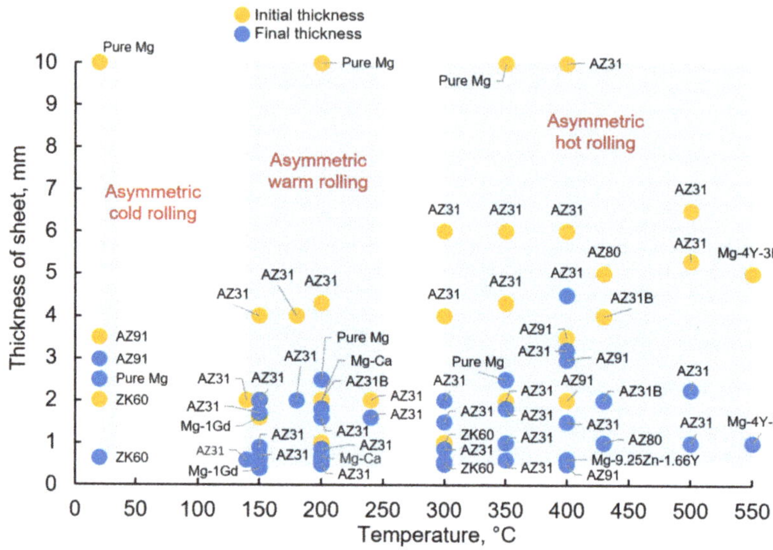

Figure 34. Different Mg alloys, which were processed by asymmetric (hot, warm, cold) rolling (based on data from Tables 2–4).

In 2007, Kim et al. [120] first reported about AZ61 sheet processed by severe plastic deformation using differential speed rolling with a high speed ratio ($S_R = 3$) and the AZ61 sheets contained ultrafine grains 0.3-0.5 μm in size after single-pass rolling at 200 °C (70% thickness reduction). The sheets exhibited good low-temperature superplasticity at 200 and 250 °C, including a maximum elongation of 850% at 250 °C and 3×10^{-4} s^{-1}.

Based on the analysis of publications related to asymmetric (hot, warm, cold) rolling of Mg alloys, it was found that exceptionally high strength in a Mg-3Al-1Zn (AZ31) alloy was achieved by Kim et al. in 2011 [116]. After high-ratio differential speed rolling ($S_R = 2$) at 150 °C followed by immediate water quenching, a grain size of 0.6 μm was obtained. Figure 34 shows the TEM micrograph of the AZ31 sheet. Some grain boundaries were sharp while others were not, indicating a mix of high- and low-angle boundaries. There were many fringes (Figure 35a) and network structures (Figure 35b) within the grains, which corresponded to a high density of dislocations arranged into cell and subgrain boundaries or dislocation networks in the grains. Few twins were spotted. Figure 35b shows many very small spherical or rod-shaped particles homogeneously dispersed over the matrices of AZ31. Energy-dispersive spectroscopy elemental mapping indicated that most of the particles were Al–Mn phase. The Al/Mn atomic ratios were between 1.6 and 1.8, implying that the particles were most likely to be Al_8Mn_5 [116]. As well as the Al–Mn particles, β-$Mg_{17}Al_{12}$ particles were also occasionally encountered. They had a spherical morphology, with sizes of 100–150 nm, and were located mainly on the grain boundaries (Figure 35c). The ultrafine-grained AZ31 exhibited a very high yield stress of 382 MPa, ultimate tensile strength of 401 MPa and total elongation of 6.8% [116].

In 2014, W.Y. Kim and W.J. Kim [117] presented the continuous high-ratio differential speed rolling (HRDSR) technique (Figure 36) for the continuous production of ultrafine-grained AZ31 alloy sheets with enhanced room temperature mechanical properties compared to commercial Mg sheets.

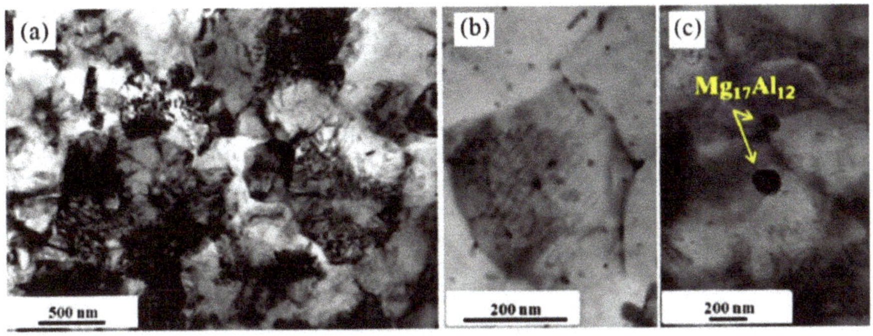

Figure 35. (**a**) TEM micrograph of AZ31 sheet obtained after high-ratio differential speed rolling ($S_R = 2$) at 150 °C followed by immediate water quenching. (**b**) Al–Mn and (**c**) β-$Mg_{17}Al_{12}$ particles in the matrix, reproduced from [116], with permission from Elsevier, 2011.

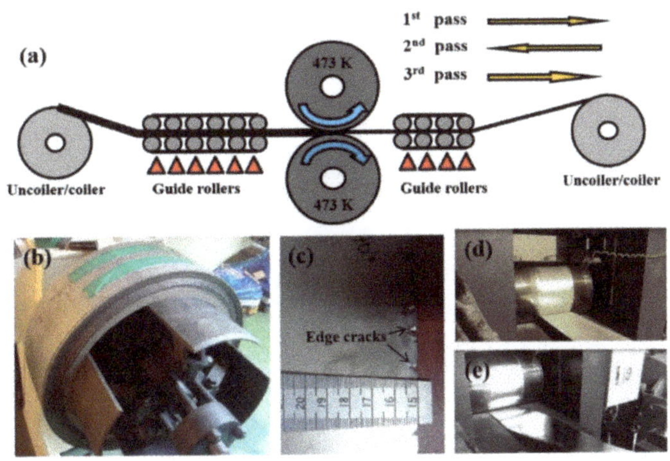

Figure 36. (**a**) The layout of the continuous HRDSR mill. (**b**) The appearance of the HRDSR-processed sheets in coil form. (**c**) The appearance of the HRDSR-processed sheets. Snap shots of the sheets coming out of the rollers during the second pass under (**d**) HRDSR and (**e**) conventional rolling conditions, reproduced from [117], with permission from Elsevier, 2014.

5. Asymmetric (Hot, Warm, Cold, Cryo) Rolling of Al Alloys

Aluminum and aluminum alloys are well known for their high mechanical property anisotropy (a so-called earing behavior) upon a deep drawing process [62]. It was recognized that the main determinant of such behavior is a {100}<100> cubic crystallographic texture formed in a fully annealed state [62]. On the other hand, it was proposed by Lequeu and Jonas [123] that formation of the undesired {100}<100> recrystallization texture component may be prominently inhibited through an application of shear strain prior to a heat treatment. Therefore, a number of works have been devoted to the development of asymmetric rolling-based processing techniques that allow for the fabrication of aluminum alloy sheets with enhanced formability [62]. Engler et al. [124] reported that the most efficient method of formability improvement is to introduce {111} textures (composed of crystallographic orientations that are characterized by {111} crystallographic planes parallel to a rolling plane). Since these orientations are normally found in BCC metals and alloys (and are responsible for an excellent drawability of low-carbon steels), in the case of FCC metals, they may be produced only by shear strain. Jin and Lloyd [125] proved that the recrystallization texture of AA5754 aluminum alloy is randomized (the {001}<100>

component is prominently reduced), when a high-ratio differential speed rolling ($S_R = 1.5$ and $S_R = 2$) is applied before the annealing treatment. The {111} shear strain texture is maintained in the material after annealing that allows for the lowering the so-called planar anisotropy (that characterizes an alteration of mechanical properties in different directions lying in the rolling plane). Analogous results were also obtained by Sakai et al. [39], who showed that 5052 aluminum alloy cold deformed with a 75% thickness reduction in a two-pass asymmetric rolling process followed by recrystallization annealing at a temperature of 310–460 °C exhibits almost perfectly isotropic mechanical behavior (values of the planar anisotropy coefficient were reduced to nearly zero) [62].

Results of an extensive study on various Al alloys (Tables 5–8 and Figure 37) showed that (hot, warm, cold, cryo) differential speed rolling has a great impact on the grain size and mechanical properties of these materials.

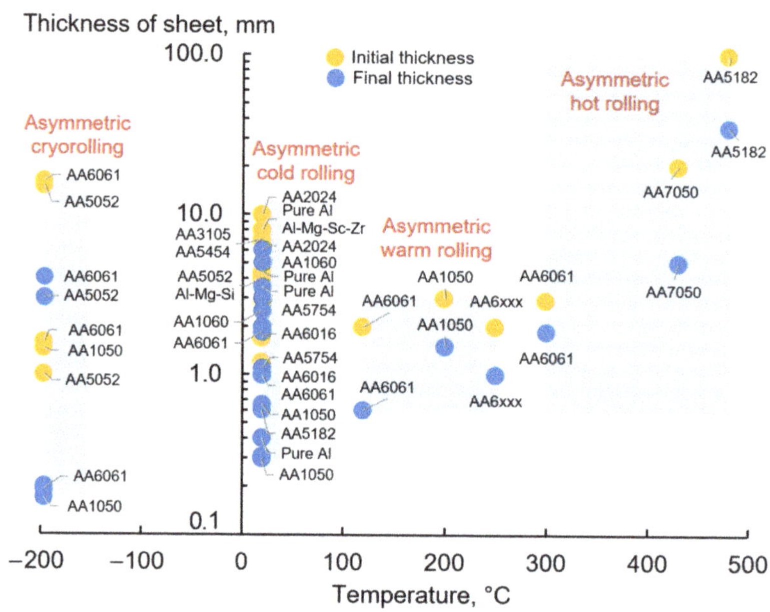

Figure 37. Different Al alloys, which were processed by asymmetric (hot, warm, cold) rolling (based on data from Tables 5–8).

Table 5. Experimental data related to asymmetric hot rolling of Al alloys.

Material	Initial/Final Sheet Thickness, mm	Top/Bottom Work Roll Diameters, mm	Speed Ratio S_R	Total Reduction/Number of Passes	T, °C	Final Grain Size, µm	YS, MPa	UTS, MPa	EL, %	Ref.
AA5182	100/35	450/480	1.07	65%/6	480	-	113.3	272.7	30.5	[55]
							112.8	271.8	31.2	
							113.7	273.3	30.1	
	100/35	400/480	1.2	65%/6	480	-	113.3	273.3	29.2	
							112	272.2	30.5	
							115.7	275	28.7	
AA7050	20/5	400/400	1.25	75%/4	430	-	445	568	22.8	[126]
AA7050	20/5	400/400	1.25	75%/1	430	18.2	459	543	11.6	[72]
							475	539	5.2	
							467	548	12.2	
	20/5	400/400	1.25	75%/2	430	14.6	467	548	12.4	
							468	534	9.6	
							428	540	13.2	
	20/5	400/400	1.25	75%/4	430	12.7	466	550	16.4	
							434	559	14.4	
							411	548	18	
	20/5	400/400	1.25	75%/10	430	13.2	478	553	17.6	
							499	564	18	
							474	562	18	
AA6061	2.9/1.85	180/180	1.5	36%/1	300	-	-	-	-	[127]

Table 6. Experimental data related to asymmetric warm rolling of Al alloys.

Material	Initial/Final Sheet Thickness, mm	Top/Bottom Work Roll Diameters, mm	Speed Ratio S_R	Total Reduction/Number of Passes	T, °C	Final Grain Size, µm	YS, MPa	UTS, MPa	EL, %	Ref.
AA1050	3.0/1.5	130/130	1.5	50%/1	200	-	-	-	-	[37]
AA6061	2.0/0.6	300/300	3.0	70%/1	120	0.37	455	489	7.4	[19]
AA6xxx	2.0/1.0	-	2.0	30%/2	250	-	134	261	37	[128]

Table 7. Experimental data related to asymmetric cold rolling of Al alloys.

Material	Initial/Final Sheet Thickness, mm	Top/Bottom Work Roll Diameters, mm	Speed Ratio S_R	Total Reduction/Number of Passes	T, °C	Final Grain Size, μm	YS, MPa	UTS, MPa	EL, %	Ref.
AA5xxx	2.0/1.0	-	2.0	50%/2	20	-	-	306	33.6	[129]
AA1060	5.0/2.5	100/100	1.02	50%/1	20	12.49	131.6	141	8.88	[130]
AA2024	10/6	150/150	-	40%/-	20	-	-	-	-	[131, 132]
AA3105	7/-	-	1.18	-	20	-	-	-	-	[133]
AA5052	4.0/1.0	220/220	4.0	75%/4	20	0.7	380	390	4.2	[134]
AA5052	4.0/1.0	220/220	4.0	75%/2	20	0.5	-	-	-	[22]
AA5052	4.0/1.0	220/220	4.0	75%/4	20	0.7	380	-	4	[135]
AA5182	1.2/0.6	180/180	3.0	50%/2	20	-	-	-	-	[136]
AA5182	3.0/2.25	180/180	1.36	25%/1	20	33	230	380	13.5	[137]
	3.0/1.5	180/180	1.36	50%/2	20	20	280	410	10.5	
	3.0/1.0	180/180	1.36	66%/4	20	-	310	420	10	
	3.0/0.3	180/180	1.36	90%/6	20	0.3	370	450	7.5	
AA5454	6.7/-	295/295	1.5	33%/1	20	17.7	-	280	-	[138]
	6.7/-	295/295	2	33%/1	20	17.5	235	280	-	
	6.7/-	295/295	1.5	44%/1	20	18	-	293	-	
	6.7/-	295/295	2	44%/1	20	17.5	253	291	-	
AA5754	-	161.5/161.5	2.0	-	20	-	-	-	-	[139]
AA5754	2.5/1.1	161.5/161.5	2.0	56%/2	20	-	-	-	-	[125]
AA6016	1.7/1.0	-	1.5	41%/2	20	-	-	-	-	[140]
AA6061	4.0/1.0	-	4.0	75%/4	20	0.8	322	-	-	[141]
AA6061	2.9/1.85	180/180	1.5	36%/1	20	4.0	-	-	-	[130, 142]
AA6111	-	161.5/161.5	4.0	50%/2	20	30	-	-	-	[143]
Pure Al	2.0/1.0	126/157.5	1.25	50%/1	20	-	-	-	-	[82]
	2.0/1.0	126/189	1.5	50%/1	20	-	-	-	-	
AA1050	2.92/0.3	300/300	1.5	90%/4	20	-	-	-	-	[144]
Pure Al	8.0/3.0	180/180	3.0	62.5%/6	20	-	108	121	-	[145]
AA1050	7.0/3.5	-	2.0	50%/8	20	-	-	-	-	[146]
AA6016	1.22/1.0	310/460	1.5	18%/1	20	-	-	-	-	[147]
AA1050	8.0/0.65	180/180	1.36	92%/15	20	-	-	-	-	[148]
	8.0/0.65	180/180	1.36	92%/4	20	-	-	-	-	
Pure Al	4.0/0.4	130/130	-	90%/22	20	1.0	250	-	4	[149]
Pure Al	4.0/1.2	130/130	1.5	70%/11	20	0.5	-	-	-	[59]
AA1050	4.4/0.3	248/128	1.94	93%/4	20	1.4	-	-	-	[79,86]
	4.4/0.3	248/128	1.94	93%/52	20	1.5	-	-	-	
Al-Mg-Sc-Zr	8.0/2.0	100/100	1.3	75%/10	20	1.5	-	-	3200 *	[150]
Al-Mg-Mn-Sc-Zr	8.0/2.0	100/100	1.5	75%/10	20	1.1	-	-	3170 *	[151]
Al-Mg-Si	4.0/1.0	-	4.0	75%/4	20	-	-	312	7.4	[152]

* Superplasticity at 500 °C and 5×10^{-2} s^{-1}.

Table 8. Experimental data related to asymmetric cryorolling of Al alloys.

Material	Initial/Final Sheet Thickness, mm	Top/Bottom Work Roll Diameters, mm	Speed Ratio S_R	Total Reduction/Number of Passes	T, °C	Final Grain Size, µm	YS, MPa	UTS, MPa	EL, %	Ref.
AA1050	1.45/0.17	-	1.1	88%/7	−196	0.36	154	160	-	[24,153]
	1.45/0.17	-	1.2	88%/7	−196	-	159	179	-	
	1.45/0.17	-	1.3	88%/7	−196	-	160	182	-	
	1.45/0.17	-	1.4	88%/7	−196	0.21	162	196	-	
AA6061	1.5/0.19	120/120	1.1	87.3%/7	−196	0.235	-	235	-	[25]
AA6061	1.5/0.19	-	1.1	87.3%/7	−196	-	-	-	-	[26]
AA5052	1.0/0.2	-	1.4	80%/3	−196	0.35	-	343	3.3	[154]
AA5052	15/3	130/130	1.2	80%/15	−196	-	267	295	6.8	[155]
	15/3	130/130	1.5	80%/15	−196	-	252	286	9.2	
AA6061	16/4	-	1.5	75%/13	−196	-				[156]

In 2009, Jiang et al. [149] showed that ultrafine-grained pure Al with an average grain size of ~1 μm (Figure 38a) could be prepared by asymmetric cold rolling. The sheets were rolled with a total thickness reduction of 90% from 4 mm to 0.4 mm after 22 rolling passes. The grains of the material were nearly equiaxed and the microstructure was homogeneous. Moreover, the high-angle grain boundaries were predominant. The fraction of high-angle grain boundaries (>15°) was about 50%. Before rolling, the original pure Al was of very low yield stress (~100 MPa) and high ductility (~17%). The yield strength of asymmetrically rolled Al increased to 250 MPa, but the corresponding elongation decreased to 4% [149].

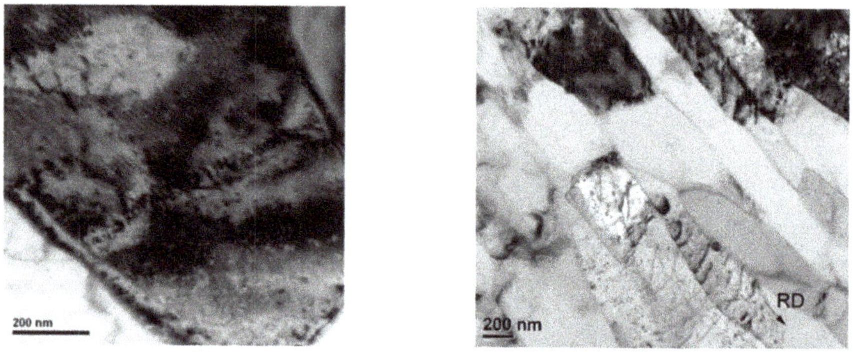

(**a**) (**b**)

Figure 38. (**a**) TEM micrograph of pure Al after asymmetric cold rolling, reproduced from [149], with permission from Elsevier, 2009; (**b**) TEM micrograph of Al 1050 after asymmetric cryorolling [24] (RD—rolling direction).

In 2012, Yu et al. [24] reported about nanostructural Al 1050 sheets with a grain size of 0.211 μm (Figure 38b). Sheets were manufactured using a novel method of asymmetric cryorolling. Asymmetric cryorolling ($S_R = 1.4$) was performed by dipping the sheets into liquid nitrogen for at least 8 min before each rolling pass. The sheets were rolled with a total thickness reduction of 88.3% from 1.45 mm to 0.17 mm after seven rolling passes.

In 2009, Kim et al. [19] showed that ultrafine-grained Al-Mg-Si alloy (AA6061) sheets could be fabricated by SPD using high-ratio differential speed rolling (HRDSR) ($S_R = 3$) and subsequent low-temperature aging. The asymmetric warm rolling was conducted on preheated specimens (120 °C) and the work roll surfaces were maintained at 140 °C throughout the process. Sheets with a width of 100 mm were rolled from 2 mm to 0.6 mm (70% thickness reduction) by a single pass. Processed sheets exhibited an ultra-high strength (yield stress: 455 MPa, ultimate tensile strength: 489 MPa, total elongation: 7.4%). The strengthening effect was impressive compared with the results obtained by using other SPD techniques. The high strength of the AA6061 could be attributed to a significant decrease in grain size (0.37 μm) and increased Hall-Petch constant. The additional strengthening gained after the low-temperature aging was due to the precipitation of nanosized β' particles. Figure 39 shows the microstructures of the AA6061 observed by TEM [19].

In 2014, Loorentz and Y.G. Ko [134] reported about nanostructured AA5052 (with grains of 0.7 μm) manufactured by differential speed rolling ($S_R = 4$). The sheets were rolled with a total thickness reduction of 75% from 4 mm to 1 mm after four rolling passes. The yield strength, ultimate tensile strength and total elongation of the initial sheets were 65 ± 5 MPa, 137 ± 10 MPa and 32 ± 2%, respectively. The yield strength and ultimate tensile strength of the nanostructured sheets were several times higher than that of the initial coarse counterpart: 380 ± 10 MPa and 390 ± 10 MPa, respectively. However, total elongation was decreased to 4.2 ± 0.5% [134].

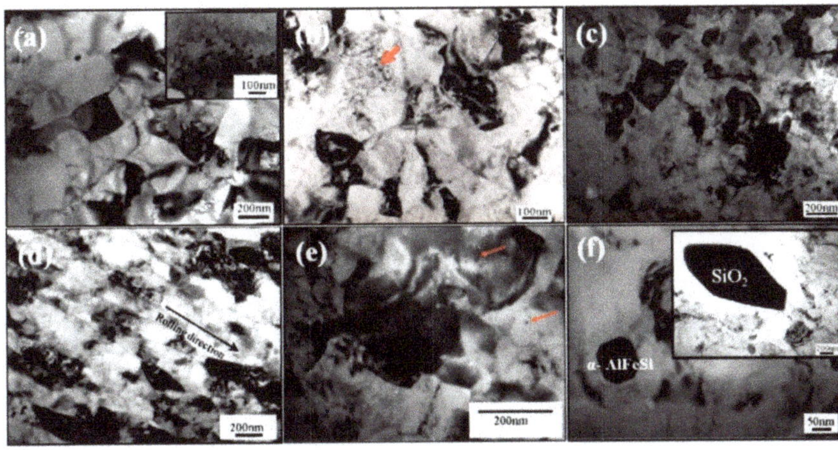

Figure 39. TEM micrographs of HRDSR AA6061 after thickness reduction of 70%: (**a**) furnace-cooled (FC) HRDSR Al, (**b**) water-quenched (WQ) HRDSR Al and (**c**) WQ-HRDSR Al after aging at 100 °C for 48 h. (**d**) WQ-HRDSR Al after aging at 100 °C for 48 h (taken on RD–ND plane) (**e**) β'' or β' precipitates in the aged WQ-HRDSR Al (**f**) SiO_2 (>200 nm) and α-AlFeSi particles (~50 nm) commonly detected in all the HRDSR 6061 Al, reproduced from [19], with permission from Elsevier, 2009.

6. Asymmetric (Warm, Cold, Cryo) Rolling of Ti Alloys

Results of an extensive study (Tables 9–11 and Figure 40) showed that (warm, cold, cryo) differential speed rolling has a great impact on the grain size and mechanical properties of pure Ti and Ti-6Al-4V alloy. In 2010, Kim et al. [20] reported that high-strength CP-Ti (ASTM grade 2) sheets with ultrafine grains of 0.1–0.3 μm and ultimate tensile strength of 895–915 MPa could be fabricated by single-pass high-ratio differential speed rolling (HRDSR) ($S_R = 3$) at room temperature (thickness reduction of 63% from 2 mm to 0.74 mm).

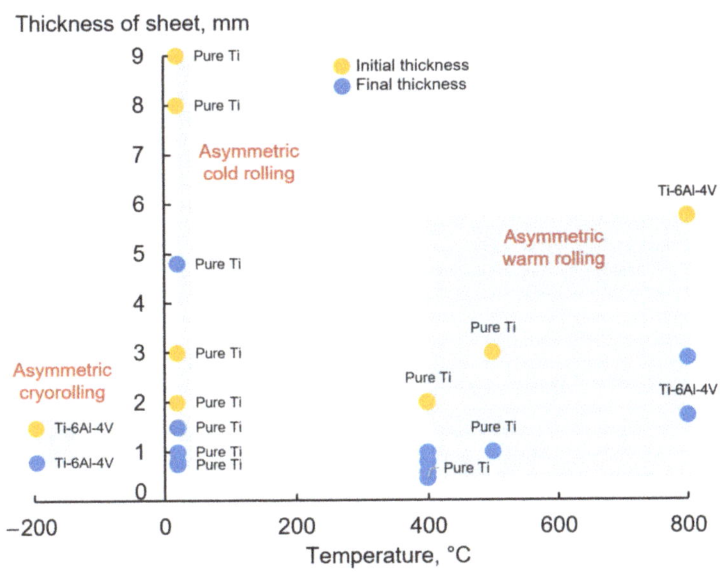

Figure 40. Different Ti alloys, which were processed by asymmetric (warm, cold, cryo) rolling (based on data from Tables 9–11).

Table 9. Experimental data related to asymmetric warm rolling of Ti alloys.

Material	Initial/Final Sheet Thickness, mm	Top/Bottom Work Roll Diameters, mm	Speed Ratio S_R	Total Reduction/Number of Passes	T, °C	Final Grain Size, μm	Direction	YS, MPa	UTS, MPa	EL, %	Ref.
Pure Ti	2.0/1.18	300/300	1	41%/1	400	-	(0°)	451	507	25.1	[21]
							(45°)	437	476	26.9	
							(90°)	425	477	23.5	
	2.0/0.98	300/300	2	51%/1	400	-	(0°)	525	580	19.6	
							(45°)	456	511	23.1	
							(90°)	511	559	20.4	
	2.0/0.78	300/300	3	61%/1	400	-	(0°)	532	633	19.1	
							(45°)	507	550	20	
							(90°)	557	637	11.2	
	2.0/0.58	300/300	4	71%/1	400	-	(0°)	558	693	17.8	
							(45°)	502	562	15.3	
							(90°)	631	691	10.7	
	2.0/0.46	300/300	5	77%/1	400	-	(0°)	555	693	16.5	
							(45°)	494	550	17	
							(90°)	626	737	10.2	
Ti-6Al-4V	5.75/2.88	-	1.67	50%/-	800	0.227	-	1205	1273	18.7	[157]
	5.75/1.72	-	1.67	70%/-	800	0.229	-	1231	1365	22.8	
Pure Ti	3.0/1.0	-	1.36	66.7%/12	500	-	(0°)	232	385	35.4	[158]
							(45°)	247	356	44.7	
							(90°)	270	367	41	

Table 10. Experimental data related to asymmetric cold rolling of Ti alloys.

Material	Initial/Final Sheet Thickness, mm	Top/Bottom Work Roll Diameters, mm	Speed Ratio S_R	Total Reduction/Number of Passes	T, °C	Final Grain Size, μm	Direction	YS, MPa	UTS, MPa	EL, %	Ref.
Pure Ti	9.0/1.5	130/130	1.5	83%/15	20	0.13	-	652	780	9.8	[159]
Pure Ti	3.0/1.0	-	1.36	66.7/12	20	-	(0°)	236	397	39.8	[158]
							(45°)	278	372	42.4	
							(90°)	333	403	38.7	
Ti-6Al-4V	1.5/0.8	50/50	1.2	46.7%/-	20	-	-	-	1046	11.2	[160]
Pure Ti	2.0/0.74	400/400	3.0	63%/1	20	0.1-0.3	-	780	895	-	[20]
Pure Ti	8.0/4.8	180/180	1.5	40%/-	20	-	-	-	-	-	[161]

Table 11. Experimental data related to asymmetric cryorolling of Ti alloys.

Material	Initial/Final Sheet Thickness, mm	Top/Bottom Work Roll Diameters, mm	Speed Ratio S_R	Total Reduction/Number of Passes	T, °C	Final Grain Size, μm	YS, MPa	UTS, MPa	EL, %	Ref.
Ti-6Al-4V	1.5/0.8	50/50	1.2	46.7%/-	−196	-	-	1113	10.1	[160]

The microstructure of the HRDSR-processed Ti was composed of two types of region: the "U" region, where shear bands were densely populated, and the "F" region, where some fragments of the original grains remained. The U region was clearly the dominant area fraction. The formation of nearly unidirectional shear bands is a characteristic of the HRDSR process, where large shear strain is imposed on the sheet. The interval of the lamellar boundaries in the U region was between 0.9 and 1.2 μm. Very fine and equiaxed grains resided in shear bands and their sizes were comparable to or smaller than those of the lamellar spacing, indicating that the lamellar structure was related to the formation mechanism of ultrafine grains. Figure 41 shows TEM micrographs and the selected area electron diffraction (SAED) pattern of the HRDSR-processed Ti, taken at two places showing different microstructural features. There are two regions distinguished in the U region. One region (Figure 41a) is composed of elongated or nearly equiaxed subgrains with thick cell walls (0.2–0.3 μm in size), inside of which the dislocation density is high. In the other region (Figure 41b), grains are more equiaxed and smaller. Comparison of the SAED pattern between the two regions indicates that the grains in the latter have higher angle grain boundaries. The SAED patterns in both regions can be indexed as the close-packed hexagonal structure of α-Ti, indicating that no phase transformation occurred during rolling. The "A" region marked in Figure 41a indicates that the ultrafine grains were formed via a dynamic recovery process, by which the high density of dislocations forms equiaxed subgrains with low angle boundaries between the lamellar boundaries, which were then converted into equiaxed crystallites with high-angle boundaries via dynamic continuous recrystallization. As the local temperature in the shear bands can be greatly increased due to intense shear banding, dynamic recovery and recrystallization processes will be effectively promoted in the bands, even though the material is deformed at room temperature.

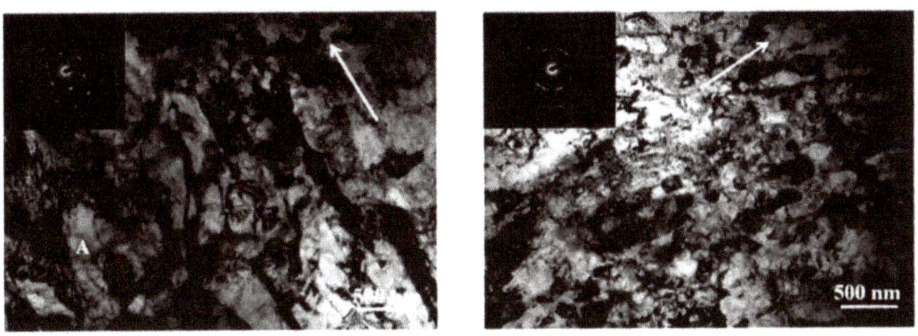

(a) (b)

Figure 41. TEM micrographs and the selected area electron diffraction (SAED) pattern of the HRDSR-processed CP-Ti (ASTM grade 2). (**a**) Region with elongated or nearly equiaxed subgrains. (**b**) Region with equiaxed and smaller grains. The arrows indicate the rolling direction, reproduced from [20], with permission from Elsevier, 2010.

Extensive formation of high-density shear bands over the entire thickness of the section and their conversion to equiaxed crystallites with high-angle boundaries via dynamic recovery or dynamic continuous recrystallization were important in achieving ultrafine grains with a size of 0.1–0.3 μm [20]. The resultant mechanical properties of pure Ti (Figure 42) exceeded those reported from tests using multiple passes of ECAP at elevated temperatures [162] or room temperature [163] and multiple cycles of accumulative roll bonding at room temperature [164].

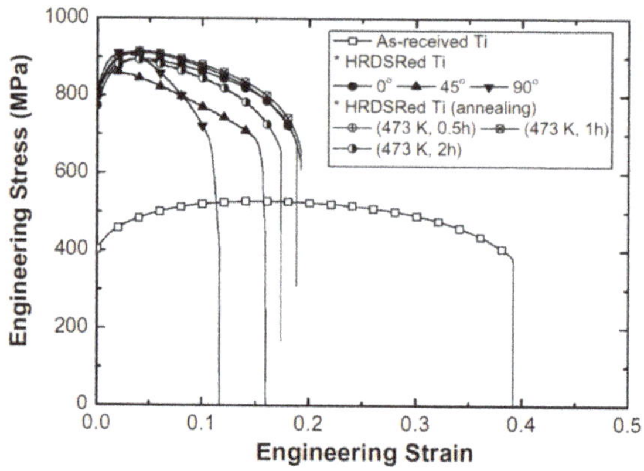

Figure 42. Stress–strain curves of the as-received and the HRDSR-processed Ti sheets, reproduced from [20], with permission from Elsevier, 2010.

A superior balance of strength (YS = 1231 MPa, UTS = 1365 MPa) and ductility (uniform elongation = 4.93%, total elongation = 22.8%) in Ti-6Al-4V alloy was achieved in 2016 by Chao et al. [157] through warm asymmetric rolling by a pair of rolls with different diameters with a ratio of 5:3 ($S_R = 1.67$). The sheets were rolled with a total thickness reduction of 70% from 5.75 mm to 1.72 mm at 800 °C. The application of asymmetric rolling at 800 °C led to a higher strength–ductility balance, reaching the highest UTS × TE of 30,000 MPa%. The exceptional mechanical properties were ascribed to the formation of an ultrafine-grained structure (0.229 μm), texture characteristics and transformation of the β phase upon straining.

7. Conclusions

Differential speed rolling, i.e., rolling with different angular speeds of the work rolls when both rolls are independently driven by two motors, is the most suitable way to implement asymmetric rolling in industry. A relatively low speed ratio of the work rolls ($S_R \leq 1.5$) can be used for reducing rolling force, improving sheet flatness, minimizing the ski effect and obtaining thinner sheets. A high speed ratio of the work rolls ($S_R = 2 \ldots 4$) can be used as an SPD method for grain refinement and improvement of the texture and mechanical properties of large-scale sheets of Mg, Al and Ti pure metals and alloys.

The mechanics of the differential speed rolling process as an SPD method are based on simultaneous pure and simple shear, which combine the advantages of simple shear (rotation of the material as with ECAP) and the advantages of pure shear (compression and elongation of the material as with symmetric rolling). The mechanism for creating shear strain during differential speed rolling is the presence of the cross-shear zone in the deformation zone, in which the forces of contact friction are oppositely directed. The equivalent strain $\bar{\varepsilon} \geq 3 \ldots 4$ can be obtained by single-pass differential speed rolling without lubrication (at high contact friction), when shear angle φ is no less than 80° due to the high speed ratio of the work rolls ($S_R = 2 \ldots 4$), large work roll diameter (D $\geq$ 300 mm), high thickness reduction per pass ($\varepsilon \geq 50\%$) and thinner initial thickness of the sheet ($h_0 \leq 2$ mm).

Based on the analysis of publications related to asymmetric (hot, warm, cold, cryo) rolling, it was found that a superior balance of strength and ductility of Mg, Al and Ti alloys could be achieved:

a. ultrafine-grained (0.6 μm) AZ31 sheets with YS of 382 MPa, UTS of 401 MPa and TE of 6.8% could be fabricated by differential speed rolling with high speed ratio of work rolls ($S_R = 2$) at 150 °C followed by immediate water quenching;

b. ultrafine-grained (0.37 μm) AA6061 sheets with YS of 455 MPa, UTS of 489 MPa, TE of 7.4% could be fabricated by differential speed rolling with high speed ratio of work rolls ($S_R = 3$) at 120 … 140 °C and subsequent low temperature aging;
c. ultrafine-grained (0.7 μm) AA5052 sheets with YS of 380 MPa, UTS of 390 MPa, TE of 4.2% could be fabricated by differential speed rolling with high speed ratio of work rolls ($S_R = 4$) at room temperature;
d. ultrafine-grained (0.1–0.3 μm) CP-Ti (ASTM grade 2) sheets with YS of 780 MPa, UTS of 895 MPa could be fabricated by differential speed rolling with high speed ratio of work rolls ($S_R = 3$) at room temperature.

It should be noted, and it is very important, that the resulting mechanical properties of Mg, Al and Ti alloys are much better than those of conventionally cold-rolled materials and at least not worse than those of counterparts subjected to conventional SPD methods, such as ECAP, while having an undeniable advantage in terms of the possibility of the production of large-scale sheets. Future prospects for the development of differential speed rolling technologies lie in the optimization of process parameters, as well as in the industrial application of these technologies to a wider range of processed materials.

Author Contributions: Conceptualization, D.P.; formal analysis, A.P., P.T.; investigation, D.P.; writing—original draft preparation, D.P.; writing—review and editing, A.P., P.T.; supervision, A.P.; project administration, A.P.; funding acquisition, A.P. All authors have read and agreed to the published version of the manuscript.

Funding: The study was carried out within the framework of the implementation of the Resolution of the Government of the Russian Federation of 9 April 2010, No. 220 (Contract No. 075-15-2019-869 from 12 May 2019) and by a grant of the Russian Science Foundation (project No. 20-69-46042 of 20 May 2020).

Institutional Review Board Statement: Not applicable.

Informed Consent Statement: Not applicable.

Data Availability Statement: The data presented in this study are available on request from the corresponding author.

Conflicts of Interest: The authors declare no conflict of interest.

References

1. Siebel, E. Zur Theorie des Walzvorganges bei ungleich angetriebenen Walzen. *Arch. Eisenhüttenwesen* **1941**, *15*, 125–128. [CrossRef]
2. Sachs, G.; Klinger, L.J. The Flow of Metals through Tools of Circular Contour. *J. Appl. Mech.* **1947**, *14*, 88–98. [CrossRef]
3. Razuvaev, D.S. A Method of Rolling Metal. USSR Patent 63448, 19 March 1940. (In Russian)
4. Chekmarev, A.P.; Nefedov, A.A. Rolling with unequal diameter rolls (complex cases of rolling). *Obrab. Met. Davl.* **1956**, *4*, 2–15. (In Russian)
5. Holbrook, R.L.; Zorowski, C.F. Effects of Nonsymmetry in Strip Rolling on Single-Roll Drive Mills. *J. Eng. Ind.* **1966**, *88*, 401–408. [CrossRef]
6. Johnson, W.; Needham, G. Further experiments in asymmetrical rolling. *Int. J. Mech. Sci.* **1966**, *8*, 443–455. [CrossRef]
7. Dewhurst, P.; Collins, I.F.; Johnson, W. A theoretical and experimental investigation into asymmetrical hot rolling. *Int. J. Mech. Sci.* **1974**, *16*, 389–397. [CrossRef]
8. Collins, I.F.; Dewhurst, P. A slipline field analysis of asymmetrical hot rolling. *Int. J. Mech. Sci.* **1975**, *17*, 643–651. [CrossRef]
9. Pan, D.; Sansome, D.H. An experimental study of the effect of roll-speed mismatch on the rolling load during the cold rolling of thin strip. *J. Mech. Work. Technol.* **1982**, *6*, 361–377. [CrossRef]
10. Dyja, H.; Korczak, P.; Pilarczyk, J.W.; Grzybowski, J. Theoretical and experimental analysis of plates asymmetric rolling. *J. Mater. Process. Technol.* **1994**, *45*, 167–172. [CrossRef]
11. Dyja, H.; Markowski, J.; Stoiniski, D. Asymmetry of the roll gap as a factor improving work of the hydraulic gauge control in the plate rolling mill. *J. Mater. Process. Technol.* **1996**, *60*, 73–80. [CrossRef]
12. Richelsen, A.B. Elastic-plastic analysis of the stress and strain distributions in asymmetric rolling. *Int. J. Mech. Sci.* **1997**, *39*, 1199–1211. [CrossRef]
13. Salganik, V.M.; Pesin, A.M. *Asimmetrichnaya Tonkolistovaya Prokatka: Razvitiye Teorii, Tekhnologii i Novyye Resheniya [Asymmetric Sheet Rolling: Development of Theory, Technology and New Solutions]*; MISIS Publisher: Moscow, Russia, 1997; p. 192. (In Russian)

14. Choi, C.-H.; Kim, K.-H.; Lee, D.N. The effect of shear texture development on the formability in rolled aluminum alloy sheets. *Mater. Sci. Forum* **1998**, *273–275*, 391–396. [CrossRef]
15. Cui, Q.; Ohori, K. Grain refinement of high purity aluminium by asymmetric rolling. *Mater. Sci. Technol.* **2000**, *16*, 1095–1101. [CrossRef]
16. Kim, W.J.; Lee, J.B.; Kim, W.Y.; Jeong, H.T.; Jeong, H.G. Microstructure and mechanical properties of Mg-Al-Zn alloy sheets severely deformed by asymmetrical rolling. *Scr. Mater.* **2007**, *56*, 309–312. [CrossRef]
17. Ji, Y.H.; Park, J.J.; Kim, W.J. Finite element analysis of severe deformation in Mg-3Al-1Zn sheets through differential-speed rolling with a high-speed ratio. *Mater. Sci. Eng. A* **2007**, *454–455*, 570–574. [CrossRef]
18. Kim, W.J.; Kim, M.J.; Wang, J.Y. Superplastic behavior of a fine-grained ZK60 magnesium alloy processed by high-ratio differential speed rolling. *Mater. Sci. Eng. A* **2009**, *527*, 322–327. [CrossRef]
19. Kim, W.J.; Wang, J.Y.; Choi, S.O.; Choi, H.J.; Sohn, H.T. Synthesis of ultra high strength Al-Mg-Si alloy sheets by differential speed rolling. *Mater. Sci. Eng. A* **2009**, *520*, 23–28. [CrossRef]
20. Kim, W.J.; Yoo, S.J.; Lee, J.B. Microstructure and mechanical properties of pure Ti processed by high-ratio differential speed rolling at room temperature. *Scr. Mater.* **2010**, *62*, 451–454. [CrossRef]
21. Kim, W.J.; Yoo, S.J.; Jeong, H.T.; Kim, D.M.; Choe, B.H.; Lee, J.B. Effect of the speed ratio on grain refinement and texture development in pure Ti during differential speed rolling. *Scr. Mater.* **2011**, *64*, 49–52. [CrossRef]
22. Loorentz; Ko, Y.G. Microstructure evolution and mechanical properties of severely deformed Al alloy processed by differential speed rolling. *J. Alloy. Compd.* **2012**, *536*, S122–S125. [CrossRef]
23. Polkowski, W.; Jozwik, P.; Polanski, M.; Bojar, Z. Microstructure and texture evolution of copper processed by differential speed rolling with various speed asymmetry coefficient. *Mater. Sci. Eng. A* **2013**, *564*, 289–297. [CrossRef]
24. Yu, H.L.; Lu, C.; Tieu, K.; Liu, X.H.; Sun, Y.; Yu, Q.B.; Kong, C. Asymmetric cryorolling for fabrication of nanostructural aluminum sheets. *Sci. Rep.* **2012**, *2*, 772. [CrossRef] [PubMed]
25. Yu, H.L.; Tieu, K.; Lu, C.; Liu, X.H.; Godbole, A.; Kong, C. Mechanical properties of Al-Mg-Si alloy sheets produced using asymmetric cryorolling and ageing treatment. *Mater. Sci. Eng. A* **2013**, *568*, 212–218. [CrossRef]
26. Yu, H.L.; Tieu, K.; Lu, C.; Lou, Y.S.; Liu, X.H.; Godbole, A.; Kong, C. Tensile fracture of ultrafine grained aluminum 6061 sheets by asymmetric cryorolling for microforming. *Int. J. Damage Mech.* **2014**, *23*, 1077–1095. [CrossRef]
27. Vincze, G.; Simoes, F.J.P.; Butuc, M.C. Asymmetrical Rolling of Aluminum Alloys and Steels: A Review. *Metals* **2020**, *10*, 1126. [CrossRef]
28. Kraner, J.; Smolar, T.; Volsak, D.; Cvahte, P.; Godec, M.; Paulin, I. A review of asymmetric rolling. *Mater. Technol.* **2020**, *54*, 731–743. [CrossRef]
29. Philipp, M.; Schwenzfeier, W.; Fischer, F.D.; Wodlinger, R.; Fischer, C. Front end bending in plate rolling influenced by circumferential speed mismatch and geometry. *J. Mater. Process. Technol.* **2007**, *184*, 224–232. [CrossRef]
30. Anders, D.; Münker, T.; Artel, J.; Weinberg, K. A dimensional analysis of front-end bending in plate rolling applications. *J. Mater. Process. Technol.* **2012**, *212*, 1387–1398. [CrossRef]
31. Balgabekov, T.K.; Azbanbayev, E.M.; Isagulov, A.Z.; Isagulova, D.A.; Zakariya, N.B.; Yermaganbetov, N.Z.; Kenzhekeeva, A.R.; Konirova, Z.A. Effect of Asymmetric Rolling with Cone-Shaped Rolls on Microstructure and Tensile Properties of Low-Carbon Steel. *Adv. Mater. Res.* **2015**, *1105*, 149–153. [CrossRef]
32. Azbanbayev, E.M. Research and development of production technology of metals with nanostructure and high mechanical properties during the rolling in the back-taper rolls. Ph.D. Thesis, Karaganda State Technical University, Karaganda, Kazakhstan, December 2015. (In Russian)
33. Hartung, H.G.; Klamma, K.; Rohde, W.; Seidel, J. Roll Stand Comprising a Crown-Variable-Control (CVC) Roll Pair. US Patent 7,059,163, 13 June 2006.
34. Kiyoshi, N.; Hori, Y.; Ogawa, S.; Mizutani, Y.; Kojima, A. High reduction rolling technology on pair cross mill. *Nippon. Steel Technol. Rep.* **1997**, *75*, 9–23.
35. Wronski, S.; Wierzbanowski, K.; Wronski, M.; Bacroix, B. Three-dimensional analysis of asymmetric rolling with flat and inclined entry. *Arch. Metall. Mater.* **2014**, *59*, 585–591. [CrossRef]
36. Gao, H.; Ramalingam, S.C.; Barber, G.C.; Chen, G. Analysis of asymmetrical cold rolling with varying coefficients of friction. *J. Mater. Process. Technol.* **2002**, *124*, 178–182. [CrossRef]
37. Utsunomiya, H.; Ueno, T.; Sakai, T. Improvement in the r-value of aluminum sheets by differential-friction rolling. *Scr. Mater.* **2007**, *57*, 1109–1112. [CrossRef]
38. Kasai, D.; Komori, A.; Ishii, A.; Yamada, K.; Ogawa, S. Strip Warpage Behavior and Mechanism in Single Roll Driven Rolling. *ISIJ Int.* **2016**, *56*, 1815–1824. [CrossRef]
39. Sakai, T.; Hamada, S.; Saito, Y. Improvement of the r-value in 5052 aluminum alloy sheets having through-thickness shear texture by 2-pass single-roll drive unidirectional shear rolling. *Scr. Mater.* **2001**, *44*, 2569–2573. [CrossRef]
40. Pesin, A. Modeling and Development of Asymmetric Deformation Processes to Improve the Efficiency of Sheet Rolling. Ph.D. Thesis, Nosov Magnitogorsk State Technical University, Magnitogorsk, Russia, 2003. (In Russian)
41. Potapkin, V.F.; Fedorinov, V.A.; Morozov, I.A.; Ravich, A.S. Klet' Dlya Deformatsii Metalla Mezhdu Nepodvizhnym i Rabochim Privodnymi Valkami [Stand for Metal Deformation Between Stationary and Driven Work Rolls]. USSR Patent 871855, 30 September 1979. (In Russian)

42. Pesin, A. Regularities of Asymmetric Deformation and Increasing the Efficiency of the Cold Sheet Rolling Process. Ph.D. Thesis, Nosov Magnitogorsk State Technical University, Magnitogorsk, Russia, 1989. (In Russian)
43. Vydrin, V.N. *Dinamika Prokatnykh Stanov [The Dynamics of Rolling Mills]*; Metallurgizdat Publisher: Sverdlovsk, Russia, 1960; p. 255. (In Russian)
44. Vydrin, V.N.; Ageyev, L.M. Printsipial'nyye i teoreticheskiye osnovy novogo protsessa "prokatka-volocheniye" [Fundamental and theoretical foundations of the new process "rolling-drawing"]. *Teoriya i tekhnologiya prokatki Chelyabinsk* **1971**, *76*, 3–21. (In Russian)
45. Shimoyama, K.; Yokoyama, S.; Kaneko, S.; Fujita, F. Effect of grooved roll profiles on microstructure evolutions of AZ31 sheets in Periodical Straining Rolling process. *Mater. Sci. Eng. A* **2014**, *611*, 58–68. [CrossRef]
46. Li, L.; Nagai, K.; Yin, F.X. Progress in Cold Roll Bonding of Metals. *Sci. Technol. Adv. Mater.* **2008**, *9*, 1–11. [CrossRef]
47. Li, X.; Zu, G.; Ding, M.; Mu, Y.; Wang, P. Interfacial microstructure and mechanical properties of Cu/Al clad sheet fabricated by asymmetrical roll bonding and annealing. *Mater. Sci. Eng. A* **2011**, *529*, 485–491. [CrossRef]
48. Wang, T.; Wang, Y.; Bian, L.; Huang, Q. Microstructural evolution and mechanical behavior of Mg/Al laminated composite sheet by novel corrugated rolling and flat rolling. *Mater. Sci. Eng. A* **2019**, *765*, 138318. [CrossRef]
49. Wang, T.; Li, S.; Ren, Z.; Jia, Y.; Fu, W.; Guo, M.; Ma, X.; Han, J. Microstructure Characterization and Mechanical Property of Mg/Al Laminated Composite Prepared by the Novel Approach: Corrugated + Flat Rolling (CFR). *Metals* **2019**, *9*, 690. [CrossRef]
50. Wang, T.; Liu, W.; Liu, Y.; Wang, Z.; Ignatov, A.V.; Huang, Q. Formation mechanism of dynamic multi-neutral points and cross shear zones in corrugated rolling of Cu/Al laminated composite. *J. Mater. Process. Technol.* **2021**, *295*, 117157. [CrossRef]
51. Pesin, A.; Pustovoitov, D.; Biryukova, O.; Ilyina, N. FEM simulation of fabrication of Al-steel layered composites with mechanical bonding through the interfacial concavo-convex lock effect. *Procedia Manuf.* **2020**, *50*, 579–583. [CrossRef]
52. Hirt, G.; Thome, M. Rolling of functional metallic surface structures. *CIRP Ann.* **2008**, *57*, 317–320. [CrossRef]
53. Hirt, G.; Thome, M. Large area rolling of functional metallic micro structures. *Prod. Eng. Res. Devel.* **2007**, *1*, 351–356. [CrossRef]
54. Stille, S.; Beck, T.; Singheiser, L. Influence of riblet geometry on fatigue life of surface structured AA 2024 thin sheets. *Int. J. Fatigue* **2014**, *68*, 48–55. [CrossRef]
55. Zuo, Y.; Fu, X.; Cui, J.; Tang, X.; Mao, L.; Li, L.; Zhu, Q. Shear deformation and plate shape control of hot-rolled aluminium alloy thick plate prepared by asymmetric rolling process. *Trans. Nonferrous Met. Soc. China* **2014**, *24*, 2220–2225. [CrossRef]
56. Tian, Y.; Guo, Y.C.; Wang, Z.-D.; Wang, G.-D. Analysis of rolling pressure in asymmetrical rolling process by slab method. *J. Iron Steel Res. Int.* **2009**, *16*, 22–26. [CrossRef]
57. Park, J.-J. Finite-element analysis of severe plastic deformation in differential-speed rolling. *Comput. Mater. Sci.* **2015**, *100*, 61–66. [CrossRef]
58. Pesin, A.M.; Pustovoytov, D.O.; Shveeva, T.V.; Steblyanko, V.L.; Fedoseev, S.A. Simulation of nonmonotonic metal flow during asymmetric sheet rolling with different velocities of the rolls. *Vestn. Nosov Magnitogorsk. State Tech. Univ.* **2017**, *15*, 56–63. [CrossRef]
59. Zuo, F.; Jiang, J.; Shan, A.; Fang, J.; Zhang, X. Shear deformation and grain refinement in pure Al by asymmetric rolling. *Trans. Nonferrous Met. Soc. China* **2008**, *18*, 774–777. [CrossRef]
60. Sidor, J.; Miroux, A.; Petrov, R.; Kestens, L. Microstructural and crystallographic aspects of conventional and asymmetric rolling processes. *Acta Mater.* **2008**, *56*, 2495–2507. [CrossRef]
61. Pesin, A.; Pustovoytov, D. Influence of process parameters on distribution of shear strain through sheet thickness in asymmetric rolling. *Key Eng. Mater.* **2014**, *622*, 929–935. [CrossRef]
62. Polkowski, W. *Differential Speed Rolling: A New Method for a Fabrication of Metallic Sheets with Enhanced Mechanical Properties, Progress in Metallic Alloys, Vadim Glebovsky*; IntechOpen: London, UK, 2016. Available online: https://www.intechopen.com/books/progress-in-metallic-alloys/differential-speed-rolling-a-new-method-for-a-fabrication-of-metallic-sheets-with-enhanced-mechanica (accessed on 20 April 2021). [CrossRef]
63. Shrivastava, S.; Ghosh, C.; Jonas, J.J. A comparison of the von Mises and Hencky equivalent strains for use in simple shear experiments. *Philos. Mag.* **2012**, *92*, 779–786. [CrossRef]
64. Shrivastava, S.C.; Jonas, J.J.; Canova, G. Equivalent strain in large deformation torsion testing: Theoretical and practical considerations. *J. Mech. Phys. Solids* **1982**, *30*, 75–90. [CrossRef]
65. Onaka, S. Equivalent strain in simple shear deformation described by using the Hencky strain. *Philos. Mag. Lett.* **2010**, *90*, 633–639. [CrossRef]
66. Polakowski, N.H.; Ripling, E.J. *Strength and Structure of Engineering Materials*; Prentice-Hall: Hoboken, NJ, USA, 1966; pp. 381–383.
67. Beygelzimer, Y. Equivalent strain in simple shear of metals. *Phys. Technol. High Press.* **2015**, *25*, 90–97.
68. Saito, Y.; Sakai, T.; Maeda, F.; Kato, K. Deformation and recrystallization behavior of ferritic stainless steel in high speed hot rolling. *Tetsu Hagane (ISIJ)* **1986**, *72*, 799–806. [CrossRef]
69. Sakai, T.; Saito, Y.; Hirano, K.; Kato, K. Deformation and recrystallization behavior of low carbon steel in high speed hot rolling. *Trans. ISIJ* **1988**, *28*, 1028–1035. [CrossRef]
70. Angella, G.; Esfandiar Jahromi, B.; Vedani, M. A comparison between equal channel angular pressing and asymmetric rolling of silver in the severe plastic deformation regime. *Mater. Sci. Eng. A* **2013**, *559*, 742–750. [CrossRef]
71. Kang, S.-B.; Min, B.-K.; Kim, H.-W.; Wilkinson, D.S.; Kang, J. Effect of asymmetric rolling on the texture and mechanical properties of AA6111-aluminum sheet. *Metall. Mater. Trans. A* **2005**, *36*, 3141–3149. [CrossRef]

72. Ma, C.; Hou, L.; Zhang, J.; Zhuang, L. Influence of thickness reduction per pass on strain, microstructures and mechanical properties of 7050 Al alloy sheet processed by asymmetric rolling. *Mater. Sci. Eng. A* **2016**, *650*, 454–468. [CrossRef]
73. Pesin, A.; Pustovoytov, D.; Shveyova, T.; Vafin, R. Finite element simulation and comparison of a shear strain and equivalent strain during ECAP and asymmetric rolling. *IOP Conf. Ser. Mater. Sci. Eng.* **2017**, *293*, 012007. [CrossRef]
74. Megantoro, R.B.; Loorentz; Ko, Y.G. Temperature rise during differential speed rolling. *J. Alloy. Compd.* **2014**, *586*, S254–S257. [CrossRef]
75. Pesin, A.M.; Pustovoitov, D.O.; Vafin, R.K. Simulation of temperature fields in a deformation zone during asymmetric rolling of aluminum alloys. *Vestn. Nosov Magnitogorsk. State Tech. Univ.* **2015**, *4*, 75–81.
76. Pesin, A.; Pustovoytov, D. Heat transfer modeling in asymmetrical sheet rolling of aluminium alloys with ultra high shear strain. *MATEC Web Conf.* **2016**, *80*, 04005. [CrossRef]
77. Simões, F.J.P. Asymmetrical rolling of an aluminum alloy 1050. Ph.D. Thesis, Universidade de Aveiro, Aveiro, Portugal, 2008; p. 152. Available online: http://hdl.handle.net/10773/2560 (accessed on 20 April 2021).
78. Lee, S.H.; Lee, D.N. Analysis of deformation textures of asymmetrically rolled steel sheets. *Int. J. Mech. Sci.* **2001**, *43*, 1997–2015. [CrossRef]
79. Lee, J.K.; Lee, D.N. Texture control and grain refinement of AA1050 Al alloy sheets by asymmetric rolling. *Int. J. Mech. Sci.* **2008**, *50*, 869–887. [CrossRef]
80. Bobor, K.; Hegedus, Z.; Gubicza, J.; Barkai, I.; Pekker, P.; Krallics, G. Microstructure and mechanical properties of Al 7075 alloy processed by differential speed rolling. *Period. Polytech. Mech. Eng.* **2012**, *56*, 111–115. [CrossRef]
81. Pesin, A.; Pustovoytov, D. Finite Element Simulation of Shear Strain during High-Ratio Differential Speed Rolling of Aluminum Alloy 5083. *Key Eng. Mater.* **2016**, *716*, 700–707. [CrossRef]
82. Kim, K.-H.; Lee, D.N. Analysis of deformation textures of asymmetrically rolled aluminum sheets. *Acta Mater.* **2001**, *49*, 2583–2595. [CrossRef]
83. Lee, D.N. Asymmetric Rolling as Means of Texture and Ridging Control and Grain Refinement of Aluminum Alloy and Steel Sheets. *Mater. Sci. Forum* **2004**, *449–452*, 1–6. [CrossRef]
84. Hamad, K.; Ko, Y.G. A cross-shear deformation for optimizing the strength and ductility of AZ31 magnesium alloys. *Sci. Rep.* **2016**, *6*, 29954. [CrossRef] [PubMed]
85. Ko, Y.G.; Hamad, K. Structural features and mechanical properties of AZ31 Mg alloy warm-deformed by differential speed rolling. *J. Alloy. Compd.* **2018**, *744*, 96–103. [CrossRef]
86. Lee, J.K.; Lee, D.N. Shear Texture Development and Grain Refinement in Asymmetrically Rolled Aluminum Alloy Sheets by Varied Reduction per Pass. *Mater. Sci. Forum* **2002**, *408–412*, 1419–1424. [CrossRef]
87. Beausir, B.; Biswas, S.; Kim, D.; Toth, L.S.; Suwas, S. Analysis of microstructure and texture evolution in pure magnesium during symmetric and asymmetric rolling. *Acta Mater.* **2009**, *57*, 5061–5077. [CrossRef]
88. Biswas, S.; Kim, D.; Suwas, S. Asymmetric and symmetric rolling of magnesium: Evolution of microstructure, texture and mechanical properties. *Mater. Sci. Eng. A* **2012**, *550*, 19–30. [CrossRef]
89. Seong, J.W.; Kim, W.J. Development of biodegradable Mg-Ca alloy sheets with enhanced strength and corrosion properties through the refinement and uniform dispersion of the Mg2Ca phase by high-ratio differential speed rolling. *Acta Biomater.* **2015**, *11*, 531–542. [CrossRef]
90. Gong, X.; Kang, S.B.; Cho, J.H.; Li, S. Effect of annealing on microstructure and mechanical properties of ZK60 magnesium alloy sheets processed by twin-roll cast and differential speed rolling. *Mater. Charact.* **2014**, *97*, 183–188. [CrossRef]
91. Ucuncuoglu, S.; Ekerim, A.; Secgin, G.O.; Duygulu, O. Effect of asymmetric rolling process on the microstructure, mechanical properties and texture of AZ31 magnesium alloys sheets produced by twin roll casting technique. *J. Magnes. Alloy.* **2014**, *2*, 92–98. [CrossRef]
92. Kaseem, M.; Chung, B.K.; Yang, H.W.; Hamad, K.; Ko, Y.G. Effect of Deformation Temperature on Microstructure and Mechanical Properties of AZ31 Mg Alloy Processed by Differential-Speed Rolling. *J. Mater. Sci. Technol.* **2015**, *31*, 498–503. [CrossRef]
93. Luo, D.; Wang, H.-Y.; Zhao, L.-G.; Wang, C.; Liu, G.-J.; Liu, Y.; Jiang, Q.-C. Effect of differential speed rolling on the room and elevated temperature tensile properties of rolled AZ31 Mg alloy sheets. *Mater. Charact.* **2017**, *124*, 223–228. [CrossRef]
94. Watanabe, H.; Mukai, T.; Ishikawa, K. Effect of temperature of differential speed rolling on room temperature mechanical properties and texture in an AZ31 magnesium alloy. *J. Mater. Process. Technol.* **2007**, *182*, 644–647. [CrossRef]
95. Tolouie, E.; Jamaati, R. Effect of β-Mg17Al12 phase on microstructure, texture and mechanical properties of AZ91 alloy processed by asymmetric hot rolling. *Mater. Sci. Eng. A* **2018**, *738*, 81–89. [CrossRef]
96. Huang, X.; Suzuki, K.; Watazu, A.; Shigematsu, I.; Saito, N. Effects of thickness reduction per pass on microstructure and texture of Mg–3Al–1Zn alloy sheet processed by differential speed rolling. *Scr. Mater.* **2009**, *60*, 964–967. [CrossRef]
97. Gong, X.; Kang, S.B.; Li, S.; Cho, J.H. Enhanced plasticity of twin-roll cast ZK60 magnesium alloy through differential speed rolling. *Mater. Des.* **2009**, *30*, 3345–3350. [CrossRef]
98. Huang, X.; Suzuki, K.; Watazu, A.; Shigematsu, I.; Saito, N. Improvement of formability of Mg-Al-Zn alloy sheet at low temperatures using differential speed rolling. *J. Alloy. Compd.* **2009**, *470*, 263–268. [CrossRef]
99. Huang, X.; Suzuki, K.; Watazu, A.; Shigematsu, I.; Saito, N. Mechanical properties of Mg-Al-Zn alloy with a tilted basal texture obtained by differential speed rolling. *Mater. Sci. Eng. A* **2008**, *488*, 214–220. [CrossRef]

100. Huang, X.; Suzuki, K.; Watazu, A.; Shigematsu, I.; Saito, N. Microstructural and textural evolution of AZ31 magnesium alloy during differential speed rolling. *J. Alloys Compd.* **2009**, *479*, 726–731. [CrossRef]
101. Chang, L.L.; Cho, J.H.; Kang, S.B. Microstructure and mechanical properties of AM31 magnesium alloys processed by differential speed rolling. *J. Mater. Process. Technol.* **2011**, *211*, 1527–1533. [CrossRef]
102. Xia, W.; Chen, Z.; Chen, D.; Zhu, S. Microstructure and mechanical properties of AZ31 magnesium alloy sheets produced by differential speed rolling. *J. Mater. Process. Technol.* **2009**, *209*, 26–31. [CrossRef]
103. Huang, X.; Suzuki, K.; Saito, N. Microstructure and mechanical properties of AZ80 magnesium alloy sheet processed by differential speed rolling. *Mater. Sci. Eng. A* **2009**, *508*, 226–233. [CrossRef]
104. Chang, L.L.; Cho, J.H.; Kang, S.K. Microstructure and mechanical properties of twin roll cast AM31 magnesium alloy sheet processed by differential speed rolling. *Mater. Des.* **2012**, *34*, 746–752. [CrossRef]
105. Kim, Y.S.; Kim, W.J. Microstructure and superplasticity of the as-cast Mg-9Al-1Zn magnesium alloy after high-ratio differential speed rolling. *Mater. Sci. Eng. A* **2016**, *677*, 332–339. [CrossRef]
106. Huang, X.; Suzuki, K.; Watazu, A.; Shigematsu, I.; Saito, N. Microstructure and texture of Mg-Al-Zn alloy processed by differential speed rolling. *J. Alloy. Compd.* **2008**, *457*, 408–412. [CrossRef]
107. Kwak, T.Y.; Lim, H.K.; Han, S.H.; Kim, W.J. Refinement of the icosahedral quasicrystalline phase and the grain size of Mg-9.25Zn-1.66Y alloy by high-ratio differential speed rolling. *Scr. Mater.* **2015**, *103*, 49–52. [CrossRef]
108. Huang, X.; Suzuki, K.; Chino, Y. Static recrystallization and mechanical properties of Mg-4Y-3RE magnesium alloy sheet processed by differential speed rolling at 823 K. *Mater. Sci. Eng. A* **2012**, *538*, 281–287. [CrossRef]
109. Kwak, T.Y.; Kim, W.J. Superplastic behavior of an ultrafine-grained Mg-13Zn-1.55Y alloy with a high volume fraction of icosahedral phases prepared by high-ratio differential speed rolling. *J. Mater. Sci. Technol.* **2017**, *33*, 919–925. [CrossRef]
110. Kim, S.-H.; You, B.-S.; Yim, C.D.; Seo, Y.-M. Texture and microstructure changes in asymmetrically hot rolled AZ31 magnesium alloy sheets. *Mater. Lett.* **2005**, *59*, 3876–3880. [CrossRef]
111. Ji, Y.H.; Park, J.J. Analysis of thermo-mechanical process occurred in magnesium alloy AZ31 sheet during differential speed rolling. *Mater. Sci. Eng. A* **2008**, *485*, 299–304. [CrossRef]
112. Cho, J.-H.; Kim, H.-W.; Kang, S.-B.; Han, T.-S. Bending behavior, and evolution of texture and microstructure during differential speed warm rolling of AZ31B magnesium alloys. *Acta Mater.* **2011**, *59*, 5638–5651. [CrossRef]
113. Li, X.; Li, X.; Ye, Y.; Zhang, R.; Kure-Chu, S.-Z.; Tang, G. Deformation mechanisms and recrystallization behavior of Mg-3Al-1Zn and Mg-1Gd alloys deformed by electroplastic-asymmetric rolling. *Mater. Sci. Eng. A* **2019**, *742*, 722–733. [CrossRef]
114. Lee, J.B.; Konno, T.J.; Jeong, H.G. Effect of differential speed rolling on the anisotropy of mechanical properties and texture evolution of AZ31 Mg alloys. *J. Alloy. Compd.* **2010**, *499*, 273–277. [CrossRef]
115. Kim, W.J.; Hwang, B.G.; Lee, M.J.; Park, Y.B. Effect of speed-ratio on microstructure, and mechanical properties of Mg-3Al-1Zn alloy, in differential speed rolling. *J. Alloy. Compd.* **2011**, *509*, 8510–8517. [CrossRef]
116. Kim, W.J.; Lee, Y.G.; Lee, M.J.; Wang, J.Y.; Park, Y.B. Exceptionally high strength in Mg-3Al-1Zn alloy processed by high-ratio differential speed rolling. *Scr. Mater.* **2011**, *65*, 1105–1108. [CrossRef]
117. Kim, W.Y.; Kim, W.J. Fabrication of ultrafine-grained Mg-3Al-1Zn magnesium alloy sheets using a continuous high-ratio differential speed rolling technique. *Mater. Sci. Eng. A* **2014**, *594*, 189–192. [CrossRef]
118. Lee, J.B.; Konno, T.J.; Jeong, H.G. Grain refinement and texture evolution in AZ31 Mg alloys sheet processed by differential speed rolling. *Mater. Sci. Eng. B* **2009**, *161*, 166–169. [CrossRef]
119. Hamad, K.; Chung, B.K.; Ko, Y.G. Microstructure and mechanical properties of severely deformed Mg-3%Al-1%Zn alloy via isothermal differential speed rolling at 453 K. *J. Alloy. Compd.* **2014**, *615*, S590–S594. [CrossRef]
120. Kim, W.J.; Park, J.D.; Wang, J.Y.; Yoon, W.S. Realization of low-temperature superplasticity in Mg-Al-Zn alloy sheets processed by differential speed rolling. *Scr. Mater.* **2007**, *57*, 755–758. [CrossRef]
121. Tolouie, E.; Jamaati, R. Asymmetric cold rolling: A technique for achieving non-basal textures in AZ91 alloy. *Mater. Lett.* **2019**, *249*, 143–146. [CrossRef]
122. Kim, W.J.; Lee, B.H.; Lee, J.B.; Lee, M.J.; Park, Y.B. Synthesis of high-strain-rate superplastic magnesium alloy sheets using a high-ratio differential speed rolling technique. *Scr. Mater.* **2010**, *63*, 772–775. [CrossRef]
123. Lequeu, P.H.; Jonas, J.J. Modeling of the plastic anisotropy of textured sheet. *Metall. Mater. Trans.* **1988**, *19*, 105–120. [CrossRef]
124. Engler, O.; Kim, H.C.; Huh, M.Y. Formation of {111} fibre texture in recrystallized aluminium sheet. *Mater. Sci. Technol.* **2001**, *17*, 75–86. [CrossRef]
125. Jin, H.; Lloyd, D.J. The reduction of planar anisotropy by texture modification through asymmetric rolling and annealing in AA5754. *Mater. Sci. Eng. A* **2005**, *399*, 358–367. [CrossRef]
126. Ma, C.; Hou, L.; Zhang, J.; Zhuang, L. Effect of deformation routes on the microstructures and mechanical properties of the asymmetrical rolled 7050 Aluminum alloy plates. *Mater. Sci. Eng. A* **2018**, *733*, 307–315. [CrossRef]
127. Wronski, S.; Ghilianu, B.; Chauveau, T.; Bacroix, B. Analysis of textures heterogeneity in cold and warm asymmetrically rolled aluminium. *Mater. Charact.* **2011**, *62*, 22–34. [CrossRef]
128. Kim, H.-K.; Kim, H.-W.; Cho, J.-H.; Lee, J.-C. High-formability Al alloy sheet produced by asymmetric rolling of strip-cast sheet. *Mater. Sci. Eng. A* **2013**, *574*, 31–36. [CrossRef]
129. Cheon, B.-H.; Kim, H.-W.; Lee, J.-C. Asymmetric rolling of strip-cast Al-5.5Mg-0.3Cu alloy sheet: Effects on the formability and mechanical properties. *Mater. Sci. Eng. A* **2011**, *528*, 5223–5227. [CrossRef]

130. Li, S.; Qin, N.; Liu, J.; Zhang, X. Microstructure, texture and mechanical properties of AA1060 aluminum plate processed by snake rolling. *Mater. Des.* **2016**, *90*, 1010–1017. [CrossRef]
131. Goli, F.; Jamaati, R. Asymmetric cross rolling (ACR): A novel technique for enhancement of Goss/Brass texture ratio in Al-Cu-Mg alloy. *Mater. Charact.* **2018**, *142*, 352–364. [CrossRef]
132. Goli, F.; Jamaati, R. Intensifying Goss/Brass texture ratio in AA2024 by asymmetric cold rolling. *Mater. Lett.* **2018**, *219*, 229–232. [CrossRef]
133. Diniz, S.B.; Benatti, E.A.; Paula, A.S.; Bolmaro, R.E.; Silva, L.C.; Meirelles, B.G. Microstructural evaluation of an asymmetrically rolled and recrystallized 3105 aluminum alloy. *J. Mater. Res. Technol.* **2016**, *5*, 183–189. [CrossRef]
134. Loorentz; Ko, Y.G. Effect of differential speed rolling strain on microstructure and mechanical properties of nanostructured 5052 Al alloy. *J. Alloy. Compd.* **2014**, *586*, S205–S209. [CrossRef]
135. Ko, Y.G.; Hamad, K. Microstructure stability and mechanical properties of ultrafine grained 5052 Al alloy fabricated by differential speed rolling. *Mater. Sci. Eng. A* **2018**, *733*, 24–27. [CrossRef]
136. Tamimi, S.; Correia, J.P.; Lopes, A.B.; Ahzi, S.; Barlat, F.; Gracio, J.J. Asymmetric rolling of thin AA-5182 sheets: Modelling and experiments. *Mater. Sci. Eng. A* **2014**, *603*, 150–159. [CrossRef]
137. Bintu, A.; Vincze, G.; Picu, R.C.; Lopes, A.B. Effect of symmetric and asymmetric rolling on the mechanical properties of AA5182. *Mater. Des.* **2016**, *100*, 151–156. [CrossRef]
138. Kraner, J.; Fajfar, P.; Palkowski, H.; Kugler, G.; Godec, M.; Paulin, I. Microstructure and Texture Evolution with Relation to Mechanical Properties of Compared Symmetrically and Asymmetrically Cold Rolled Aluminum Alloy. *Metals* **2020**, *10*, 156. [CrossRef]
139. Jin, H.; Lloyd, D.J. Effect of a duplex grain size on the tensile ductility of an ultra-fine grained Al-Mg alloy, AA5754, produced by asymmetric rolling and annealing. *Scr. Mater.* **2004**, *50*, 1319–1323. [CrossRef]
140. Sidor, J.; Petrov, R.H.; Kestens, L.A.I. Deformation, recrystallization and plastic anisotropy of asymmetrically rolled aluminum sheets. *Mater. Sci. Eng. A* **2010**, *528*, 413–424. [CrossRef]
141. Ko, Y.G.; Chaudry, U.M.; Hamad, K. Microstructure and mechanical properties of AA6061 alloy deformed by differential speed rolling. *Mater. Lett.* **2020**, *259*, 126870. [CrossRef]
142. Wronski, S.; Bacroix, B. Microstructure evolution and grain refinement in asymmetrically rolled aluminium. *Acta Mater.* **2014**, *76*, 404–412. [CrossRef]
143. Jin, H.; Lloyd, D.J. Evolution of texture in AA6111 aluminum alloy after asymmetric rolling with various velocity ratios between top and bottom rolls. *Mater. Sci. Eng. A* **2007**, *465*, 267–273. [CrossRef]
144. Lee, J.-H.; Kim, G.-H.; Nam, S.K.; Kim, I.; Lee, D.N. Calculation of plastic strain ratio of AA1050 Al alloy sheet processed by heavy asymmetric rolling-annealing followed by light rolling-annealing. *Comput. Mater. Sci.* **2015**, *100*, 45–51. [CrossRef]
145. Dutta, S.; Kaiser, M.S. Effect of asymmetric rolling on formability of pure aluminium. *J. Mech. Eng.* **2015**, *44*, 94–99. [CrossRef]
146. Zanchetta, B.D.; Da Silva, V.K.; Sordi, V.L.; Rubert, J.B.; Kliauga, A.M. Effect of asymmetric rolling under high friction coefficient on recrystallization texture and plastic anisotropy of AA1050 alloy. *Trans. Nonferrous Met. Soc. China* **2019**, *29*, 2262–2272. [CrossRef]
147. Shore, D.; Kestens, L.A.I.; Sidor, J.; Houtte, P.V.; Bael, A.V. Process parameter influence on texture heterogeneity in asymmetric rolling of aluminium sheet alloys. *Int. J. Mater. Form.* **2018**, *11*, 297–309. [CrossRef]
148. Simoes, F.J.P.; Sousa, R.J.A.; Gracio, J.J.A.; Barlat, F.; Yoon, J.W. Mechanical behavior of an asymmetrically rolled and annealed 1050-O sheet. *Int. J. Mech. Sci.* **2008**, *50*, 1372–1380. [CrossRef]
149. Jiang, J.; Ding, Y.; Zuo, F.; Shan, A. Mechanical properties and microstructures of ultrafine-grained pure aluminum by asymmetric rolling. *Scr. Mater.* **2009**, *60*, 905–908. [CrossRef]
150. Xu, G.; Cao, X.; Zhang, T.; Duan, Y.; Peng, X.; Deng, Y.; Yin, Z. Achieving high strain rate superplasticity of an Al-Mg-Sc-Zr alloy by a new asymmetrical rolling technology. *Mater. Sci. Eng. A* **2016**, *672*, 98–107. [CrossRef]
151. Duan, Y.; Xu, G.; Tang, L.; Liu, Y.; Xu, J.; Deng, Y.; Yin, Z. Excellent high strain rate superplasticity of Al-Mg-Sc-Zr alloy sheet produced by an improved asymmetrical rolling process. *J. Alloy. Compd.* **2017**, *715*, 311–321. [CrossRef]
152. Yang, H.W.; Widiantara, I.P.; Ko, Y.G. Effect of deformation path on texture and tension properties of submicrocrystalline Al-Mg-Si alloy fabricated by differential speed rolling. *Mater. Lett.* **2018**, *213*, 54–57. [CrossRef]
153. Yu, H.; Lu, C.; Tieu, A.K.; Li, H.J.; Godbole, A.; Zhang, S.H. Special Rolling Techniques for Improvement of Mechanical Properties of Ultrafine-Grained Metal Sheets: A Review. *Adv. Eng. Mater.* **2016**, *18*, 754–769. [CrossRef]
154. Yu, H.; Tieu, A.K.; Lu, C.; Liu, X.; Liu, M.; Godbole, A.; Kong, C.; Qin, Q. A new insight into ductile fracture of ultrafine-grained Al-Mg alloys. *Sci. Rep.* **2015**, *5*, 9568. [CrossRef]
155. Shi, J.; Hou, L.; Ma, C.; Zuo, J.; Cui, H.; Zhuang, L.; Zhang, J. Mechanical Properties and Microstructures of 5052 Al Alloy Processed by Asymmetric Cryorolling. *Mater. Sci. Forum* **2016**, *850*, 823–828. [CrossRef]
156. Magalhaes, D.C.C.; Kliauga, A.M.; Ferrante, M.; Sordi, V.L. Asymmetric cryorolling of AA6061 Al alloy: Strain distribution, texture and age hardening behavior. *Mater. Sci. Eng. A* **2018**, *736*, 53–60. [CrossRef]
157. Chao, Q.; Cizek, P.; Wang, J.; Hodgson, P.D.; Beladi, H. Enhanced mechanical response of an ultrafine grained Ti-6Al-4V alloy produced through warm symmetric and asymmetric rolling. *Mater. Sci. Eng. A* **2016**, *650*, 404–413. [CrossRef]
158. Huang, X.; Suzuki, K.; Chino, Y. Improvement of stretch formability of pure titanium sheet by differential speed rolling. *Scr. Mater.* **2010**, *63*, 473–476. [CrossRef]

159. Li, Z.; Fu, L.; Fu, B.; Shan, A. Effects of annealing on microstructure and mechanical properties of nano-grained titanium produced by combination of asymmetric and symmetric rolling. *Mater. Sci. Eng. A* **2012**, *558*, 309–318. [CrossRef]
160. Yu, H.; Yan, M.; Li, J.; Godbole, A.; Lu, C.; Tieu, K.; Li, H.; Kong, C. Mechanical properties and microstructure of a Ti-6Al-4V alloy subjected to cold rolling, asymmetric rolling and asymmetric cryorolling. *Mater. Sci. Eng. A* **2018**, *710*, 10–16. [CrossRef]
161. Wronski, M.; Wierzbanowski, K.; Wronski, S.; Bacroix, B.; Lipinski, P. Texture variation in asymmetrically rolled titanium. Study by Finite Element Method with implemented crystalline model. *Int. J. Mech. Sci.* **2014**, *87*, 258–267. [CrossRef]
162. Stolyarov, V.V.; Zeipper, L.; Mingler, B.; Zehetbauer, M. Influence of post-deformation on CP-Ti processed by equal channel angular pressing. *Mater. Sci. Eng. A* **2008**, *476*, 98–105. [CrossRef]
163. Zhao, X.; Fu, W.; Yang, X.; Langdon, T.G. Microstructure and properties of pure titanium processed by equal-channel angular pressing at room temperature. *Scr. Mater.* **2008**, *59*, 542–545. [CrossRef]
164. Terada, D.; Inoue, S.; Tsuji, N. Microstructure and mechanical properties of commercial purity titanium severely deformed by ARB process. *J. Mater. Sci.* **2007**, *42*, 1673–1681. [CrossRef]

MDPI
St. Alban-Anlage 66
4052 Basel
Switzerland
Tel. +41 61 683 77 34
Fax +41 61 302 89 18
www.mdpi.com

Metals Editorial Office
E-mail: metals@mdpi.com
www.mdpi.com/journal/metals

www.ingramcontent.com/pod-product-compliance
Lightning Source LLC
Chambersburg PA
CBHW061442210926
51850CB00008B/1581
* 9 7 8 3 0 3 6 5 8 0 7 4 6 *